007# The Colorado Plateau IV

The Colorado Plateau IV

SHAPING CONSERVATION THROUGH
SCIENCE AND MANAGEMENT

Edited by Charles van Riper III,
Brian F. Wakeling, and Thomas D. Sisk

The University of Arizona Press
Tucson

This volume is based on research presented at the Ninth Biennial Conference of Reseaerch on the Colorado Plateau held at Northern Arizona University, Flagstaff, Arizona, and hosted by the U.S. Geological Survey Southwest Biological Science Center Colorado Plateau Research Station, the Merriam-Powell Center for Environmental Resreaech, Bureau of Land Management, National Park Service, Diablo Trust, and the Center for Sustainable Environments at Northern Arizona University.

The University of Arizona Press
(c) 2010 The Arizona Board of Regents
All rights reserved

ISBN 978-0-8165-2914-8
Libary of Congress Control Number: 2010922393

Manufactured in the United States of America on acid-free archival-quality paper containing a minimum of 30% post-consumer waste and procssed chlorine free.

15 14 13 12 11 10   6 5 4 3 2 1

This book was published from formatted electronic copy that was edited and typeset by the volume editors.

CONTENTS

Foreword	vii
Dedication	ix
In Memoriam	xiii
Introduction and Acknowledgments	xv

CONSERVATION VISIONS AND FRAMEWORKS

1. The Legacy and Future Visions of Conservation Biology on the Colorado Plateau — 1
   Allison L. Jones, Ethan Aumack, Jan Balsom, Paul Beier, Jayne Belnap, James Catlin, Thomas L. Fleischner, Ed Grumbine, David J. Mattson, and Charles van Riper III

2. Downscaling Climate Projections in Topographically Diverse Landscapes of the Colorado Plateau in the Arid Southwestern United States — 21
   Gregg M. Garfin, Jon K. Eischeid, Melanie T. Lenart, Kenneth L. Cole, Kirsten Ironside, and Neil Cobb

3. Integrating Restoration and Conservation Objectives at the Landscape Scale: The Kane and Two Mile Ranch Project — 45
   Thomas D. Sisk, Christine Albano, Ethan Aumack, Eli J. Bernstein, Timothy E. Crews, Brett G. Dickson, Steve Fluck, Melissa McMaster, Andi S. Rogers, Steven S. Rosenstock, David Schlosberg, Ron Sieg, and Andrea Thode

ASSESSING MONITORING FRAMEWORKS AND SYSTEMS

4. Mapping Ecological Sites for Long-term Monitoring in National Parks — 69
   Steven L. Garman, Dana Witwicki, and Aneth Wight

5. Quantifying Sample Bias in Long-term Monitoring Programs: A Case Study at Bandelier National Monument, New Mexico — 89
   Chris L. Lauver, Jodi Norris, Lisa Thomas, and Jim DeCoster

6. Wisardnet Field-to-Desktop: Building a Wireless Cyberinfrastructure for Environmental Monitoring — 101
   Kenji Yamamoto, Yuxin He, Paul Heinrich, Alex Orange, Bill Ruggeri, Holland Wilberger, and Paul Flikkema

7. Finding Gaps in the Protected Area Network in the Colorado Plateau: A Case Study Using Vascular Plant Taxa in Utah — 109
   Walter Fertig

8. Fire and Fire Surrogate Treatment Impacts on Soil Moisture Condition in Southwest Ponderosa Pine Forests 121
Boris Poff, Daniel G. Neary, and Aregai Tecle

WILDLIFE SURVEYS AS A CONSERVATION FRAMEWORK

9. Milksnakes at Petrified Forest National Park, Arizona: Adaptive Monitoring of Rare Vertebrates 133
Erika M. Nowak and Trevor B. Persons

10. Mormon Cricket Control in Utah's West Desert: Evaluation of Impacts of the Pesticide Diflubenzuron on Non-target Arthropod Communities 151
Tim B. Graham, Anne M. D. Brasher, and Rebecca N. Close

11. Yellow-billed Cuckoo Distribution and Habitat Associations in Arizona 1998 - 1999: Future Monitoring and Research Implications 197
Matthew J. Johnson, Robert T. Magill, and Charles van Riper III

12. A Historical Assessment of Changes in Avian Community Composition from Montezuma Castle National Monument, with Observations from the Camp Verde Region of Arizona 213
Charles van Riper III, Mark K. Sogge, and Matthew J. Johnson

LARGE MAMMAL CONSERVATION AND MANAGEMENT

13. Mule Deer Antler Growth and Hunting Management on the North Kaibab, Arizona 271
Brian F. Wakeling

14. Female Elk Habitat Use after the Rodeo-Chediski Fire in Northeast Arizona 277
Kirby Bristow and Stan Cunningham

15. Scent-station Surveys: Indexing Relative Abundance of Mesopredators in Arizona 293
Ted McKinney and Thorry W. Smith

16. Mountain Lion Depredation Harvests in Arizona, 1976 - 2005 303
Ted McKinney, Brian F. Wakeling, and Johnathan C. O'Dell

SYNTHESIS

17. Shaping Conservation through the Integration of Research with Resources Management on The Colorado Plateau: A Synthesis 315
Charles van Riper III, Brian F. Wakeling, and Thomas D. Sisk

List of Contributors 323

Index 329

FOREWORD

Roughly centered on the Four Corners region of the southwestern United States, the Colorado Plateau covers some 130,000 square miles of sparsely vegetated plateaus, mesas, canyons, arches, and cliffs in Arizona, Utah, Colorado, and New Mexico. With elevations ranging from 3,000 to 14,000 feet, the natural systems found within the plateau are dramatically varied, from desert to alpine conditions. The place names that dot the Colorado Plateau are evocative of its rugged natural beauty and diverse peoples—Paradox Basin, Grand Canyon, El Malpais National Monument, Salt Wash, and Paiute Wilderness, to name a few. The region's diverse and distinctive ecosystems have resulted in high rates of plant endemism and species richness for vertebrates and invertebrates alike. Creatures as distinct as the canyon tree frog and Uncompahgre fritillary butterfly grace the Colorado Plateau.

A tribute to its natural and cultural wonders, the Colorado Plateau boasts 10 national parks, 17 national monuments, and millions of acres of national forests. The natural beauty and rich natural and cultural resources of the Colorado Plateau have attracted growing numbers of visitors and recreationists. The Colorado River, which drains some 90 percent of the plateau, has been extensively harnessed to produce hydropower and to provide water to the West's ever-growing communities. The region is also rich in coal, oil, natural gas, uranium, and oil shale. However, as growing human populations alter natural systems to meet their needs, it becomes imperative to better understand the relationships that are at work among the plants, animals, people, and the environment on the Colorado Plateau, particularly in the face of climate change.

Developing a better understanding of the living things and natural systems of the Colorado Plateau was the genesis for the first Biennial Conference for Research on the Colorado Plateau, which took place in 1991 on the campus of Northern Arizona University in Flagstaff, Arizona. Over the past two decades the Colorado Plateau Biennial Conferences have been made possible by a partnership between Northern Arizona University (NAU) and the U.S. Geological Survey (USGS). The series of books produced as a result of these conferences focuses on providing information to NAU and USGS partners, particularly land managers on the Colorado Plateau. Today, well into the new millennium, the need for sound scientific research that can be used to guide management activities is more urgent than ever.

This book presents scientific findings that were shared during the ninth biennial conference, containing 16 chapters of papers and presentations contributed by federal, state, and private sector researchers investigating the cultural, biological, and physical resources of the Colorado Plateau. We are all indebted to the thoughtful and capable scientists who come together with land managers every other year to share scientific information supporting land management activities on the Colorado Plateau. The value of this forum in which information about current research and resource management issues are integrated cannot be overstated.

As climate change and other factors increasingly alter physical, cultural, and biological resources on the lands of the Colorado Plateau, and throughout the world, science can serve as the compass that guides decision-making in the face of change. Science can help elected officials, resource managers, and citizens chart a path to a future that can sustain us and those who follow us—a future that meets not only our physical needs but also feeds our minds and imaginations with the bounty of the Colorado Plateau's natural beauty and cultural heritage.

KATE KITCHELL
U.S. Geological Survey
Acting Center Director
Southwest Biological Science Center
Flagstaff, Arizona

DEDICATION

We are pleased to dedicate this ninth volume of The Colorado Plateau conference proceeding to three people who have been instrumental in advancing the integration of conservation biology and natural resource management in Arizona, particularly on the Colorado Plateau—Laura Huenneke, Vice President for Research at Northern Arizona University, and Duane L. Shroufe and Larry D. Voyles, past and current directors, respectively, of the Arizona Game and Fish Department. All three have been staunch supporters of the Biennial Conference of Research on the Colorado Plateau during their tenure. With Laura's support, Northern Arizona University has been instrumental in both encouraging research on the plateau and in the realization of the biennial conferences. Under the direction of Shroufe and Voyles, the Arizona Game and Fish Department has been repeatedly recognized as a leader in the nation and the state, receiving honors from the Arizona Quality Alliance, including Showcase in Excellence Awards for the Endangered Species Recovery Program, Urban Fish Stocking Program, and the Hunt Recommendations Process. Resource management and research have long been integrated in this agency, and its work has contributed significantly to understanding issues on the Colorado Plateau.

Laura F. Huenneke

Dr. Laura Huenneke has worked tirelessly to promote research on the Colorado Plateau's rich biological and cultural diversity since joining Northern Arizona University (NAU) in 2003. Bringing an international reputation in ecosystem ecology and expertise in the role of biological diversity on ecosystem structure and function, Dr. Huenneke has helped to strengthen links between the research community and those managing Western lands and natural resources. In so doing, she has expanded the collaborative relationships that link universities with state and federal agencies, non-governmental organizations, and local citizens.

Laura attended the University of Missouri, Columbia, obtaining her Bachelor's degree in Biological Sciences with departmental honors, *summa cum laude*. She earned her Ph.D. in Ecology and Evolutionary Biology from Cornell University, where she developed a lifelong commitment to engaging students in research, fostering experiences fundamental to a modern science education. Before coming to NAU, Laura spent 16 years on the faculty at New Mexico State University in Las Cruces, where she developed an interdisciplinary curriculum in environmental biology, served five years as department chair, and received the honor of being named Regents Professor. In 1999 she was selected as a Fellow of the Aldo Leopold Leadership Program. She came to NAU as Dean of the College of Arts and Sciences and

Professor of Biological Sciences in 2003. Subsequently, she served three years as Dean of Engineering and Natural Sciences before becoming Vice President for Research.

Laura has served on the boards of the Las Cruces Natural History Museum, the Chihuahuan Desert Nature Park, the Arboretum at Flagstaff, and the University of Arizona's Biosphere II. She has served on several editorial boards for ecological research journals and on National Science Foundation and other review panels, and has been elected twice to the governing board of the Ecological Society of America (most recently as Vice President for Public Affairs). She is also a member of the Steering Committee for the Arizona Bioscience Roadmap and of the Northern Arizona Economic Development Advisory Council.

Under the leadership of Duane Shroufe and Larry Voyles, the Arizona Game and Fish Department (Department) has set the bar for excellence in state wildlife agencies. In addition to Arizona Quality Alliance awards for specific programs and process, the Department also received an Arizona Pioneer in Quality Award for establishing fundamental quality programs, an essential foundation for resource management and conservation research. The collective leadership of Shroufe and Voyles has resulted in the Department taking a primary role in conservation and restoration efforts for many wildlife species—Gould's turkey, black-footed ferret, black-tailed prairie dog, bald eagle, Mexican gray wolf, California condor, and Apache trout. They have repeatedly focused on connecting the public with the natural world through their emphasis on recruiting and retaining wildlife recreationists, including hunters, anglers, recreational shooters, and wildlife-viewing enthusiasts. They have served as champions for the North American Model of Wildlife Conservation (including the Colorado Plateau).

Duane L. Shroufe

Duane Shroufe was born in Jackson, Michigan. He received a Bachelors of Science in Wildlife Biology from Michigan State University and was accredited as a Certified Wildlife Biologist by The Wildlife Society. Prior to coming to Arizona, Duane served the Indiana Department of Natural Resources' Division of Fish and Wildlife for 17 years as a property manager, property staff specialist, and chief of wildlife. He moved to Arizona in 1984 when he accepted a job as Assistant Director of Arizona Game and Fish Department. He became Director there 1989. When he retired in 2008, Duane was the second-longest-tenured director of any state wildlife agency in the United States.

As director, Duane made major contributions to wildlife conservation on state, national, and international scales. He supported and encouraged international collaboration with Mexico. Under his leadership, the Arizona Game and Fish Department maintained a strong field operations program, while nurturing a robust research and nongame program. At the same time, a proactive habitat program was developed. Duane was instrumental in the implementation of the State Heritage Program, providing funding for the protection of sensitive species. He chaired the North American Wetland Conservation Council and the Pacific Flyway Council, and served as president of the Colorado River Fish and Wildlife Council. He has served as president of the Western Association of Fish and Wildlife Agencies and the Association of Fish and Wildlife Agencies.

Duane has been recognized with numerous awards during his career, including the International Association of Fish and Wildlife Agencies' Ernest Thompson Seton Award in 1996 and the Seth Gordon Award in 2001. In 1997, he was recognized with the Wildlife Professional Award from the Western Association of Fish and Wildlife Agencies. In 2000, he received the Promoting Outdoor Ethics Award from the Boone and Crockett Club. Humbly, he always asserted that his recognition was a direct result of the men and women who worked for the Department. In retirement, Duane and his wife, Linda, continue to enjoy hunting and angling.

Larry D. Voyles

Larry Voyles is a third generation native to Arizona. His paternal grandfather's family settled near the confluence of the Salt and Gila Rivers, where they farmed in what is now part of Phoenix, Arizona. Larry was born and raised in Tempe.

Larry attended Mesa College and the University of Arizona before transferring to Arizona State University, where he graduated with a Bachelor of Science degree and soon landed his first job with Arizona Game and Fish Department, capturing and banding doves in the Arlington area during the summer. In September 1974, Larry was hired as a banding Wildlife Manager. Larry worked as Wildlife Manager in Welton, Wickenburg, and Prescott before moving to the Phoenix Headquarters to fill the new Law Enforcement Training Manager position. In 1988, he was promoted to Regional Supervisor for the Yuma Region, where he remained until January 2008, when he was selected as the new director for the Department. Larry and his wife Donna have one grown daughter, Leena. Larry enjoys many outdoor activities with his family and friends, such as hunting, fishing, and horseback riding.

As Director of the Arizona Game and Fish Department, Larry has articulated an ambitious vision. He sees the agency's center of gravity as its credibility, which is supported by the quality of the workforce, the quality of the information, the Commission system, and the public's passion for wildlife. Larry recognizes the need for geospatial planning to prepare for and address the challenges that will present themselves in the decades to come. Under his leadership, the Department established two new branches—Wildlife Recreation Recruitment and Retention, and Shooting Sports—to garner broader involvement and encourage wider participation in Arizona's outdoor recreation, which in turn fosters greater relevancy for wildlife in the new century.

Laura Huenneke, Duane Shroufe, and Larry Voyles have been and will continue to be key figures in the integration of research and resources management on the Colorado Plateau, each providing a unique, and equally important, perspective. We are grateful to their insight and their support for the biennial conferences of research on the Colorado Plateau. Their commitment and their continued support for identifying and facing the challenges of sustaining our natural resources will be critical in coming years.

IN MEMORIAM

Ted D. McKinney
July 16, 1937 – October 22, 2008

Ted McKinney was born in Miami, Arizona, and passed at his home in Mesa, Arizona, at the age of 71. Passionate about research and wildlife, he worked tirelessly with the Arizona Game and Fish Department's Research Branch to the end of his days. Ted's drive to learn and discover was matched only by his desire to mentor. Scores of young and veteran biologists alike benefited from the time they spent alongside this man.

He enjoyed a wide array of experiences and held multiple professions during his lifetime: a cowboy in his youth, bank loan officer, gold miner in Costa Rica, assistant and associate professor in Oklahoma and Texas, outfitter and guide in Colorado, self-employed environmental consultant, and wildlife research biologist in Arizona. Ted's formal education started at the University of Arizona, where he earned his B.S. in range management, followed by an M.S. in wildlife management from Colorado State University. At Virginia Polytechnic Institute, he obtained his Ph.D. in wildlife ecology and then followed up with a post–doctoral fellowship at the National Institutes of Health, Albert Einstein Medical Center in Philadelphia, Pennsylvania.

The depth and breadth of Ted's contribution to wildlife science is remarkable. He is widely published in topics ranging from the ecology of the flannelmouth sucker and rainbow trout in the Colorado River, to dietary overlap in mesocarnivore populations, to the identification of trans-highway movement corridors for desert bighorn sheep, to the ecology of mountain lions. Perhaps his greatest contribution to the conservation of bovids is the monograph

Ted D. McKinney

he authored entitled, "Evaluation of factors potentially influencing a desert bighorn sheep population." At the time of his passing, Ted was embarking on an ambitious study to determine the effect of mountain lion predation on stable, increasing, and decreasing desert bighorn sheep populations.

Our challenge is to gain the intuition of Ted's unique perspective, earned through a lifetime of awareness and perseverance, and to continue on from where he left us. Ted dedicated his life to the pursuit of knowledge through research for the betterment of wildlife. As a research scientist for the Arizona Game and Fish Department, he never considered retirement. He followed the tracks of wildlife across the landscape. May we follow in his.

BRIAN F. WAKELING
Arizona Game and Fish Department
Game Branch
Phoenix, Arizona

INTRODUCTION AND ACKNOWLEDGEMENTS

This is the ninth in a series of books that focus on research and resource management issues across the Colorado Plateau. Of the eight previous volumes, the last three have been published by the University of Arizona Press (van Riper and Cole 2004, van Riper and Mattson 2005, van Riper and Sogge 2008), while the earlier volumes were published by the U.S. Government Printing Office (Rowlands et al. 1993, van Riper 1995, van Riper and Deshler 1997, van Riper and Stuart 1999, van Riper et al. 2001). These books have highlighted efforts at integrating research findings into management of natural, cultural, and physical resources within the biogeographic province of the Colorado Plateau.

This ninth volume also highlights aspects of biological, cultural, and physical research, combined with a series of chapters explaining collaborative tools and decision-making processes that can be used to better manage resources at broader spatial and temporal scales, and over a a larger geographic region. The mix of chapters covering many diverse research and resource management subjects also addresses how scientists and land managers can better interface when dealing with Colorado Plateau management issues.

The 16 chapters that constitute this book were selected from 177 research papers, panel sessions, and posters presented at the Ninth Biennial Conference of Research on the Colorado Plateau. Held at Northern Arizona University from 29 October to 1 November 2007 in Flagstaff, Arizona, the conference was hosted by the U.S. Geological Survey Southwest Biological Science Center's Colorado Plateau Research Station; Northern Arizona University's Center for Sustainable Environments, Merriam-Powell Center for Environmental Research, and Ecological Monitoring and Assessment Program and Foundation; and the National Park Service, Diablo Trust, and Colorado Plateau Cooperative Ecosystem Studies Unit (CESU). The theme of this conference revolved around research, inventory, and monitoring of lands over the Colorado Plateau, with a focus on tools that can be useful in the integration of biophysical and socioeconomic research into land management decision-making and actions.

No scientific work is a single effort, and this book is no exception; it is a direct result of assistance by many individuals. We would especially like to thank the following scientific peer reviewers: Eli Bernstein, George Billingsley, Chad Bishop, Ken Boykin, Bethany Bradley, Bill Burger, Neil Cobb, Troy Corman, Ron Day, James C. de Vos, Jr., Charles Drost, Andrew Ellis, Don Falk, Peter Fflolliott, Dave Freddy, Peter Fulé, Jim Graham, Michael F. Ingraldi, Brian Jacobs, Tom Jones, Tom Keegan, Paul Krausman, Rick Langley, Scott Lerich, Matthew Loeser, Kate Maher, Marc Mazerolle, Chris Menges, Mark Miller, Fiona Nagle, Pamela Nagler, Chantal O'Brien, Louis Pech, Ed Pfeiffer, Barb Phillips, Andi Rogers, Cecil Schwalbe, Steve Sesnie, Harley Shaw, Vashti Supplee, Don Swann, Kathryn Thomas, Ron Thompson, John Tinsley, and Gary White—all who unselfishly devoted their time and effort to improving each chapter that they reviewed.

This series of books has received continued financial support from the U.S. Geological Survey (USGS) and its Southwest Biological Science Center. Andrea Alpine, Anne Kinsinger, Kate Kitchell, Michael Schulters, and Sue Haseltine all provided encouragement and/or financial assistance for this publication. We also very much appreciate the support of Northern Arizona University (NAU) and especially Laura Huenneke and John Haeger for providing encouragement and/or financial assistance. The dedicated USGS Colorado Plateau Research Station staff (S. Adson, T. Arundel, K. Cole, C. Drost, J. Hart, J. Holmes, M. Johnson, D. Mattson, E. Nowak, M. Saul, and especially Mark K. Sogge and Scott Durst), volunteers and staff from NAU, and the conference co-sponsors, as well as the audio-visual staff of Chris Calvo and Chris Taesali, provided much needed assistance during this Ninth Biennial Conference. We also appreciate the assistance of Neil Cobb, J. J. Wynne, Ron Heibert, Matthew J. Johnson, Eli Bernstein, Kirsten Ironside, Ken Cole, Michele James, Diana Anderson, Amy Whipple, Judy Springer, and Susan Nyoka, who organized numerous special sessions at the conference. Linda Brewer, in addition to her editorial and graphic duties, helped in many ways, and without her tenacity this book would never have been a reality. We would also particularly like to thank Kimberly A. van Riper for the line drawings throughout the book. Finally, we express deep appreciation to our families for their support and understanding during the time that this book was in production.

CHARLES VAN RIPER III
U.S. Geological Survey
Southwest Biological Science Center
Sonoran Desert Research Station
University of Arizona
Tucson, Arizona

BRIAN F. WAKELING
Arizona Game and Fish Department
Game Branch
Phoenix, Arizona

THOMAS D. SISK
Center for Sustainable Environments
Northern Arizona University
Flagstaff, Arizona

LITERATURE CITED

Rowlands, P. G., C. van Riper III, and M. K. Sogge editors. 1993. Proceedings of the First Biennial Conference on Research in Colorado Plateau National Parks. Transaction and Proceedings Series NPS/NRNAU/NRTP-93/10. U.S. Department of the Interior, Washington, D.C.

van Riper, C., III, editor. 1995. Proceedings of the Second Biennial Conference on Research in Colorado Plateau National Parks. Transaction and Proceedings Series NPS/NRNAU/NRTP-95/11. U.S. Department of the Interior, Washington, D.C.

van Riper, C., III., and K. A. Cole, editors. 2004. The Colorado Plateau: Cultural, Biological, and Physical Research. University of Arizona Press, Tucson.

van Riper, C., III, and E. T. Deshler, editors. 1997. Proceedings of the Third Biennial Conference of Research on the Colorado Plateau. Transactions and Proceedings Series NPS/NRNAU/NRTP-97/12. U.S. Department of the Interior, Washington, D.C.

van Riper, C., III., and D. J. Mattson, editors. 2005. The Colorado Plateau II: Biophysical, Socioeconomic, and Cultural Research. University of Arizona Press, Tucson.

van Riper, C., III., and M. K. Sogge, editors. 2008. The Colorado Plateau III: Integrating Research and Resources Management for Effective Conservation. The University of Arizona Press, Tucson.

van Riper, C., III., and M. A. Stuart, editors. 1999. Proceedings of the Fourth Biennial Conference of Research on the Colorado Plateau. U.S. Geological Survey Forest and Rangeland Ecosystem Science Center CPFS/99/16 Rep. Ser., Flagstaff, AZ.

van Riper, C., III, K. A. Thomas, and M. A. Stuart, editors. 2001. Proceedings of the Fifth Conference of Research on the Colorado Plateau. U.S. Geological Survey/ Forest and Rangeland Ecosystem Science Center USGSFRESC/COPL/2001/21 Rep. Ser., Flagstaff, Arizona.

Conservation Visions and Frameworks

THE LEGACY AND FUTURE VISIONS OF CONSERVATION BIOLOGY ON THE COLORADO PLATEAU

Allison L. Jones, Ethan N. Aumack, Jan Balsom, Paul Beier, Jayne Belnap, James C. Catlin, Thomas L. Fleischner, Ed Grumbine, David J. Mattson, and Charles van Riper III

ABSTRACT

This paper summarizes a half–day panel discussion and larger audience discussion on a topic that doesn't receive enough attention at the scientific conferences we all attend: collectively reflecting on how our individual research and efforts have served the Colorado Plateau for the past several decades. And, what do we need to do more of in the future to assure the preservation of biological diversity in this magnificent region? During the 2007 USGS conference at Northern Arizona University, the Colorado Plateau Chapter of the Society for Conservation Biology held a two-part session, discussing first with our distinguished panelists what strategies have either forwarded, or hindered, the understanding and protection of biodiversity on the Plateau. In the second session, we hypothetically traveled forward in time to develop visions of a future for conservation biology on the Plateau within the context of current (and future) challenges such as a changing climate and shifting human demands for natural resources. Our panelists, all leaders in the field of conservation biology, gave us many valuable "take-home lessons," ranging from the benefits of collaboration with non-scientists, to how to deal with the influence of political processes in the scientific realm, to the power of hope to sustain conservationists through tough times. Overall, a common thread among the presenters was one of using beneficial lessons learned to proactively and effectively work towards a positive future for our Colorado Plateau landscapes and wildlife.

INTRODUCTION

The Colorado Plateau is home to a thriving network of practitioners, researchers, and academics in the field of conservation biology. Many of them are well-recognized leaders in this discipline, having guided its evolution over the last three decades. They have cut their professional teeth on issues across North America, from grizzlies in Yellowstone, to cougars in Southern California, to legal battles for indigenous water rights, but for many years they have rooted their work here, on the Colorado Plateau. They are affiliated with universities, federal agencies, and research organizations in four states and many watersheds, and their individual work is as unique as the region. Unfortunately, many of those involved in conservation on the Plateau rarely avail themselves of the opportunity to draw on the wisdom and experiences of these leaders. Similarly, how often do we collectively reflect on how our work has served this place we all call home? In short, how has conservation biology served the Colorado Plateau in the past, and how can it be most effective in the future?

During the ninth annual Conference for Research on the Colorado Plateau, the

Colorado Plateau Chapter of the Society for Conservation Biology held a two-part panel session. This session provided an opportunity to engage a suite of conservation leaders around these questions in an open panel setting. In the first session, we educated ourselves about the recent history of conservation efforts across the Plateau, identifying strategies that have forwarded or hindered the protection of structural, compositional, and functional biodiversity. Our distinguished panelists in this first gathering included Jan Balsom, Deputy Chief of Science and Resource Management at Grand Canyon National Park, Paul Beier, professor and researcher at Northern Arizona University, James Catlin, Executive Director of the Wild Utah Project, and Charles van Riper III, research scientist and professor at the USGS Southwest Biological Science Center Sonoran Desert Research Station.

In the second session, we looked to the future to develop visions of a future for conservation biology within the context of current challenges such as a changing climate and shifting human demands for natural resources. Special emphasis was paid to "the passing of the torch," as seminal thinkers in the field of conservation biology helped us explore what skills the future leaders of conservation on the Plateau will require to face these challenges. Our distinguished panelists in this second discussion included Thomas L. Fleischner of Prescott College, Ethan Aumack with the Grand Canyon Trust, Jayne Belnap with the USGS Southwest Biological Science Center Canyonlands Research Station, Ed Grumbine with Prescott College, and David J. Mattson with the USGS Southwest Biological Science Center Colorado Plateau Research Station.

In an attempt to capture the ideas presented during that afternoon of discussion, we asked each panelist to provide a written summary of their presentation. Those are presented here, collected under the heading of each panel. We hope these essays will both inform and inspire current and future conservation biologists here on the Colorado Plateau and around the world. In addition, some of the best insight comes from open audience discussion that follows the presentations of deep thinkers. As such, the highlights of the audience discussion follow the panelist presentations.

PANEL 1 PRESENTATIONS
LEGACY OF CONSERVATION BIOLOGY ON THE COLORADO PLATEAU

Paul Beier

My experience conserving wildlife corridors began in 1988, when I started a study of mountain lions in Southern California. My first scientific discovery was grim. Due to unrelenting development, mountain lions were on the road to extinction in every southern California mountain range. As the encirclement of each mountain range became complete, each mountain lion population would wink out, one by one (Beier 1993). But my other discovery was that mountain lions were still moving between some mountain ranges, and often using narrow, highly disturbed corridors through urban areas in order to do it. If they could continue to do so, they could survive in every linked mountain range (Beier 1995). Imagine how successful a corridor would be if we designed them to facilitate movement by animals.

As recipient of these scientific insights, I felt obliged to bring them to the attention of managers. I published scientific papers on my findings, but I knew that wasn't enough. So for several years, I did the only thing I could think to do. I fought against proposed highways and housing developments that would sever corridors in my study area. Through this, I learned that reacting to proposals to destroy connectivity is like fighting a one-way ratchet; sometimes you'll stop a bad action, sometimes you'll reduce the impact on connectivity, but you'll never permanently protect or restore a corridor. The significance of this, in turn, was that I learned that we could only win if we moved beyond reacting to bad proposals and put forward a positive proposal for a connected landscape.

This is the first lesson I'm sharing with you all today. Here are the others…

Lesson 2: Scientists cannot lead by asking others to follow us. Lead by serving. My first effort at presenting a linkage plan was a collaboration of 15 conservation biologists. It was a fine positive vision that collected dust. The problem was not its vision, but the fact that the authors were 15 PhDs who wanted to help "the befuddled management agencies." If they had been part of the process, they might have agreed with our priorities, but instead they'd been handed a map and told to "make it happen." I had failed. Instead of asking managers to "trust me – I'm a scientist," I needed to say "I trust you to identify needs for connectivity; I'm ready to work with you to develop sound and fair ways to meet those needs."

Shortly afterwards a nascent NGO called South Coast Wildlands gave me the opportunity to go back and do it right. South Coast Wildlands and five co-sponsors invited 250 people to map a connected conservation landscape for the state of California. The workshop included a few scientists like me, but the meeting was dominated by managers, all of whom were concerned about the increasing isolation of the lands and populations they managed. These people knew more than I did about what was important. They were passionate about creating a landscape more than the sum of its parts, because they owned the parts. Our report instantly became the primary reference book on connectivity issues for agencies, consultants, and corridor advocates in California. On the release date, almost every daily newspaper in the state carried a positive front-page story on the report.

Lesson 3: Leadership means engaging diverse people to develop comprehensive solutions to difficult problems. As a scientist, it took me a while to embrace the idea of inviting non-scientists to participate in scientific issues. But then I came to realize that no assumption or logical chain in ecology is so esoteric that a practitioner can't understand it. A scientist who wants to be effective simply must invite managers, other practitioners and NGOs to participate in our science. Our science is improved by having managers challenge our assumptions and offer alternative evidence and alternative interpretations of the evidence. In every case the result was better than what I could've invented on my own. Invite appropriate skeptics, real-estate developers, and other potential opponents to your workshops and meetings. This kind of openness demonstrates that your process is transparent, honest, and inclusive. You have nothing to lose and much to gain by inviting anyone who wants to advance the scientific rigor of the conservation plan and its implementation. Sometimes I walk out of a meeting thinking nobody agreed with me. A few months later, I find my harshest critics have become allies. I now see that my role in conservation is to insist on scientific rigor, consistency, and honesty, and to trust our partners to grapple with the scientific issues and set the agenda.

Jan Balsom

When I was first approached about participating in the panel organized by folks involved in conservation biology, I was sure they had made a mistake. After all, I am an archaeologist by training and have not been involved in conservation biology. I was told there was no mistake in the invitation, so I saved the date for the panel on my calendar and hoped I would have an idea by the time of the session.

As I read through the questions to prepare for the panel, I was struck by the central notion of extraction as a theme, resource extraction specifically. As a resource manager at Grand Canyon National Park (GCNP), conservation and preservation are fundamental missions for all our resources. Resource extraction is not central to the work of the National Park Service. However, extraction of water is a type of resource extraction I am familiar with so I decided to focus my comments on the effect of water extraction and Glen Canyon Dam for the panel discussion.

Grand Canyon National Park is deeply affected by the extraction and exploitation of the Colorado River as it passes through the Park. The Colorado River is a desert river; it went from being a seasonally fluctuating, sediment laden system to a highly controlled, clear and cold river, totally changing the ecology of the system. Glen Canyon Dam was authorized in 1956 and completed in 1963. Its primary purpose was flood control and water storage, but it has primarily been used as a cash register dam, with water releases timed for peak power generation. The National Park Service (NPS) has found itself trying to care for a world heritage site with plugs at each end and an altered ecosystem in between. Our challenge continues to be how to conserve all the resources along the Colorado River within NPS jurisdiction in a way that will preserve them for future generations.

The jurisdictional issues and cultural barriers that are paramount in this system have relevance for hydrological systems and their management across the Colorado Plateau. The water in the Colorado River was divided between the seven basin states in 1922 with the signing of the Colorado River Compact. The water was divided between these users based upon the wettest years on record (at that time), so the river water has always been over allocated. Never has the needs of the national park, or the resources dependent on the water, been considered. However, in 1992, Congress passed the Grand Canyon Protection Act, an act we thought finally put the resources of the Grand Canyon National Park on par with the water and power interests that dictated how the water flows through the canyon. The Act specifically told the Secretary of the Interior that Glen Canyon Dam was to be managed to improve the conditions for which Grand Canyon National Park and Glen Canyon National Recreation Area were established, including, but not limited to, natural and cultural resources and recreational use. The law of the river was not the only body of law dictating how the water was released from the dam; resource concerns also had to be taken into account.

Once the act was passed, we thought that the playing field was finally leveled and that the resources of the Park would be considered in the long-term management of Glen Canyon Dam. A 25-member stakeholder group was established to recommend management of the dam to the Secretary of the Interior. This group was made up of the 7 basin states, power and water users, environmental interests, recreational groups, and the federal agencies that have responsibilities for either the physical resources, the dam, or the power. The cultures of these different entities have become the institutional barriers to achieving an affective adaptive management program. The NPS has a singular mission: to conserve the resources for future generations. The Bureau of Reclamation has a singular mission: to operate the dam in the most efficient way possible, recognizing all of the laws regulating the Colorado River and the water storage and flood control mission of the original act authorizing the dam. Western Area Power Administration has a singular mission: to market the electricity generated by the Reclamation at the greatest economic benefit. The problem is that these three federal agencies, all deeply involved in the management of Glen Canyon Dam, have three very different missions, all of them righteous and potentially in conflict. The basin states have their missions too; for the upper basin states, they want to keep as much of their water for beneficial uses in the upper basin (above Glen Canyon Dam) for as long as possible. The lower basin states have their mission to receive their allocation of Colorado River water as soon as possible for the same beneficial use. There is an inherent struggle in how the water is used, by whom and when.

The NPS is a conservation agency. Our only mission is to conserve the resources of the park. The challenge we face is how best to conserve the resources when you have all the different entities deciding what is best to conserve their own interests, which don't seem to have anything to do with park resources and values.

One of the best examples of the challenge is represented by our interest in having periodic, special high flows released from the dam. A "beach-habitat-building flow" is a controlled high flow above power plant capacity intended to use sediment stored on the bed of the river to rebuild sandbars and backwater areas and put sand back on the beaches for a variety of resource benefits. There has been an amazing amount of science done in the Canyon since 1982, including the original Glen Canyon Environmental Studies and the current program run by the Grand Canyon Monitoring and Research Center, an office of the USGS. Tens of millions of dollars of scientific research has been done in the canyon to provide the policy makers information on how to best manage the system. Yet, with all this research, the resources of the canyon continue to degrade, huge silt banks and tamarisk invade the lower end of the canyon as the river reaches Lake Mead, and the decision about flow seems to rest more with politics than the conservation of the resources.

The NPS is committed to science-based decisions, yet the decision makers in this process are tied by legal constructs, in some cases created over 80 years ago at a time when conservation of resources or the benefits of the natural system were not a consideration. These institutional constructs keep us from being able to use the science to manage this river system. Conservation of the resource is a compelling story and we can hope that the information we provide will be used in the future to help make decisions about this world heritage resource. Some of you in the audience, especially younger conservation biologists just entering the field, may be helping us make these decisions, since this management challenge will likely be with us for some time to come.

James Catlin

We all wish to address the global crisis concerning the loss of biodiversity. As we review our collective work over past few decades, both as an international society and as a regional community here on the Colorado Plateau, we can identify many important successes. However, our collective effort has still, on the whole, failed to significantly slow (and has certainly not reversed) the overall trend of species loss. As I pondered the question, "where have we been; what's worked?" as I prepared for this panel today, I just kept going back to another problem that has been a major part of this conference —climate change.

Recognition of the climate change crisis has grown enormously in the past few years. Today most of the public and its leaders recognize this crisis and many are actively working to reverse the emissions that are a primary cause. Unfortunately, we have not seen a similar recognition and commitment concerning the biodiversity crisis. Why is this? Is there something that we might learn from the climate change campaign that would benefit our work on restoring and maintaining biodiversity?

The negative effects of climate change and its anthropogenic causes have been well known by scientists for decades. However, climate change is complex and hard to understand and teach to the general public. How can changing the temperature a few degrees be important? Scientists can go into great detail about the complexity of climate, and it is very easy to become lost in the jargon of climate change (such as measuring carbon dioxide concentrations in parts per million, the make-up of components of the atmosphere by weight, tons of CO2 emitted, etc.). With information like this, it is no wonder that the public had trouble grasping the enormity of the crisis on climate change.

But by the 1990s things began to change. The Intergovernmental Panel on Climate Change (IPCC) was formed. Scientists, including many government scientists like NASA's James Hansen, found the courage to go to the public about this crisis. And, importantly, scientists simplified the complexity of the issue by choosing one metric in order to make this crisis understandable and progress possible…they flatly stated that we need to limit the CO2 level in the atmosphere to 450 to 500 parts per million. In a nutshell, the scientific community had come together on climate change, developed a key focus and a common understandable goal for resolving this issue. Subsequently, governments have come to understand the climate change crisis and are now developing specific actions in order to meet this goal. Within the past decade there has been a major shift in the public attitude on climate change. More importantly, the public sees what they can do in their own actions to help address this issue.

We have not been able to achieve this level of commitment and change in order to address the biodiversity crisis. While our community may have a clear understanding of the loss of biodiversity and its causes, for the most part this understanding has not reached the general public and few elected officials see this as an issue requiring action.

We are at that moment today that climate scientists reached 20 years ago. We can now develop the focus and commitment to replicate what they did in order to address the biodiversity crisis. Here are a few of the lessons learned from the climate change campaign that may help us:

1. Scientists established an official, international panel under the United Nations to focus on this issue.

2. The results of this collaboration developed a common metric that simplified a very complex system.

3. Specific goals that address the climate change crisis were developed that are understandable to the general public.

4. The consequences of climate change to people's lives is widely publicized and generally understood.

5. Required changes in human actions are identified. These changes have precision that describes both the kind and amount of change needed to reach a goal.

6. An achievable plan on how to change human actions to achieve these goals was developed.

7. Elected officials, starting in local communities, adopted a plan to reduce climate-changing emissions. These plans call for radical changes in land use and transportation.

Today, this issue is on the short list of significant issues that elected officials need to address.

All of this amounts to a meaningful start, however, climate changing emissions are still on the rise. We have a long way still to go.

There is generally broad agreement among the scientific community that biodiversity faces a crisis. The Convention on Biological Diversity stated in April 2002 that the Parties to the Convention committed themselves to achieve by 2010 a significant reduction of the current rate of biodiversity loss at the global, regional and national level as a contribution to poverty alleviation and to the benefit of all life on Earth. While important, this policy leaves unclear how biodiversity will be restored or protected. Again, perhaps some sort of a simple measure is needed that can be applied at many scales to determine whether ecosystems are functioning adequately in order to sustain biodiversity. We have agreed on goals, such as to sustain populations of native species. However, we lack the system connection to identify the actual ecological conditions needed to sustain species. We also lack clear causal links between human actions to ecosystems needed to ensure sustainable populations.

Broad public understanding of the biodiversity loss crisis is substantially less than that of climate change. While almost all scientists are familiar with the term biodiversity, a relatively small percentage of the population is (Christie et al. 2005). Thus, in spite of significant work by a large number of scientists and nongovernmental organizations, this issue has not received much recognition in the policy-setting arena (Loreau et al. 2006). Without public understanding of this issue, policy makers will be slow or less able to act.

The IPCC was established by the United Nations in 1988 to "to assess on a comprehensive, objective, open and transparent basis the scientific, technical and socio-economic information relevant to understanding the scientific basis of risk of human-induced climate change, its potential impacts and options for adaptation and mitigation." With these words, scientists and policy makers began a process that has lead to the climate crisis being accepted by the public and policy makers as one of the most significant issues that society faces today. We must learn from this campaign in order to achieve the same level of acceptance and recognition of need for action for the biodiversity crisis.

Charles van Riper III

Our challenge today, as members of Panel I for the Society for Conservation Biology, Colorado Plateau Chapter, is to provide you with observed legacies of conservation biology issues on the Colorado Plateau. This includes our experiences of what has been successful, and what has not worked when addressing strategies for the protection of biodiversity over the Colorado Plateau. I have spent the past two decades dealing with conservation issues on the Plateau, functioning as a federal Department of Interior scientist within the university and government research environments. My conservation efforts have, of course, been influenced by where I have conducted research (primarily in national parks and on other DOI lands). In order to fully document these conservation efforts, we have produced a series of books that document the results of conservation research over the past 20 years on the Colorado Plateau (e.g., Rowlands et al. 1993, van Riper 1995, van Riper and Deshler 1997, van Riper and Stuart 1999, van Riper et al. 2001, van Riper and Cole 2004, van Riper and Mattson 2005, van Riper and Sogge 2008). Hopefully these products, along with the experiences that I will share with you today, will be portable and applicable to all of you who deal with the conservation of resources throughout the many federal, state, and private management units over the Colorado Plateau.

Prior to dealing with past conservation successes and failures, I believe that it is important to recognize several things about the Colorado Plateau. First, *from a national perspective, we do not have any conservation problems.* Although each of us in this room are cognizant of the multitude of conservation issues that presently exist over the Colorado Plateau, most people (including decision makers) view the "Plateau" as isolated, beautiful, and without major conservation problems. Second is that the geographic region of the Colorado Plateau encompasses portions of four states, thus four very different political entities. Third, we must recognize that the sparse human population over the Colorado

Plateau, although beneficial for conservation, does not result in many votes for political decision makers.

My presentation thus far has included heavy political overtones. One reason for this is because when I reflected on my past experiences in dealing with conservation "victories" and "failures" over the past several decades, the single unifying feature I could find was a political thread. Each conservation "success" that I have observed has been closely tied to strong political support at four levels (local, regional, state and federal), while efforts that have not resulted in adequately conserving resources seem often to lack support at one or more of those political levels. Adequate political support also seems to be closely tied to financial support, and that very often assures continuance and success of a conservation effort.

Now I will expand on this "political" theme, sharing with you several instances of what I believe have been Colorado Plateau conservation successes, and then several instances where I feel that conservation could have been better served by more fully involving people at all political levels. What I consider to be one of the greatest conservation successes on the Colorado Plateau over the last 20 years has been the formation of new Bureau of Land Management (BLM) National Monuments. This conservation success was achieved by local scientists and conservation advocates, working within the political system to secure these lands as national monuments. The process utilized Section 2 of the Act of June 8, 1906 (34 Stat. 225, 16 U.S.C. 431), which authorizes the President to declare by public proclamation historic landmarks, historic and prehistoric structures, and other objects of historic or scientific interest that are situated upon the lands owned or controlled by the Government of the United States to be national monuments, and to reserve as a part thereof parcels of land,… objects to be protected. Of key importance, in September of 1996 President Clinton issued Presidential Proclamation #6920, establishing the Grand Staircase Escalante National Monument. This proclamation resulted in a true conservation success, preserving almost two million acres on the Colorado Plateau. I argue that this was achieved principally through engaging necessary parties in the political process, and coupling that effort with sound scientific information.

Several other conservation successes that I have noted have also involved the creation of additional natural areas on the Colorado Plateau. One was the expansion of Petrified Forest National Park, when on 3 December 2004 the U.S. President signed a bill that authorized expanded boundaries, increasing park acreage from 93,533 to approximately 218,533 acres of land in Apache and Navajo Counties, Arizona. This was an increase of 125,000 acres, more than doubling the size of Petrified Forest National Park. This expansion conserved important paleontological resources, and was scientifically based, as the boundaries were set by known geological substrates of the Chinle and Bidahochi Formations. We have also seen a recent expansion to the boundaries of Walnut Canyon National Monument which preserved significant archeological resources, adding the "first" and "fifth" Anasazi Forts to that park.

I would also consider the development of several land-partnerships, or Community-Based Collaboratives (CBCs), over the past decade as real conservation success stories. The one CBC that created a benchmark, and is the present standard for the Colorado Plateau, is the Diablo Trust. A background on this collaborative can be found in previous chapters of the Colorado Plateau book series (e.g., Sisk et al. 1999, Loeser et al. 2001). Initially founded in 1993 by two ranches, the Bar-T-Bar and Flying M, the Diablo Trust CBC was created to link private and public values under one holistic goal: "to create sustainable rangeland management that maintains the tradition of working ranches and provides for economic viability while managing for ecosystem health." The focus area of this CBC is east of Flagstaff, Arizona, encompassing checker-boarded private and state lands, augmented with U.S. Forest Service summer grazing allotments. Collaborators of the Diablo Trust now include local ranchers, state and federal agencies, scientists, environmentalists, and other interested stakeholders.

Another very successful conservation collaborative, although not focused entirely on the plateau, has been a "grass-banking" effort by members of the Malpai borderlands ecosystem management project. This 800,000-acre planning area includes about 57% private land, 20% state trust lands, 11% National Forest, and 7% BLM administered land. Included in the Malpai Borderlands effort are a number of partners from the private sector, and all local state and federal land management agencies. The Malpai CBC grass-banking effort would certainly seem to be applicable to areas on the Colorado Plateau. In all CBC efforts, Tilt et al. (2008) point out that conservation success is often realized only after partners recognize three parameters. First, all collaborative processes face issues where the problem and its solution are poorly understood, where there are few scientific data and little understanding of what that information means, and where personnel and financial resources are small or nonexistent. Next, they point out that conflicting values confuse the process and innovation is often viewed as risky and expensive. Not only must collaborations bring together a diverse and representative group of stakeholders, but they must also recognize the amount of time, effort, and funding that is necessary for creating and sustaining a successful collaborative process. Finally, the authors point out the importance of gaining the trust of the stakeholders and outside interest groups, which is accomplished by maintaining an open and transparent process that incorporates research and monitoring protocols that will properly evaluate the CBC goals.

The final conservation success that I would like to deal with is the establishment of the Ecological Restoration Institute (ERI) at Northern Arizona University (NAU). Through the efforts of W. Wallace Covington, the ERI was formally established by the Arizona Board of Regents in 1997 and by federal legislation in 2004. Working at all political levels, Covington recognized a conservation need (reducing forest fuel loads) and educated political entities at the local, state, and national levels, to that need. This resulted in the ERI being funded by a combination of programmatic state and federal programs. At the national level, line-item funding introduced by Arizona Senator Kyl, through the BLM to NAU, assured continued success of this effort. The ERI is now nationally recognized for mobilizing the unique assets of a university to help solve the conservation problem of unnaturally severe wildfire and degraded forest health. The ERI has continued its success through working with and helping land management agencies and communities with comprehensive focused studies, monitoring and evaluation research, and technical support. This Colorado Plateau conservation success story serves as an example of the importance in educating people at all political levels.

There are several instances in the past where I feel that conservation on the Plateau could have been better served by more fully involving entities at all political levels. This is by no means meant to be a disparaging to any single effort, but is noted merely to point out where we might pay more attention to aspects of conservation efforts in the future. In 1958 the Flagstaff City Council passed the world's first lighting ordinance banning advertising search lights that were making it difficult for professional astronomers to do their work. This year (2008) marked the fiftieth anniversary of that historic event—the beginning of the dark skies movement that is now gaining attention across the planet. Unfortunately, the conservation of dark sky has not been embraced by all areas of the Colorado Plateau. Our region has one of the last true expanses of dark sky in the United States, yet the conservation of this resource has not been completely recognized at the local and regional levels.

The Land Use History of North America project (LUHNA) had a pilot project on the Colorado Plateau. Through the efforts of Thomas D. Sisk, working with John Grahame, the Center for Sustainable Environments at NAU created the COPL LUHNA web site *http://www.cpluhna.nau.edu*. This is still one of the premier locations for people to access information about the background of conservation issues and concerns over the Colorado Plateau. Unfortunately, due to political changes at the government level, funding for this effort ended. Without current and updated web-based information available to people at all four political levels of our country, Colorado Plateau conservation efforts will be hindered and greatly hampered by outdated information. At present, the bibliographic citations are not even available for this web site. Here is an example of local levels recognizing the importance of the conservation effort, but that importance not being recognized at the national level.

The National Park Service Inventory and Monitoring Program that was begun in the 1990s on the Colorado Plateau can certainly be viewed as a success (e.g., Persons et al. 2008). However, many of the other recent inventory and monitoring initiatives that we began to create information databases for (e.g., BLM, Native American Tribal Lands), have not fully integrated with other conservation efforts and thus have not achieved true conservation goals. Much of this rests with the fact that at all political levels, and especially in this case the local level, managers did not visualize benefits that the integrated information would have provided. As the Colorado Plateau landscape is further subdivided into smaller and smaller pieces, if we truly want to conserve resources we must begin managing lands on a much larger scale. No single land owner/manager, no matter how large the piece of land, is going to be able to alone conserve all resources.

The Southwest Strategy is another conservation effort that never really achieved its stated conservation goals. The overarching goal of the Southwest Strategy was to maintain and restore cultural, economic, and environmental quality of life in Arizona and New Mexico. The strategy attempted to address community development and natural resources conservation and management within the jurisdictions of all involved federal agencies, and in a manner that was scientifically based, legally defensible, and implementable. Why did the Southwest Strategy fail to meet their desired goals? I argue that it was due in large part to mid-level managers not embracing the effort, thus the gradual erosion of support. There also was never sufficient funding provided to the process, as was the case with the Colorado Plateau LUHNA effort described above.

In summary, there have been many conservation issues that have arisen over the past several decades on the Colorado Plateau. Some of the efforts have been successful, while others have not worked quite as well. I have provided herein examples of what I have seen work and situations where I believe that conservation efforts could have been improved. But, hopefully, all of these experiences will be useful in recognizing directions to move in helping to conserve resources throughout the many federal, state, and private management units over the Colorado. In all instances, we must recognize that, from a national perspective most people do not believe that we have any conservation problems on the Colorado Plateau. In working with conservation efforts we must also recognize that the geographic region of the Colorado Plateau encompasses portions of four states, thus four very different political entities. The lack of numerous high population centers on the Plateau, although a blessing, does not bode well when weighing political decisions for conservation funding initiatives. Each of the conservation "success" stories that I have addressed has been closely tied to strong political support at the four levels of government: the local, regional, state, and federal levels. Conversely, those efforts that I discussed that did not result in conserving a resource seem to have lacked political support at one or more of those four levels. Adequate political support also seems to be closely tied to financial support, and that very often assures continuance and success of a conservation effort. I hope that the material in this presentation will provide conservation practitioners and land managers with useful information and tools, and that this information can act as a stimulus for successful future conservation of cultural, natural, and physical resources over the Colorado Plateau.

PANEL 2 PRESENTATIONS
FUTURE OF CONSERVATION BIOLOGY ON THE COLORADO PLATEAU

David J. Mattson

If current projections are any guide, environmental crises on the Colorado Plateau will almost certainly attenuate, multiply, and worsen in ways both anticipated and surprising. Environmental professionals and leaders will be increasingly confronted with critical questions. What are my obligations, and to whom? What is my role? And assuming that answers to these questions are clear, how can I be most effective?

Questions of obligations and roles bespeak questions about overarching goals. Scholars and practitioners are increasingly invoking sustainability of ecosystems and societies as central concerns. Sustainability requires resilience in both natural and human systems; an ability to absorb shocks and adaptively respond in ways that maintain key structure and function. Accumulating evidence suggests that our ability to sustain healthy ecosystems cannot be divorced from our ability to sustain healthy human societies based on equitability, fairness, openness, and mutual respect. Despotic regimes that increasingly concentrate privilege and resources—power and wealth—in the hands of the few while depriving the many have a history of not only degrading humans, but also the natural environments we depend on.

If healthy ecosystems depend on healthy human societies, then sustenance of human dignity logically becomes the concern of all who are concerned about the future of the environment on the Colorado Plateau. If so, our obligations are first and foremost to civil society and the maintenance of democratic institutions. We are also obliged to lead from whatever situations we find ourselves in.

The requirements of individuals that arise from this diagnosis may be surprising to some: capacity for self-reflection and self-knowledge; a conscious rather than opaquely acculturated ethical stance; a willingness to take calculative risks; an ability to constructively engage with others based on empathy; and skill at orienting efficiently and effectively to complex situations. I feel safe predicting that we, as a society, will much more effectively address our environmental challenges if there are more rather than fewer of us who are like this.

Ultimately, changing individual character depends in some measure on changing current paramount societal myths. If sustainability is our goal, we cannot afford to perpetuate doctrines that extol materialism and collapse individual accounting to considerations of immediate gain. Responsibility and accountability to others near and far, in space and time, need to attain a cultural status equal to that currently afforded individual rights and privileges. Our "accounting space" needs to be broad rather than narrow—a figurative tent that shelters the many rather than the few.

If culture change, enhancement of civil society, and maintenance of democratic institutions are key to addressing environmental challenges, then what role have conservation biologists and other environmental professionals to play, and how can they be most effective? For one, a paramount focus of inquiry becomes humans and human systems of governance. Information about ecosystems is important, but clearly not sufficient, especially when responses to environmental crises are seminally about people's values, attitudes, and behaviors. The history of natural resources management clearly shows that shoveling more science about biological systems at unresponsive people has little or no effect.

As a corollary, environmental professionals will likely need to embrace a new paradigm of practice if they are to be effective. This new stance prospectively entails abandoning expertise-based prerogatives and the investitures of power that have been institutionalized around them. This is no small prospect given that virtually all our current natural resource management bureaus have accrued power on the basis of doctrines that give pride of place to scientific managers with the skills purportedly needed to identify and solve our societal problems. Yet, problems are subjective constructions arising from peoples' desires and, increasingly, citizens are calling back to themselves the authority to work together to directly negotiate their interests. The hope is they produce durable common interest outcomes. In such a world, the "expert" becomes a co-equal participant, who recognizes his or her values, respects others and their skills, and is hopefully respected in turn both as an individual and for the skills that he or she brings.

Thomas L. Fleischner

In conservation biology, "conservation" comes first. Conservation biology represents constantly shifting sets of interactions between three realms: values, policy, and science.

We tend to focus too much on the science piece, but the real action and the real challenges—and where success will ultimately lie—are in the values and policy pieces of our discipline. Consequently, as we look to the future of conservation biology on the Colorado Plateau, it is important for conservation biologists to take their citizenship seriously—to understand the policies and processes, and to know the people involved, in addition to knowing places and species of concern.

The work of conservation biology should be rooted in love. Conservation then remains a personal, practical reality, not a set of academic abstractions. That's why a priority for me is to try to instill a sense of love of critters, ecology, and place in my students on the many field courses I teach. I feel it myself, after all these years, and so hopefully my own excitement is conveyed to my students, and they "catch it." This leads to strong passion for the places they end up studying and caring for, and supports the growth of these students into well-rounded conservationists, scientists, and teachers. Love is ultimately a more effective motivator than fear.

As pointed out in the first panel discussion, many of us in the Society have bemoaned what appears to be the demise of natural history in conservation biology. I too have written on this topic and I personally define natural history as "a practice of intentional, focused attentiveness and receptivity to the more-than-human world, guided by honesty and accuracy" (Fleischner, 2001 and 2005). I believe a firm grounding in natural history is essential to the practice of conservation biology, for both practical and profound reasons. Practically, natural history undergirds all good conservation work: how can we know what to save if we don't know where it is, how many exist, etc.? But on a more profound personal level, natural history continually recharges the conservation biologist's passion. Natural history is our "secret weapon"—a constant reinfusion of affirmation that helps us deflect despair.

In this age of climate change a sense of hope and empowerment is ever more critical to our long-term persistence in this work. As writer Scott Russell Sanders has pointed out, the roots of the word "hope" mean, essentially, "to leap up in expectation." So what is reasonable for us to expect? That loving the world is always a worthy endeavor. That this world will go on, in some inventive fashion. That our imaginations, coupled with our hard work, will make a difference. To quote Norman Cousins: "We don't know enough to be pessimistic." And as we put forth our best efforts, we must support our colleagues more than ever. A sense of network and support among colleagues buoys up our collective sense of hope. Lastly, we need to reach out to new and diverse, maybe even unlikely, partners. We should always seek out more voices—both human and more-than-human. We need to be vigilant about trying to represent the underrepresented.

Ed Grumbine

Any chance of conserving biodiversity on the Colorado Plateau now and into the future depends on working through two paradoxes. The first paradox is that conservation scientists, despite their professional training to do otherwise, need to advocate more actively for nature. Put simply, if we want to have success we must do more to match "conservation" with "biology." The second paradox is that in addition to the conservation work we do locally, we must pay more attention to events beyond our home region. We need to think globally and act globally. Both of these paradoxes are fed by the carbon emission trajectories of China, India, and the rest of the rapidly developing world. As these countries (and others) follow our own poor behavior around fossil fuels, global emission rates threaten to swamp any particular conservation action on the Colorado Plateau.

To resolve the first paradox, we need to recalibrate the ratio of conservation to science in our work. Gone are the days when data collection, interpretation, and publication in a science journal are our primary work. These activities remain necessary, of course, but they are no longer sufficient given the scale of pending climate change impacts on Plateau ecosystems. For example, our best science suggests that there will likely be 30% less water available in our region well before 2100. How will this affect instream flow for wildlife and what is left of riparian ecosystems? Under conditions of scarcity, will water-strapped humans leave enough for their biological kin? Publishing papers that few people read is not the answer to this problem.

To work on the conservation versus science paradox, we should take these and other climate trend data to the public and engage our fellow citizens in dialogue. Who else, after all, is in a better position to do this? At the local level, we must ask candidates for office at all political levels in 2008 to address climate change. Then we need to support those candidates with the strongest carbon-neutral positions, and encourage others to do likewise.

But our influence on the future ecological health of the Plateau does not rest with local conservation action and politics. The second paradox we face is that our best efforts will come to naught if the rest of the world continues to emulate the United States by expanding their carbon footprint. The news is unequivocal. Over the next twenty years, China alone will swamp all carbon reductions from the rest of the world if they continue to burn coal as planned. And it would be difficult to deny development to a country with a per capita carbon footprint six times less than ours. So the second paradox asks us to move beyond local conservation action to active advocacy at the federal and international levels. Until I pondered the data, I never would have imagined that China is where conservation on the Colorado Plateau will be won or lost over the next few decades (Grumbine 2007).

So how does a conservation biologist on the Colorado Plateau "act globally"? Climate science projections combine with American election cycles to yield a clue—the next U.S. President will need to engage the international community on carbon reductions in her or his first term or the window of opportunity to address this issue may close. Conservation biologists must do everything possible to help elect the U.S. presidential candidate with the strongest carbon reduction platform and the leadership skills to implement it.

Ethan Aumack

Conservation challenges across the Colorado Plateau have been described at length and are not worth relisting here. It is worth repeating, however, that our challenges are growing in scale, complexity, and consequence due to inaction and/or fragmented advocacy efforts. Climate change, rapid urban and exurban growth, increasing demand for energy and water development, and ever-declining land management budgets threaten to exacerbate protection and restoration challenges from the lowest elevation desert grasslands to alpine tundra.

Given the degree to which twenty-first century conservation challenges such as development-induced habitat fragmentation, invasive non-native species invasion and proliferating unnaturally severe wildfires are occurring at landscape scales, the need for landscape-scaled conservation science is clear. Given the need to understand future implications of current policy choices, the need for predictive science is clear. Remote sensing and spatially explicit predictive modeling approaches offer great promise for identifying optimal landscape-scale protection, restoration, and monitoring approaches.

Landscape-scale predictive conservation science is hardly simple. Given the overall methodological complexity surrounding such science, combined with the rapid pace at which modeling approaches are tested and refined, the capacity for conducting cutting-edge science of this sort is likely to reside in a relatively few academic institutions and offices of government.

Highly technical science, once cloistered, diminishes in its capacity to affect change at the rate needed to address pressing conservation challenges. Conservation scientists must ambitiously reach beyond the ivory tower into the uncomfortable realms of advocacy to ensure the credibility, legitimacy, and relevance of their work. Conservation advocates must similarly reach into the often uncomfortable realms of rigorous science to re-build credibility with the general public.

Language, behaviors, and social norms vary dramatically between the worlds of science, policy, and politics. It isn't surprising then, that individuals (and even individual organizations) capable of pushing forward policy and political discussions centered around rigorous and relevant conservation science are few and far between. Effective conservation efforts in the twenty-first century should, then, be focused around building networks of practitioners representing academia, government, NGOs, and affected communities. Such networks will be collectively conversant in science, policy and politics such that they can, in a coordinated fashion, define issue-relevant assessment, research, and monitoring agendas, build science foundations of unquestioned rigor, interpret scientific findings in a policy relevant context, and push in a coordinated and scientifically responsible fashion for ambitious social change.

This network-based, science and policy-grounded conservation advocacy strategy offers promise for bridging the gap that seems to have grown between rigorous conservation science and conservation advocacy. It offers promise as a platform for reaching across the diversity of constituencies and communities that must be involved in conservation discourse if that discourse is to be effective and have long-standing impact. It provides a bulwark against those with narrow, short-term, and often exploitative self-interests that have gained substantial ground through efficient use of wedge tactics. Most importantly, it promises to generate a sense of humility, integrity, and shared purpose which are so essential if we as a conservation community are to affect change during a time of dire need.

Jayne Belnap

The environmental issues that have faced us in the past have often seemed daunting. As time goes on and the human demand for natural resources grows with the increase in human populations, these issues will become ever more numerous and pressing. Working on environmental issues can be very disheartening, as it seems that we most often lose our battles and when we lose, it is forever. In contrast, when we win, it seems that it is only temporary. There are only a limited number of people willing to engage in environmental issues and we need everybody. We cannot afford to lose people due to disillusionment. One of the questions asked of our panel was this, "How can we make our efforts more effective and, mostly importantly, keep going in the face of constant disappointment?"

In my twenties and thirties, I truly believed that there was a universal right and wrong, that conserving natural resources was the most important goal one could strive for, and that science could show the best way to reach this goal. Based on this greatly overly-simplified belief system, I decided my strongest contribution to the larger world would be that of uncovering scientific truths on environmentally contentious issues and then communicating my results to land managers and the public in as many ways as possible. I thought that if I just showed people the data on which I based my decisions, they too would understand the solutions to the problems in the same way I did.

Over the past 25 years, I have (slowly) come to understand that the above assumptions were not valid. I can no longer find black and white right and wrongs, only a hundred shades of gray. I have found that focusing on resource conservation is not enough—that we need to find a way to balance human needs with resource conservation, and doing that may require redefining "need" in the minds of almost everyone, including myself. I have found that scientific facts are only a small part of the story. Human lives and decisions are far more complicated than can be addressed in a scientific study, and the data from those studies are only useful in decision-making once they are placed in the social and cultural contexts in which the decisions are being made. Other people may have the same facts, yet come to a different conclusion regarding what actions should, or should not, be taken. We need to learn to listen and allow ourselves to be taught about these other perspectives. This adds immense complexity to the decision-making process, as we cannot assume our values are universal nor can we assume that science is giving us all the answers we need. I believe that if we can really understand and accept this reality, our actions will be far more effective and many relationships based on conflict will turn into cooperative efforts.

Our effectiveness also depends on our ability to increase the awareness of those around us about the issues we face together. Educating our society on the current thinking of how ecosystems function, or how a specific species may be imperiled, is essential to this effort. However, we need to avoid over-simplification; instead, we need to discuss the complexity surrounding our understanding, such that we give the public the tools to integrate and process the seemingly contradictory information that they will receive. We need to tell the whole story, not just our side of it, and present both sides fairly, avoiding embellishment of our views. In the western United States, being effective requires an additional step. As the federal government owns much of the land in this region, we also need to seek out and establish working relationships with federal resource staff and land managers. This is a dedicated, interested, and overworked group of people with a huge task before them. They need help in addressing the many issues they face daily.

And lastly, how do we deal with the many disappointments we all face while working in the environmental arena? This is probably the most important issue each of us will have to address. Many people give up along the way because it is just too discouraging to be confronted with repeated failure on issues that evoke such strong feelings. I spent many years on this roller coaster: incredible joy when something went my way and deep depression when it did not. At some point, it became clear that I could not continue in this way. I then discovered advice that had been around for thousands of years: Do what you do because it is your passion, because you couldn't imagine not doing it, because it makes you want to get up in the morning. Take action because you believe in the action, not because you think it will attain a certain outcome. Because there is no way you or anyone else can know the ultimate outcome of a given act, do the act because it is what you feel compelled by your passion to do. It may be years before the outcome becomes manifest, and you may never see the result. In addition, the outcome we are attached to may not be the best outcome that is possible. So do what you do because you think it is right and then let go. If you hang on, your life will be a roller coaster of extreme ups and downs, with little peace of mind along the way.

Basically, this means giving up the concept of hope. I don't mean in a negative way, I mean going beyond hope. Hope is a funny thing; it is something we often use to keep up our optimism. But hoping for something means staying attached to a specific outcome, as we "hope" things will go our way. It keeps one living in the future rather than the present, as you "hope" something will be different than it is now. And it implies powerlessness; if you had power, you would not just hope for the change, you would make it happen. These downsides to hope, especially the way it keeps us attached to outcome, need to be closely examined in our search for better ways to do what we do. And there is a powerful consequence to giving up hope in this way; the equanimity that comes with it can help quell the animosity that so often characterizes the interactions among people working on different sides of an environmental issue.

Maybe we won't save the world, but that does not mean we should give up trying. And even if we can't save the world, we can make a huge difference in the quality of life for all beings, including ourselves, during our time on this planet.

HIGHLIGHTS OF AUDIENCE DISCUSSION
The Interface of Science and Politics

As we moved from panelist presentations into audience discussion, the boundaries between politics and science was the first discussion topic to be broached. We conservation biologists cannot be naïve about, or let ourselves be caught off guard by, the role of politics in affecting conservation policy, and even conservation research, on the Colorado Plateau. Politics affect who gets hired, what areas and issues become management priorities, and what gets funded. Some of us have experienced situations where science blends into politics to such a degree that sometimes there is actually not a firm line between one and the other. It is a reality that we must all operate within. And it probably will not change significantly with a new administration.

One of the upshots of science performed in a political environment is "competing science." This is not a phenomenon unique to conservation biology. An example is a suite of studies that show that livestock grazing can have negative ecological impacts in southwestern shrublands, and another suite of studies that indicate that livestock grazing can have positive effects for some aspects of southwestern shrubland systems. These conflicting studies are trotted out by the advocates for or against livestock grazing, and land managers are asked to essentially pick sides. The ex-governor of Utah, Mike Levitt, once said, "Before I trust the science, I want to know who's science it is." This conveys that politicians and land managers are familiar with being put in the position of "picking sides." Unfortunately often we, the scientists, have helped to bring this situation about. The unfortunate result of this can be estrangement between the scientists who performed the disparate studies, and even between those scientists who choose to cite them. The audience concluded that the future of conservation biology on the Plateau will be best served by doing our best to not contribute to the phenomenon of "competing science."

Interfacing with the Public

Another prevalent discussion topic was built around the question of whether Colorado Plateau conservation biologists and practitioners are interacting in the most productive way possible with the public. The public can be the best ally for conservation and good science. Having a hundred constituents approach a legislator or a hundred written comments on a proposed project are more effective than one scientific report on the same issue.

At the same time, we, as scientists, can all do a better job both getting our good science into the hands of those citizen-activists and motivating both scientists and non-scientists to become activists. The key in both cases is getting people excited about science and natural resource issues. Probably, most of us are not doing enough of that. We now have a strong new motivator…climate change. The Colorado Plateau and the Southwest are predicted by the IPCC to experience drier and warmer conditions than almost anywhere else on the planet. If we can get the people of the Plateau concerned enough to get engaged, angry even, and demand change of our governments, we can help guide messaging towards better conservation policy and water management, in addition to the obvious need for renewable energy portfolios in all the Colorado Plateau states.

Interfacing with the Agencies

Perhaps no thread was more common to the panelists' presentations than the need to work directly, and productively, with a host of committed and diverse stakeholders to solve conservation problems…especially land management agencies. Among the group of conservation biologists assembled for this panel presentation and discussion, we can see many different ways that we collaborate with and affect the land management agencies. It can, and needs to be, a two-way street. We see examples of agencies reaching out to individual scientists in academia, and science-based agencies such as The Nature Conservancy, asking for help to inform and improve agency analyses and management. And there are certainly examples of independent scientists reaching out to the agencies, often with other stakeholders, to join Community-Based Collaboratives. These sort of agency-scientist partnerships have grown and matured in the last couple of decades, including on the Colorado Plateau. In a new administration we may enjoy even more productive relationships.

As we approach a new conservation future on the Colorado Plateau, we will sometimes find that the agencies are not even aware that they need scientific assistance. An example is panelist Jim Catlin's work with the Bureau of Land Management in Utah. Jim discovered that the BLM was not using scientifically robust methods to assess ecosystem health, in particular, methods used to set cattle-stocking levels on public lands. These methods are generally not ecologically based, but are a throwback to earlier times when livestock grazing was seen as more of a "sustainable agriculture" pursuit. We have learned from this example that we may need to ask the agencies how we can help them use science to assure they are effectively meeting their conservation mandates. In the example above, we could really use independent scientific validation of some of the methods that have become entrenched in agency management and monitoring. Agency protocols should be driven by strong science and a sociallycollaborative framework. The question is simple: Can we help managers ascertain whether those practices and methods accepted by managers are actually helping agencies reach their ecological goals? There are now indications that agency staff "closer to the ground," who are actually conducting agency monitoring, would welcome this sort of method testing from us.

Tribal Relations

One reason the Colorado Plateau is a unique place for conservation biologists to work is the large presence of native tribes on the Plateau. The audience and panel asked themselves, what has the conservation community learned over the years from the tribal biologists, and what lessons have we learned about working effectively with the tribes?

We have learned is that it is often difficult to gain the trust of the tribal leaders and elders. This could be partly because of the higher rates of staff turnover in non-tribal agencies. Although tribal leaders and tribal resource staff tend to stay in one place, we cannot say the same for the non-tribal agency resource people, supervisors, and superintendents; they are constantly turning over. Another thing we've learned is that if you help a tribe with some problem that they need assistance with, they are far more likely to trust and work with you on your conservation priority. Therefore, the challenge is to develop continuity and generosity in our tribal partnerships.

CONCLUSIONS

The Colorado Plateau Chapter of the Society for Conservation Biology's panel presentation and discussion at the 2007 Biennial Conference recharged, invigorated, and unified the participants as we embark on a new conservation future on the Colorado Plateau. We hope it does the same for readers.

We have learned many valuable lessons from our collective decades of experience in this amazing and biologically diverse part of the Southwest. We have experienced the benefits of "inviting the skeptics" to join in, and critique, our work. We now see the campaign to reduce emissions and combat climate change resonate with the American public like no recent other environmental initiative. We could achieve much by applying this successful model to a campaign to save biodiversity in the face of a changing climate.

We have learned that our best successes come from earnest work with diverse groups of stakeholders. This group should include all levels of government, from local to national, to really have hope for success. These collaborative efforts cannot just pay lip service to compromise. If we truly understand and empathize with the interests of other stakeholders, express our preferences in terms of our interest in conservation rather than a "bottom line" outcome, and seek solutions that meet everyone's interests, we can create the most important ingredient for conservation success—political capital. As the Colorado Plateau has a meager population base to rally around our cause, it is even more critical that we achieve this capital through genuine and productive relationships with the land management agencies and land users.

With these lessons in hand, we now look to the future of conservation biology on the Colorado Plateau. With the help of impressive technological advances in remote sensing and modeling, we can create the best predictive models yet for glimpses of future resource conditions under different management scenarios. It will be up to us to help guide these uses to ensure the best future for biodiversity on the Plateau. And as we do this, we should draw not only on hope, but perhaps more importantly on simple love of nature and natural history. This is what can also give us courage to be not only scientists, but advocates. We know from experience that if done with care we can be both. After all, science needs advocates as much as advocates need science. Perhaps above all, we should keep this in mind as we go forward.

LITERATURE CITED

Alagona, P. S. 2008. Credibility. Conservation Biology 22:1365–1367.

Beier, P. 1995. Dispersal of juvenile cougars in fragmented habitat. Journal of Wildlife Management 59:228–237.

Beier, P. 1993. Determining minimum habitat areas and corridors for cougars. Conservation Biology 7:94–108.

Center for Biodiversity and Conservation, American Museum of Natural History. 1998. Louis Harris and Associates survey: Biodiversity in the Next Millennium. http://cbc.amnh.org/crisis/crisis.html, last accessed 4/30/09.

Christie, M., M. Hanley, J. Warren, K. Murphy, R. Wright, and T. Hyde. 2005. Valuing the diversity of biodiversity. Ecological Economics 58:304–317

Fleischner, T. L. 2001. Natural history and the spiral of offering. Wild Earth 11:10–13.

Fleischner, T. L. 2005. Natural history and the deep roots of resource management. Natural Resources Journal 45:1–13

Grahame, J. D., and T. D. Sisk (editors). 2002. Canyons, cultures and environmental change: An introduction to the land-use history of the Colorado Plateau. http://www.cpluhna.nau.edu, last accessed 12/31/08.

Grumbine, R. E. 2007. China's emergence and the prospects for global sustainability. BioScience 57:249–255

Loeser, M. R., T. D. Sisk, T. E. Crews, K. Olsen, C. Moran, and C. Hudenko. 2001. Reframing the grazing debate: Evaluating ecological sustainability and bioregional food production. In Proceedings of the Fifth Conference of Research on the Colorado Plateau, edited by C. van Riper III, K. A. Thomas, and M. A. Stuart, pp. 3–18. USGS Forest and Rangeland Ecosystem Science Center, Flagstaff, Arizona.

Loreau, M., A. Oteng-Yeboah, M. Arroyo, D. Babin, R. Barbault, M. Donoghue, M. Gadgil, C. Häuser, C. Heip, A. Larigauderie, K. Ma, G. Mace, H. A. Mooney, C. Perrings, P. Raven, J. Sarukhan, P. Schei, R. J. Scholes, and R. T. Watson. 2006. Diversity without representation. Nature 442:245–246.

Persons, T. B., E. M. Nowak, and D. G. Mikesic. 2008. Overview of Herpetofauna inventories in southern Colorado Plateau national parks. In The Colorado Plateau III: Integrating research and resources management for effective conservation, edited by C. van Riper III and M. K. Sogge, pp. 197–218. The University of Arizona Press, Tucson.

Rowlands, P. G., C. van Riper III, and M. K. Sogge, editors. 1993. Proceedings of the First Biennial Conference on Research in Colorado Plateau National Parks. U.S. Department of the Interior, Washington, D.C.

Sisk, T. D. and J. Palumbo. 2005. Collaborative science: making research a participatory endeavor for solving environmental challenges. The Quivira Coalition 7(1):22–27.

Sisk, T. D., T. E. Crews, R. T. Eisfeldt, M. King, and E. Stanley. 1999. Assessing impacts of alternative livestock management practices: Raging debates and a role for science. In Proceedings of the Fourth Biennial Conference of Research on the Colorado Plateau, edited by C. van Riper III, and M. A. Stuart, pp. 89–103. USGS Forest and Rangeland Ecosystem Science Center, Flagstaff, Arizona.

Tilt, W., C. Conley, M. James, J. Lynn, T. Munoz-Erickson, and P. Warren. 2008. Creating successful collaborations in the West: Lessons from the field. In The Colorado Plateau III: Integrating research and resources management for effective conservation, pp.1–22. The University of Arizona Press, Tucson.

van Riper, C. III, editor. 1995. Proceedings of the Second Biennial Conference on Research in Colorado Plateau National Parks. NPS Transaction and Proceedings Series NPS/NRNAU/NRTP-95/11.

van Riper, C. III., and K. A. Cole, editors. 2004. The Colorado Plateau: Cultural, Biological, and Physical Research. University of Arizona Press, Tucson.

van Riper, C. III, and E. T. Deshler, editors. 1997. Proceedings of the Third Biennial Conference of Research on the Colorado Plateau. NPS Transactions and Proceedings Series NPS/NRNAU/NRTP-97/12. Denver, Colorado.

van Riper, C. III., and D. J. Mattson, editors. 2005. The Colorado Plateau II: Biophysical, Socioeconomic and Cultural Research. University of Arizona Press, Tucson.

van Riper, C. III., and M. K. Sogge, editors. 2008. The Colorado Plateau III: Integrating Research and Resources Management for Effective Conservation. The University of Arizona Press, Tucson.

van Riper, C. III, and M. A. Stuart, editors. 1999. Proceedings of the Fourth Biennial Conference of Research on the Colorado Plateau. USGS Forest and Rangeland Ecosystem Science Center, Flagstaff, Arizona.

van Riper, C. III, K. A. Thomas, and M. A. Stuart, editors. 2001. Proceedings of the Fifth Conference of Research on the Colorado Plateau. USGS Forest and Rangeland Ecosystem Science Center, Flagstaff, Arizona.

DOWNSCALING CLIMATE PROJECTIONS IN TOPOGRAPHICALLY DIVERSE LANDSCAPES OF THE COLORADO PLATEAU IN THE ARID SOUTHWESTERN UNITED STATES

Gregg M. Garfin, Jon K. Eischeid, Melanie T. Lenart, Kenneth L. Cole, Kirsten Ironside, and Neil Cobb

ABSTRACT

Global Climate Models (GCMs) operate at scales much larger than the federal and state forest, range and riparian ecosystems managed by land and water professionals. Therefore, incorporating information on climate change projections into resource management plans requires GCM projections at a scale more relevant to ecosystems, especially in topographically diverse regions such as the western United States. In an effort to address this need, we developed downscaled climate projections for the Southern Colorado Plateau (SCP) (35° to 38°N, 114° to 107°W), centered on the Four Corners states. We compared twenty-two global climate models (GCMs) from the archive of model runs used in the Intergovernmental Panel on Climate Change Fourth Assessment Report, and statistically downscaled them to a 4 km grid, to accord with spatially and temporally continuous historic observations using the Parameter-elevation Regressions on Independent Slopes Model (PRISM) data set. From these results, we selected five models representing a range of plausible possible climate futures. We consider them in the context of three seasonal time frames observed to be critical for vegetation in the SCP: winter (November–March), arid foresummer (May–June), and summer monsoon (July–September). Projections for the SCP describe a warmer future, in which annual temperatures seem likely to increase by 1.5° to 3.6°C by mid-century, and 2.5° to 5.4°C by the end of the century, depending on the model chosen. Annual temperatures are projected to exceed the 1950–1999 range of variability by the 2030s. Annual precipitation changes are more equivocal. A conservative estimate, using a 22-model ensemble average, indicates that SCP annual precipitation may decrease by 6% by the end of the century. For precipitation projections, GCM agreement is greatest for the May–June arid foresummer season, and projections show SCP May–June precipitation declining by 11 to 45% during the twenty-first century. Downscaled output from this study will be used to drive vegetation change models with the intent of examining a diversity of possible outcomes so scientists can test an array of vegetation changes, and resource managers can make informed decisions in relation to the range of possible climate change scenarios.

INTRODUCTION

This paper describes the process and results of Global Climate Model (GCM) selection and statistical downscaling of climate parameters for use by ecologists examining vegetation change in the Southern Colorado Plateau (Figure 1). The downscaled climate data will eventually be used as input into

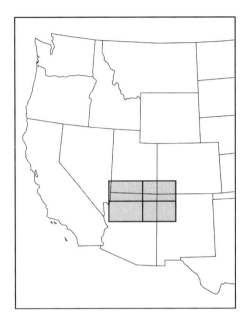

Figure 1. Western United States and Southern Colorado Plateau (shaded box) domains used in this study.

process-based, landscape-scale vegetation models. The impetus for the study is scientific and public interest regarding the recent rapid and massive forest mortality on the Southern Colorado Plateau (SCP), which affected nearly 1.4 million hectares of forest land between 2002 and 2004 (USDA 2008). Given coarse spatial scale projections of a warmer drier Southwest (Seager et al. 2007), there is concern that such changes may presage future ecosystem changes in the region. Research on pinyon pine mortality in the SCP suggests that a combination of drought and unusually high temperatures depleted soil moisture to a greater extent during a recent drought episode than during past drought episodes, thus exposing trees to "global-change-style" drought stress (Adams et al. 2009, Breshears et al. 2005). Moreover, researchers speculate that increasing temperatures also enhance insect life cycles and predispose Southwestern forests to greater risk of massive mortality in response to drought (Stephenson et al. 2006, Burkett et al. 2005, Logan et al. 2003).

Global Climate Model (GCM) grid cells, the areal unit at which parameters are projected by the models, are often larger than 1 degree latitude and longitude per side (~100 km in mid-latitudes). In a region as topographically diverse as the Colorado Plateau, having a single value for a parameter such as precipitation makes it challenging for resource managers to consider how temperature and precipitation changes will impact the considerably smaller than GCM grid cell-sized forest, range, and riparian ecosystems that they manage. For example, the dramatic topography of the San Francisco Peaks of northern Arizona would take up only a small fraction of a 1°x1° grid cell, and their effect on local climate and hydrology would be greatly attenuated. Although statistically downscaling GCM projections to finer spatial scales cannot remove so-called epistemic uncertainties (e.g., imperfect knowledge of some climate system processes), improving the spatial scale of GCM estimates to the point where the effects of regional elevation can be approximated should help improve estimates. The addition of such detail in model projections could provide guidance needed by land managers to discern projected temperature differences between mountain ranges and adjacent rangelands, and to better evaluate management strategies and climate change adaptation options (e.g., Bachelet et al. 2003). While modelers work to improve spatial resolution and dynamical processes within the next generation of models, relatively quick and cost-effective statistical downscaling efforts can allow scientists and managers to evaluate more finely detailed and plausible potential impacts based on the array of current generation GCM projections.

Climate has long been known to be important in determining the distribution of native plants on the landscape. The physiological adaptations of individual

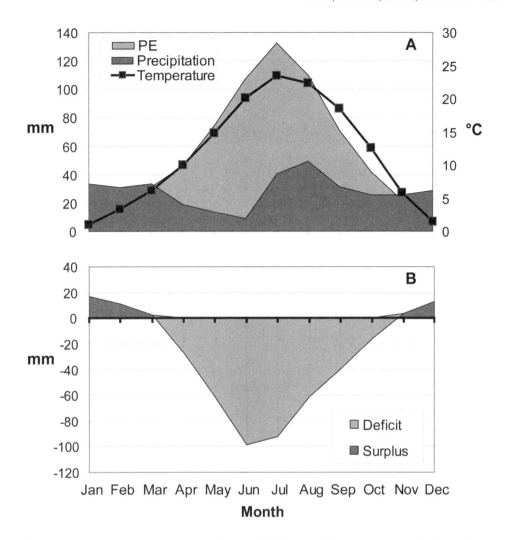

Figure 2. Southern Colorado Plateau moisture balance. (A) Mean monthly temperature, precipitation, and potential evapotranspiration (PE) for 1950–1999. (B) Precipitation minus potential evapotranspiration. Potential evapotranspiration is calculated using Hamon's method (Hamon 1961). Data: PRISM 4 km (Daly et al. 1994); PE calculations provided by Andrew Ellis, Arizona State University.

plant species allow them to take advantage of seasonal patterns in available moisture. Topographic diversity within northern Arizona creates several vegetation life zones within the region. Temperature and precipitation can vary greatly from the lower regions of the Grand Canyon to the top of the San Francisco Peaks, but the seasonality of these parameters is the same, regardless of altitude, slope, and aspect. Annual precipitation follows a bimodal distribution, split between large, spatially coherent, winter frontal storms and isolated summer monsoon convective storms.

Moisture surplus in the SCP typically occurs from November to March (Figure 2). April reflects a turning point when monthly mean temperatures rise to a point where the growing season typically begins even as precipitation decreases. By May and June, temperatures rise high enough to affect water vapor pressure. The typically sparse rainfall does little to offset high levels of potential evapotranspiration (PE). This period of time can cause severe vegetation stress and even mortality in seedlings lacking an established root system and, during extreme years, in established perennial species (Breshears et al. 2005). Many of the perennial bunch grasses common to northern Arizona remain semi-dormant during this time period. For woody plant recruitment, this is often a critical period; ponderosa pine (*Pinus ponderosa*), the dominant tree in the region, has been shown to have had cohort events occurring during anomalously cool and wet May–June periods (Savage et al. 1996). Conversely, ponderosa forests are most susceptible to damage from infestations of bark-beetle (primarily *Dendroctonus* spp. and *Ips* spp.) and wildfires when these pre-monsoon months are anomolously dry (Adams et al. 2009).

On average, Southern Colorado Plateau summer monsoon precipitation begins in mid-July (Higgins et al. 1999). Although it occurs during a time when PE and average monthly temperature are at their annual peaks, it typically brings enough precipitation to decrease the moisture deficit during this time of year (Figure 2). As average monthly temperature begins to decrease in August, monsoon precipitation reaches its regional apex, which further decreases the moisture deficit. By mid-September, monsoon precipitation typically decreases, but so does temperature. October, like April, is a transition period between annual moisture surplus and deficit.

In the rest of this chapter, we describe the process of GCM selection and statistical downscaling of climate parameters, and the implications of these results. In the introduction, we describe the GCM data, downscaling methods, and model-selection criteria. In the data and methods section we discuss the GCMs and ensemble averages selected, and we examine the spatial and temporal fidelity of the GCMs in reproducing the seasonal cycle of temperature and precipitation over the study domain, the Southern Colorado Plateau. In the section on the results of model selection and simulations of Colorado Plateau seasonal cycle, we discuss the GCM projections for the twenty-first century over the study domain. In the discussion section, we consider implications for colleagues interested in vegetation modeling and alternatives to our approach that may inform future studies. The final section contains a summary of major conclusions.

DATA AND METHODS

Historic Climate Records

To compare historic observations to GCM simulations, in order to correct model biases and implement downscaling algorithms, we used mean monthly temperature and monthly total precipitation data from the Parameter-elevation Regressions on Independent Slopes Model (PRISM) 4 km grid cell resolution dataset (Daly et al. 1994) (www.prism.oregonstate.edu). PRISM uses point data, spatial data sets, a knowledge base, and expert interaction to generate estimates of gridded monthly climatic parameters (Daly et al. 2001). A combination of linear regression and a series of rules, decisions, and calculations set weights for the station data entering the linear regression (Daly et al. 2002). The weighting function contains information about relationships between the climate field and geographic or meteorological factors. Weighting factors include measures such as elevation, distance from the predicted location, station clustering, vertical layer (to account for local inversions), topographic facet (to account for rainshadows), coastal proximity, and effective terrain weights (Daly et al. 2002). We compared PRISM estimates to simulations for the period

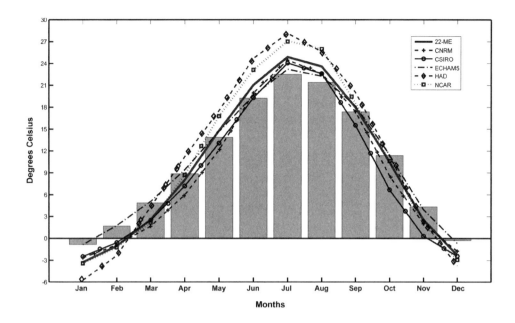

Figure 3. Southern Colorado Plateau observed (bars) and GCM simulated (lines) mean monthly temperature (°C), 1950–1999. See Table 1 for GCM acronyms. "22-ME" refers to the ensemble mean of 22 GCMs used in the IPCC Fourth Assessment Report data set (Meehl et al. 2007). Observed data: PRISM 4 km (Daly et al. 1994).

1895–2000 (used in Figures 3–9), and used 1950–1999 as the climatological average period for calculating anomalies. PRISM provides the most topographically precise, methodologically sound, quality-controlled historic climate data set available for century-long time scales; thus, PRISM is a robust choice for spatial and topographic concerns that underlie the requirements of vegetation models that will be served by downscaled GCM projections from this study.

Global Climate Model (GCM) Projections

We garnered climate model projections used in the Fourth Assessment Report (AR4) of the Intergovernmental Panel on Climate Change (IPCC) from the Program for Climate Model Diagnosis and Intercomparison (PCMDI) archive. Details on the models and their configurations are available at *http://wwwpcmdi.llnl.gov/ipcc/about ipcc.php*. These projections used coupled ocean-atmosphere models (AOGCMs) to simulate climate variations spanning the late nineteenth century to the end of the twenty-first century (see Meehl et al. 2007); this generation of models is referred to as Coupled Model Intercomparison Project version 3 (CMIP3). We analyzed, individually, 22 of these models (48 simulations; Table 1) that are forced with estimated greenhouse gas and aerosol changes from the late nineteenth century through 1999, and the IPCC Special Report on Emissions Scenarios (SRES) A1B scenario from 2000 to 2100. The A1B scenario describes a future world of rapid economic growth, with global population that peaks in mid-century then declines, and rapid introduction of new and more efficient technologies that are balanced such that no single source of energy is overly dominant (Nakicenovic et al. 2000). This scenario, sometimes referred to as "the medium non-mitigation scenario" (Moss et al. 2008), has

Table 1. GCM rankings for spatial correlation between GCM simulation and observed seasonal precipitation over land in the western United States, 1950–1999. Low rank indicates the best combined correlations with the annual cycle of precipitation. **Bold** font indicates GCM identifiers used in the text.

GCM	Modeling Center, Country	Rank
HADGEM1	Hadley Centre, UK	10
ECHAM5/MPI	Max Planck Institute, Germany	26
ECHAM4/MPI	Max Planck Institute, Germany	27
MRI-CGCM2.3.2	Meteorological Research Institute, Japan	29
GFDL-CM2.1	NOAA Geophysical Fluid Dynamics Lab, US	35
CGCM3.1	Canadian Centre for Climate Modelling & Analysis	37
GISS-EH	NASA Goddard Institute for Space Studies, US	37
HadCM3-UKMO	Hadley Centre, UK	38
MIROC3.2(medres)	Center for Climate System Research (The University of Tokyo), National Institute for Environmental Studies, and Frontier Research Center for Global Change (JAMSTEC), Japan	39
CNRM CM3	Météo-France / Centre National de Recherches Météorologiques, France	42
CSIRO MK3.0	CSIRO, Australia	42
MIROC 3.2(hires)	Center for Climate System Research (The University of Tokyo), National Institute for Environmental Studies, and Frontier Research Center for Global Change (JAMSTEC), Japan	44
MIUB ECHO-G	Meteorological Institute of the University of Bonn, Meteorological Research Institute of KMA, and Model and Data group.	50
NCAR CCSM3	National Center for Atmospheric Research, US	50
GISS-ER	NASA Goddard Institute for Space Studies, US	53
NCAR PCM1.0	National Center for Atmospheric Research, US	56
BCCR BCM2.0	Bjerknes Centre for Climate Research, Norway	58
GFDL-CM2.0	NOAA Geophysical Fluid Dynamics Lab, US	62
GISS-AOM	NASA Goddard Institute for Space Studies, US	63
IAP-FGOALS	Institute of Atmospheric Physics, People's Republc of China	67
ISPL-CM4	Institut Pierre Simon Laplace, France	71
INM-CM3.0	Institute for Numerical Mathematics, Russia	76

been favored for analyses in some situations where capturing the full range of scenario output may be too computationally intensive (*sensu* Seager et al. 2007). It is known by some as the "business as usual" scenario and, as such, is a reasonable choice of emission scenario for examining plausible futures. Further, all the scenarios yield similar output through about 2030. It is only after mid-century that the scenarios begin to diverge extensively. We analyzed total monthly precipitation and monthly mean temperature for each individual model and the 22-model ensemble mean (Hoerling et al. 2007).

Downscaling

The AR4 GCMs use a variety of grid resolutions, typically in the range of approximately 2.5° (~300 km in middle latitudes) per side of the grid box. The first step in the data treatment is to align the GCMs to a common grid, in this case to the same 4 km grid used by PRISM, using inverse distance weighting (Eischeid et al. 2000). Once the GCM estimates for each parameter have been re-gridded, we statistically downscale the GCM estimates using the method described in detail by Salathé (2005) and summarized in this section. The method uses the PRISM estimates to impose spatial structure to the GCM-simulated monthly precipitation and temperature, while preserving the atmospheric processes driving the simulations. As mentioned by Salathé (2005), Widmann et al. (2003) used a similar method, referred to as "local scaling."

To remove the bias between the large-scale simulated climate parameter and the observed climate parameter at each grid cell, we apply monthly corrections so magnitudes of the GCM simulations of the historic period conform to observations for the 1950–1999 period of overlap with the PRISM data. The twentieth-century runs used to fit each model are simulations forced by historic variations in greenhouse gases, solar output, and atmospheric aerosol loading. For each of the models presented here, the twentieth-century runs were obtained from the PCMDI archive. The aforementioned biases are presumed to be the same from year to year, because at the monthly time scale the models can resolve the large-scale weather systems that generate observed temperature and precipitation across the Colorado Plateau. The spatial biases and magnitudes are corrected independently at each grid point for each model, by multiplying the simulated parameters by monthly bias factor (for precipitation) or by taking the difference between the simulation and the bias factor (for temperature), as described with the equations adapted from Salathé (2005) below.

Let $P_{mod}(x, t)$ be the simulated monthly precipitation for the large-scale gridpoint in location x and at time t (in months); $(P_{mod})_{mth}$ is the monthly mean taken over the period of overlap between the simulated data and observations $(P_{obs})_{mth}$. The downscaled monthly mean precipitation (P_{ds}), then, can be calculated by:

$$P_{ds}(x,t) = P_{mod}(x,t) \, (P_{obs})_{mth}/(P_{mod})_{mth}$$

The fitting is performed independently for each month.

Surface air temperature is downscaled in a similar way. For temperature, the adjustment uses the difference between the mean bias and the observations. Let $T_{mod}(x, t)$ be the simulated monthly temperature, $(T_{mod})_{mth}$ be the simulated monthly mean taken over the fitting period, and $(T_{obs})_{mth}$ be the monthly mean of the observations taken over the fitting period. Then, the downscaled monthly mean surface temperature (T_{ds}) can be calculated by:

$$T_{ds}(x,t) = T_{mod}(x,t) + [(T_{obs})_{mth} - (T_{mod})_{mth}]$$

This correction assumes that the large-scale temperature predicts the local temperature, given the removal of a monthly bias in the mean. Salathé (2005) notes that this additive methodology may be thought of as a lapse-rate correction due to the elevation difference of the local gridpoint relative to the GCM grid. Like Salathé, we make no allowance for possible changes in the lapse rate as a consequence of climate change. Despite these limitations to the aforementioned methods, and the dependency of this statistical approach on the accuracy of the regional circulation patterns produced by the GCMs (CCSP 2008), the method is computationally efficient, and previous studies show a relatively high confidence in the simulations of storms and jet streams in the middle latitudes (CCSP 2008).

Ranking procedure

In order to determine the most appropriate GCMs to use in the vegetation change analyses, we ranked the models using four metrics based on the fit between the GCM climatological estimates of western U.S. seasonal precipitation during the period of fit and the observed seasonal precipitation averaged over the period 1950–1999. We compared observed parameters and GCM projections for three seasons chosen for their influence on Colorado Plateau vegetation: November–March (winter), May–June (pre-monsoon) and July–September (monsoon). We did not assess fit between simulated and observed temperature, because it is well known that there is good agreement between model temperature simulations for western North America (IPCC 2007), and the bias corrections should account for differences in magnitude. We acknowledge that, compared with temperature, spatial variations in precipitation are less well understood and that there is a greater spread between models in simulated precipitation—thus, choice of models can make a difference in the application of projections for decision-making (IPCC 2007; CCSP 2008; Brekke et al. 2008). We assume that models that simulate well the recent precipitation history of these key seasons are likely to simulate key characteristics of future climate—although Pierce et al. (2009) found no strong relationship between the score of the CMIP3 models on a set of performance metrics and the results of a detection and attribution study of western North America temperatures. In using this metric, we acknowledge that we cannot tell whether well-fitting simulations for our region produce the "right results" for the wrong (mechanistic) reasons.

We computed four values for each model, using metrics computed seasonally and then averaged, for the evaluation domain of interest, the continental United States west of 100°W. We evaluated the models based on this western-U.S. comparison rather than the much smaller domain of our study area in order to encompass a larger array of grid points in the GCMs. We reasoned that if the models cannot perform well for the West as a whole, their performance in our study area could be the result of chance rather than skill. The values we computed are as follows:

- the spatial correlation coefficient between simulated and observed (PRISM) precipitation at the 4 km scale over the domain of interest for each of the three selected seasons;
- the spatial congruence coefficient for the same;
- the mean ratio of simulated/observed area-averaged precipitation for each 4 km grid cell over the domain of interest, based on the seasonal precipitation totals in millimeters; and
- the mean ratio of simulated/observed area-averaged precipitation for each 4 km grid cell over the domain of interest, based on the seasonal precipitation expressed as a percentage of the annual total.

These metrics evaluate the GCMs' fidelity for spatial distribution of precipitation, precipitation amplitude, and seasonal cycle of precipitation. These four sets of metrics for each of the 22 models were then ranked, and the ranks summed for each model. The best rank for each metric is 1, and the worst is 22. The best possible cumulative rank is 4, i.e., a rank of 1 for each of the four metrics. For example, the HAD model produced the following ranks: 1, 3, 3, 3, which yielded an overall score of 10 (Table 1).

We acknowledge that the GCMs differ significantly in terms of basic physical and dynamical design and number of atmospheric and oceanic layers; accounting for such factors may have resulted in a different choice of models. For example, process-based measures, such as the ability of a model to reproduce the El Niño-Southern Oscillation (ENSO), are certainly appropriate for studying Colorado Plateau climate; however, a model with acceptable ENSO simulation may lack acceptable monsoon simulation. Although several studies project decreasing annual precipitation in the southwestern United States (IPCC 2007, Seager et al. 2007), we recognize the limitations of statistically downscaling GCM output to our study area. As we show below, the annual cycle of precipitation for the western United States is poorly simulated by many of these models (Pierce et al. 2009), a result also found in studies of previous generations of models (Coquard et al. 2004).

RESULTS: MODEL SELECTION AND SIMULATIONS OF COLORADO PLATEAU SEASONAL CYCLE

GCMs were ranked based upon their performance in the western half of the United States, using fidelity to seasonal and spatial statistics of precipitation as the overarching metric (Table 1). The overall score for the HAD model (10) produced the lowest rank of all models used in this analysis—that is, HAD precipitation for 1950–1999 was most faithful to the seasonal cycle and spatial distribution of precipitation in the observed record. The next closest score was 26 (ECHAM5). For each of the four ranking metrics, the order of models (best/lowest to worst/highest) did not change much (not shown). In other words, any of the above four metrics individually produce approximately the same order as that for summing the four and then ranking the models.

Rank results did not form the final basis for model selection, although all GCMs selected did rank in the top 11, accounting for ties in some of the rankings (Table 1). Along with the top two models, three other models were selected, based on subjective criteria. CSIRO (score 42) and CNRM (score 42), which tied for rank 9, were included because together they bracket the range of published projections for aridity in the southwestern United States (Seager et al. 2007). Seager and his colleagues modeled precipitation minus evaporation anomalies (P-E) using A1B-emissions scenario projections for 19 IPCC AR4 models. Mid-century (2041–2060) projections ranged from about -0.12 mm/day (CNRM) to no detectable change; CSIRO and HAD both showed near zero change or slight increases in P-E (Figure 2 of Seager et al. 2007). NCAR (score 50; rank 11) was selected because of opportunities to use this model to expand upon this work in collaboration with other investigators (e.g., Govindasamy et al. 2003, Diffenbaugh et al. 2005). None of these five models required flux corrections in order to maintain a stable climate in control runs (Kripalani et al. 2007).

A 22-model ensemble mean (22-ME) also was included for comparison with the individual models. Ensemble means generally improve the performance of climate simulations at both global (Reichler and Kim 2008) and regional scales (Pierce et al. 2009). Pierce and colleagues (2009) compared 42 performance metrics on 21 current GCMs and found multimodel ensemble means consistently outperformed individual models used for a climate change detection and attribution study. They found that a suite of at least five randomly selected models proved superior to any individual model, as long as at least 14 model runs were incorporated, because ensembles with sufficient realizations reduce the effects of internal climate variability. (N.B.: in this and other analyses, multiple realizations of models are included when available). They traced the improved performance of multimodel ensembles to the cancellation of offsetting errors in the individual models. These results

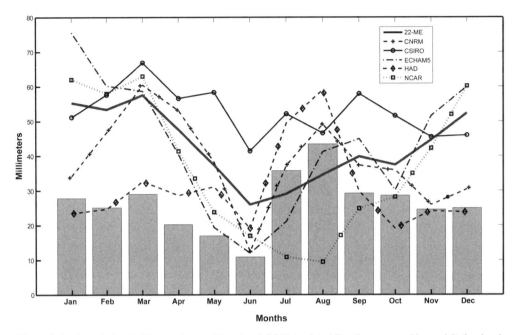

Figure 4. Southern Colorado Plateau observed (bars) and GCM simulated (lines) mean monthly precipitation (mm), 1950–1999. See Table 1 for GCM acronyms. "22-ME" refers to the ensemble mean of 22 GCMs used in the IPCC Fourth Assessment Report data set (Meehl et al. 2007). Observed data: PRISM 4 km (Daly et al. 1994).

were robust, regardless of whether the "best" models, based on comparisons with historical data, were selected. However, by defining the mean more precisely through a larger sample size, ensemble mean results display less interannual variability than individual models. For the purpose of ecological studies, which are dependent on the interannual variability in time series, it is often the extremes rather than the average conditions that define the boundaries of species distribution. Consequently, in this paper, emphasis will also be given to the individual model results.

Simulations for Southern Colorado Plateau area

The GCMs more closely simulate the seasonal cycle of 1950–1999 Colorado Plateau monthly mean temperature than they simulate the seasonal cycle of monthly mean precipitation (Figures 3 and 4). This is consistent with results from global-scale studies; Covey et al. (2003) found that historic global temperatures generated by CMIP2-coupled atmosphere-ocean Global Climate Models (AOGCMs) correlated exceedingly well with historic observed temperatures ($r > 0.93$ for each model), whereas correlations between AOGCMs and observed precipitation ranged from 0.4 to 0.7.

Temperature

The ECHAM5 model shows the closest match to the observed seasonal cycle of Colorado Plateau temperatures (Figure 3). HAD exaggerates the seasonal temperature range; HAD temperatures are too hot in summer and too cold in winter. The NCAR model shows a similar exaggeration of average monthly temperature range. The CNRM and CSIRO models exhibit seasonal cycles similar to the ensemble average—temperatures that are

Table 2. Observed and simulated temperature (top) and precipitation (bottom) for the three seasons analyzed in this study (November–March; May–June; July–September). See Table 1 for GCM acronyms. "22-ME" refers to the ensemble mean of 22 GCMs used in the IPCC Fourth Assessment Report data set (Meehl et al. 2007). Observed data: PRISM 4 km (Daly et al. 1994). Temperature is in °C. Precipitation is in mm.

TEM	PRISM	CNRM	ECHAM5	NCAR	CSIRO	HAD	22-ME
NOV-MAR	2.0	-0.3	1.8	-0.7	-0.5	-1.0	-0.3
MAY-JUN	16.6	16.0	17.2	20.0	16.2	21.1	17.9
JUL-SEP	20.4	21.5	21.2	23.9	20.7	24.6	22.1

PRECIP	PRISM	CNRM	ECHAM5	NCAR	CSIRO	HAD	22-ME
NOV-MAR	131.5	197.9	306.2	285.5	267.3	128.8	263.4
MAY-JUN	27.9	50.7	31.3	40.8	99.8	50.2	63.4
JUL-SEP	108.5	123.8	107.3	45.2	156.8	138.4	103.6

cooler than the observed average during the winter, spring, and fall months, but near the observed average during the summer months. The IPCC Fourth Assessment Report notes that Western North America temperature was underestimated by most AR4 GCMs (Christensen et al. 2007); the median bias for annual temperature was -1.3°C. The most pronounced under-estimation of seasonal temperature was for spring (March–May); the median bias was -2.0°C.

Precipitation

HAD shows the best match with the observed seasonal cycle of precipitation (Figure 4) and associated spatial distribution of precipitation (Table 3). Although HAD projections for July and August total precipitation are 38.3% and 36.0% higher, respectively, than the observed, it is the only model considered here that does not drastically overestimate Colorado Plateau winter season precipitation (Figure 4; Table 2). The NCAR model depicts a Mediterranean climate seasonal cycle of precipitation for the Southern Colorado Plateau region, whereas the CSIRO shows little variation in precipitation between months, in contrast to the observed bimodal season cycle. The CNRM, ECHAM5 and the 22-ME all show bimodal seasonal cycles, but overestimate November–March precipitation, as well as April and September precipitation.

Table 3 presents a qualitative comparison of model estimates versus observed precipitation for the Southern Colorado Plateau (SCP). For November–March, most of the models overestimate SCP mean precipitation. The HAD and 22-ME produce wetter than observed conditions in the southeastern quadrant of the SCP. The CSIRO, ECHAM5, and NCAR models all exhibit wetter than observed winter precipitation, with the ECHAM5 showing more than double the observed precipitation over most of the SCP domain. The CRNM CM3 simulates wetter than observed precipitation over the southern half of the SCP. Overestimation of winter precipitation in the U.S. West is a long-standing issue

Table 3. Qualitative assessment of GCM simulated precipitation compared to observations (1950–1999) for the Southern Colorado Plateau (SCP; see Figure 1). Each bold outlined box represents the SCP, and each quadrant of a bold outlined box represents one quadrant of the domain, corresponding to Figure 1 (clockwise from left, NW NE, SE, SW). Sign indicates the direction of projection, boldness indicates magnitude. **+** much greater than observed; + greater than observed; – less than observed; — much less than observed. Blanks indicate approximately similar total precipitation.

Model	Winter PPT (Nov-Mar)		Spring PPT (May-Jun)		Summer PPT (Jul-Sep)	
22-Model Ensemble			+	**+**	–	+
		+	+	+	–	+
UKMO-HADGEM1			+		–	+
		+	+	+	–	+
MPI-ECHAM5	**+**				–	+
	+	**+**			–	+
CSIRO-MK3	+	+	+	**+**		**+**
	+	**+**	+	**+**	–	**+**
CNRM-CM3			+		–	+
	+	**+**	+		–	**+**
NCAR-CCSM3	**+**	+			+	–	–
	+	**+**			+	—	–

among GCMs (Coquard et al. 2004), related in part to the complex topography of the region (Duffy et al. 2003). The IPCC Fourth Assessment Report notes that more than 75% of the models overestimated western North America annual and seasonal precipitation (Christensen et al. 2007). The median precipitation bias was highest for winter (93%, December–February) and lowest for summer (28%, June–August); the median annual precipitation bias was 65%.

For May–June, with the exception of the ECHAM5, all models and the 22-ME produce wetter than observed precipitation in the eastern half of the SCP. In particular, the CSIRO model produces well more than double the observed precipitation in the eastern half of the SCP. For July–September, the differences between models and observations are even more pronounced, with most models showing greater than observed precipitation in the eastern half of the SCP and less than observed precipitation in the western half of the SCP. Similar to the spring season, the CSIRO model produces double the observed summer precipitation

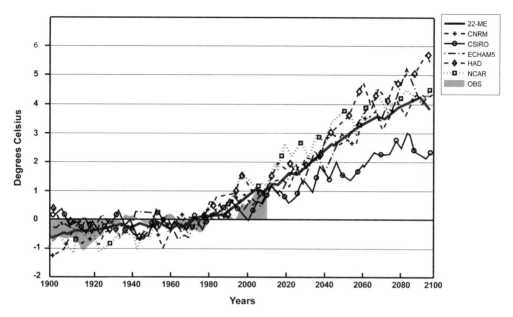

Figure 5. Southern Colorado Plateau observed (shaded) and GCM projected (lines) mean annual temperature (°C). See Table 1 for GCM acronyms. "22-ME" refers to the ensemble mean of 22 GCMs used in the IPCC Fourth Assessment Report data set (Meehl et al. 2007). Observed data: PRISM 4 km (Daly et al. 1994).

in the eastern half of the SCP; as mentioned above, the CSIRO model does not produce a strong seasonal cycle, and overestimates precipitation in every single month (Figure 4). The NCAR model produces conditions drier than observed over the entire SCP domain. A study by Lin et al. (2008) determined that most of the AR4 models overestimate precipitation in the core monsoon region and fail to show the monsoon retreat.

In the section above, we described the biases in the models and their general correspondence with the seasonal cycles of temperature and precipitation, as well as the spatial distribution of precipitation in the SCP. As mentioned in the Data and Methods section, the monthly bias between the GCM and observed estimates is removed independently at each grid cell. We assume that each model's bias, based on simulations, remains consistent in projections of future conditions; thus correcting this bias should yield more reasonable projected values.

Nevertheless, it is valuable to reflect on model bias when interpreting the downscaled projections presented in the next section.

RESULTS:
MODEL PROJECTIONS FOR THE SOUTHERN COLORADO PLATEAU

Projections for the SCP all show increasing temperatures after 1980 (Figure 5). When the 22 model projections are averaged together (22-ME), the temperature increase appears to be nearly monotonic, reaching 2.2°C above the observed average in the 2030s, and 4.0°C above the observed average by the end of the century. Individual models exhibit considerable multi-year variability within the upward trends in temperature. With the exception of the CSIRO model, which projects a considerably lower temperature increase of 2.3°C by the end of the century, individual models selected for this study show increases comparable to the

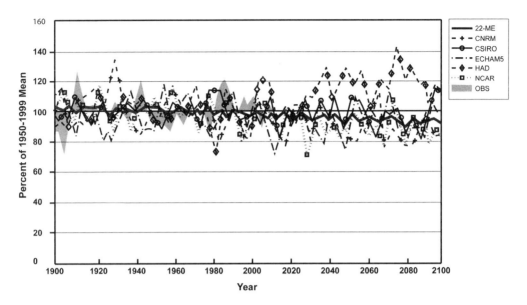

Figure 6. Southern Colorado Plateau observed (shaded) and GCM projected (lines) mean annual precipitation (mm). See Table 1 for GCM acronyms. "22-ME" refers to the ensemble mean of 22 GCMs used in the IPCC Fourth Assessment Report data set (Meehl et al. 2007). Observed data: PRISM 4 km (Daly et al. 1994).

22-ME. HAD projects the greatest annual temperature increases, reaching 5.4°C above the observed average by 2080.

Seasonal temperature projections (not shown) exhibit slightly higher rates of increase for the arid foresummer and summer seasons than for the cool season. The 22-ME projection for the SCP warm seasons reaches 2.5°C above the observed average in the 2030s, and 4.7°C higher than observed average values by the end of the century. November–March seasonal temperature projections reach 1.9°C above the observed average by 2040, and 3.6°C above the average by the end of the century. In all seasons, the CSIRO shows lower temperature increases than the other models. The HAD projects much higher winter temperature increases than the other models (6.0°C higher than average by the end of the century) and, probably due to its exceedingly high projection of July–September precipitation, less than the 22-ME mean increase during the summer. The NCAR, which characterizes the SCP as having a Mediterranean seasonal precipitation cycle for 1950–1999, projects the greatest summer season temperature increases (4.8°C by the end of the century). Timbal et al. (2008) found the CSIRO model the least sensitive (2.11°C) and the ECHAM5 most sensitive (3.69°C) models when comparing the global temperature sensitivity of 10 AR4 GCMs modeling the A1B scenario for the twenty-first century. The HAD was not among the models tested, but similar to this study, CNRM temperature sensitivity was roughly in the middle (2.81°C).

SCP precipitation projections show a wide range of possibilities, and few coherent trends. This is not surprising, and is consistent with the CCSP (2008) and IPCC (2007) statements that AOGCMs are often not reliable for simulating sub-continental scale precipitation. The 22-ME projection suggests a slight decline (6.5%) in annual precipitation for the Colorado Plateau (Figure 6). Most of the GCMs we

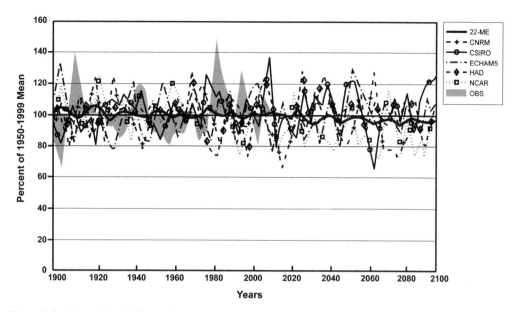

Figure 7. Southern Colorado Plateau observed (shaded) and GCM projected (lines) mean November–March total precipitation (mm). See Table 1 for GCM acronyms. "22-ME" refers to the ensemble mean of 22 GCMs used in the IPCC Fourth Assessment Report data set (Meehl et al. 2007). Observed data: PRISM 4 km (Daly et al. 1994).

selected project annual precipitation below the observed 1950–1999 average for most of the twenty-first century. Most models show decade-scale variations, with few pluvials of the magnitude seen during the twentieth century. HAD, however, projects higher than observed mean SCP precipitation for most of the twenty-first century, due to several decade-scale winter pluvials, and an overall increase in summer precipitation, averaging 50% above observed after 2040 (Figures 6–9). SCP November–March precipitation projections indicate great variability between models and no strong trends (Figure 7); the 22-ME projects a slight decline in winter precipitation of about 5% during the course of the century.

SCP May–June precipitation projections agree on mostly below-observed-average precipitation during the course of the twenty-first century, with some substantial differences in multi-decade variability and the magnitude of declining arid foresummer precipitation (Figure 8). In particular, the 22-ME declines throughout the century, averaging about 75% of climatology during the last decades of the century. The CNRM projects the greatest decline in SCP May–June precipitation (47.8%, with an average of 50% lower than climatology for the last three decades of the century). The CSIRO projects consistently below-average SCP May–June precipitation, reaching about 25% below climatology by the last few decades of the century (Figure 8).

SCP July–September precipitation projections also indicate great variability between models and no clear trends; the 22-ME projects a slight increase in summer precipitation during the course of the century, modulated by multi-decadal variability (Figure 9). The HAD model shows a clear and dramatic increase in SCP summer precipitation after 2020, with regional totals far in excess of observations. On the other hand, the ECHAM5 model projects mostly below-observed-average SCP summer precipitation during the twenty-first century.

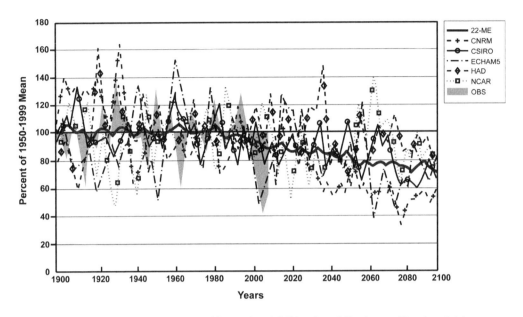

Figure 8. Southern Colorado Plateau observed (shaded) and GCM projected (lines) mean May–June total precipitation (mm). See Table 1 for GCM acronyms. "22-ME" refers to the ensemble mean of 22 GCMs used in the IPCC Fourth Assessment Report data set (Meehl et al. 2007). Observed data: PRISM 4 km (Daly et al. 1994).

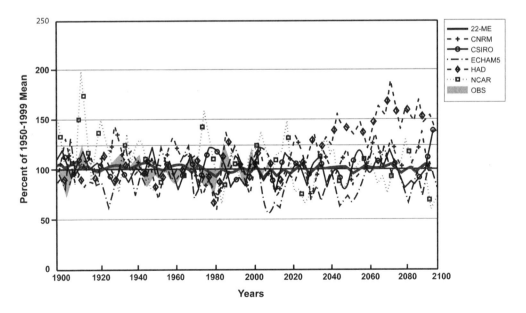

Figure 9. Southern Colorado Plateau observed (shaded) and GCM projected (lines) mean July–September total precipitation (mm). See Table 1 for GCM acronyms. "22-ME" refers to the ensemble mean of 22 GCMs used in the IPCC Fourth Assessment Report data set (Meehl et al. 2007). Observed data: PRISM 4 km (Daly et al. 1994).

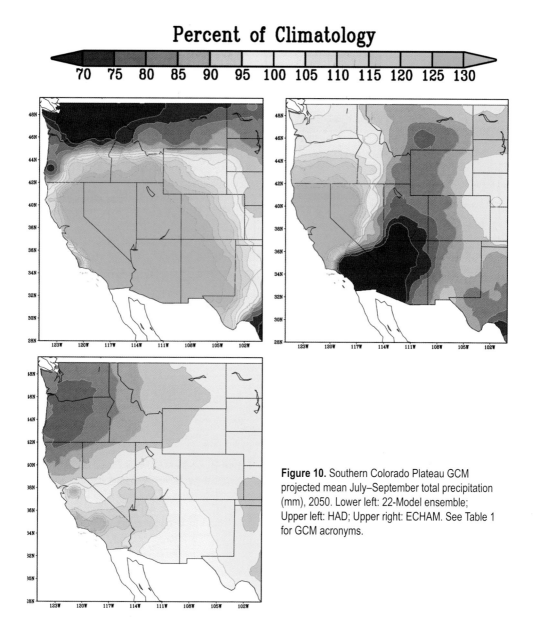

Figure 10. Southern Colorado Plateau GCM projected mean July–September total precipitation (mm), 2050. Lower left: 22-Model ensemble; Upper left: HAD; Upper right: ECHAM. See Table 1 for GCM acronyms.

In this study, the two top-ranking GCMs, based on their skill in simulating observed-average precipitation in the West (HAD and ECHAM5, Table 1), project radically different precipitation changes in the mid-century example considered; this is particularly the case for summer precipitation (Figure 10; Table 4). For mid-century July–September precipitation, the HAD projects a 30% increase for the Colorado Plateau area, whereas the ECHAM5 projects a comparable decrease for the western three-fourths of the domain (Table 4). For May–June mid-century precipitation (Table 4), the two

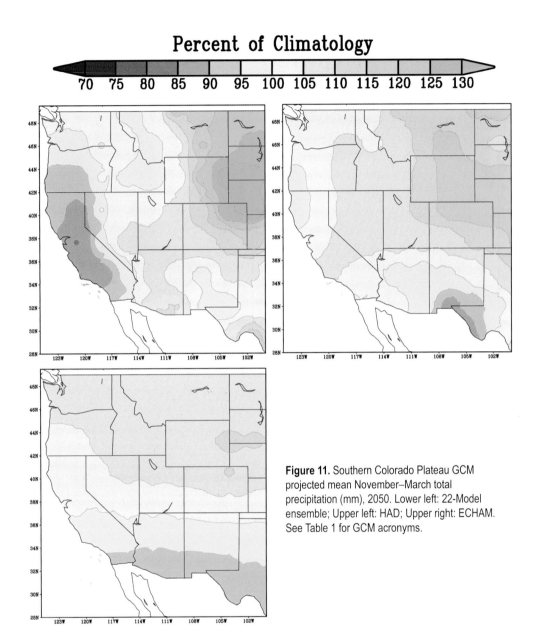

Figure 11. Southern Colorado Plateau GCM projected mean November–March total precipitation (mm), 2050. Lower left: 22-Model ensemble; Upper left: HAD; Upper right: ECHAM. See Table 1 for GCM acronyms.

models show greater agreement with each other, with prominent drying in the western two-thirds of the domain. The two models project a slight increase in May–June precipitation for the eastern part of the SCP domain, in contrast to the 22-ME projection. For November–March mid-century projections, similarly, the two models show a slight increase in precipitation for much of the SCP, while the 22-model ensemble projects a slight decrease in the southern half of the SCP (Figure 11; Table 4).

Table 4. Qualitative assessment of precipitation projections for mid-century (2050) are compared to observations (1950–1999) for the Southern Colorado Plateau (SCP; see Figure 1). Each bold outlined box represents the SCP, and each quadrant of a bold outlined box represents one quadrant of the domain, corresponding to Figure 1 (clockwise from left, NW NE, SE, SW). Sign indicates the direction of projection; boldness indicates the magnitude of the projection. **+** large increase; + increase; – decrease; **—** large decrease.

Model	Winter PPT (Nov-Mar)		Spring PPT (May-Jun)		Summer PPT (Jul-Sep)	
22-Model Ensemble	+	+	**—**	–	+	+
	–	–	**—**	–	+	+
UKMO-HADGEM1	+	+	–	–	**+**	**+**
	+	–	–	–	**+**	**+**
MPI-ECHAM5	+	+	–	–	**—**	–
	–	–	**—**	–	**—**	–
CSIRO-MK3	+	+	–	–	–	–
	+	+	**—**	–	–	–
CNRM-CM3	–	–	**—**	**—**	+	–
	–	–	**—**	**—**	+	–
NCAR-CCSM3	–	–	**—**	–	**+**	+
	–	–	–	–	**+**	+

DISCUSSION

Model Selection and Vegetation Modeling

The fact that two GCMs included in this study captured the global extremes of temperature sensitivity as tested by Timbal et al. (2008) suggests that our model-selection process succeeded in bracketing a range of temperature increases, which is important for modeling potential changes in vegetation distribution. In all seasonal precipitation projections, the ensemble projection falls about midway between the projections by the selected models, suggesting the five selected models are showing variability similar to the full set of 22 models. (N.B.: Results from a separate analysis, using the ensemble mean of the five highest ranking models [Table 1], do not differ dramatically from the major findings of the analysis presented herein). As with temperature sensitivity, the selected models span the full range of precipitation projections, which suits the goal of the selection process in capturing the range of possibilities for bracketing potential vegetation responses.

The approach used here should bracket the inputs for future vegetation change in much the same way as would a sensitivity analysis, although the results contain no information on the likelihood of a particular outcome. For instance, a sensitivity analysis might involve using all 22 models and Monte Carlo simulation or similar resampling approach to develop a suite of statistically robust projections of temperature and precipitation values for every month this century for input into a study of potential vegetation change (*sensu* Frey and Patil 2002). In the absence of that computationally extensive effort, multiple AOGCMs in ensembles can be used to provide more robust estimates of the mean climate state of the future, while projections from individual models can be used to provide estimates of annual variability and to consider the potential impacts if outliers become reality. Exploring a range of possibilities, using both climate and vegetation models, could help resource managers who are interested in developing adaptation plans that are robust to many possible futures (CCSP 2009, Dessai 2009), including those extreme possibilities that lie at the tails of probability distributions—which may have a small probability of occurrence, but a high probability of causing massive impact if realized.

Implications for the Southern Colorado Plateau

Based on the results described above, the Southern Colorado Plateau annual temperatures are projected to increase by 1.5° to 3.6°C by mid-century (22-ME = 2.9°C), and by 2.53° to 5.4°C by the end of the century (22-ME = 4.0°C), with annual temperatures exceeding the 1950–1999 range of variability by the 2030s. Annual precipitation changes are less clear. A conservative estimate, using the 22-model ensemble average, indicates that SCP annual precipitation may decrease by 6% by the end of the century. The clearest indication is that SCP May–June arid foresummer precipitation is likely to decline, by 11 to 45% (conservatively, 25%). Though a small fraction of annual precipitation falls during the arid foresummer, temperatures during this time of year may be implicated in massive forest mortality (Breshears et al. 2005, Weiss et al. 2009 [in press]). The majority of twenty-first century precipitation variations do not consistently exceed the range of historic variability.

The aforementioned results are consistent with IPCC AR4 projections, and with studies that have examined projections for northern California (Dettinger 2005) and the Upper Colorado River Basin (Christensen and Lettenmaier 2007). The hydroclimatic implications of increasing temperatures coupled with precipitation variability dominated by interannual and decadal change, but lacking trend, are well known. These include decreasing snowpack (Mote et al. 2005, Rauscher et al. 2008), early snowmelt (Stewart et al. 2004, 2005, 2009), an increased fraction of liquid winter precipitation (Knowles et al. 2006), decreased runoff (Milly et al. 2005, Ellis et al. 2008), and increased evapotranspiration (Hamlet et al. 2007).

The downscaling approach used in this study does not preserve within-month variability; consequently, ephemeral daily time-scale events such as flood-producing intense rains, frosts, extreme daily temperatures, and wilt-inducing hot, dry episodes are not captured. Studies by Diffenbaugh et al. (2005), using regional climate models, and Meehl et al. (2004) show that projected increases in mean temperature across North America are associated with the aforementioned phenomena—all of which have strong effects on vegetation. In particular, temperature-limited growth processes and the combination of increased temperatures and soil moisture deficits can affect widespread tree mortality and treeline conifer species distribution (e.g., Adams et al. 2009, van Mantgem et al. 2009, Schrag et al. 2008). Moreover, Weiss et al. (2009 [in press]) demonstrate that increased late spring temperatures and drying, consistent facets of the projections used in this study, increase evapotranspirational demand and vegetation moisture stress.

The approach used in this study did preserve annual and multidecadal climate variability, which can be important to vegetation changes. The use of individual model projections allows a more robust simulation of time-series annual and multidecadal climate fluctuations for vegetation modeling than the ensemble mean, though the statistical characteristics of multi-model ensembles are more robust for examining trends and mean values for specified time periods (Pierce et al. 2009). Annual and seasonal variability can influence many ecological processes that can affect species distribution, including seedling germination and survival, herbivore pressure, pollinator phenology, and wildfire frequency and extent. For example, dry wildfire seasons that follow relatively wet years can be associated with more area-burned than dry seasons following dry years in some cases (Swetnam and Betancourt 1998).

CONCLUSIONS

In this study, we statistically downscaled selected IPCC AR4 GCMs for the western United States. Our statistical downscaling method rapidly, and at little cost, provided simulations of future climate at a spatial scale acceptable for input to process-based landscape-scale vegetation models. GCMs were selected for their simulation of western U.S. precipitation, and for characteristics that would produce a range of future climatic conditions to drive vegetation change simulation models for the Southern Colorado Plateau. Our evaluation of selected models, based on their simulation of the seasonal cycle of precipitation and spatial correlation between GCMs and observed precipitation, revealed that all overestimate annual SCP precipitation, and that few match the observed SCP seasonal cycle of precipitation. Models better estimated the seasonal cycle of SCP temperatures, but several models exhibited biases toward warmer than observed summer temperatures and cooler than observed winter temperatures. The HAD model displayed the closest match with historic precipitation observations, but some of the projections from HAD, notably summer precipitation, are well beyond the edge of the envelope of projections of other models.

Projections of future temperature and precipitation, based on individual models and a 22-model ensemble mean, show excellent agreement with regard to projecting temperature increases for the SCP; however, differences in magnitude between GCMs spanned more than 3°C in each season, and for annual temperature. For future precipitation, the most important results are (1) the models show only slight downward trends in annual and winter precipitation and no trend in fall and summer, and (2) the selected models show a strong downward trend in May–June precipitation. In combination with increasing temperatures, lack of moisture during this time of year could increase the likelihood of massive forest mortality events, such as the die-off of Colorado Plateau conifers in the early part of the twenty-first century.

Our analysis demonstrated that the models selected for vegetation analysis produce substantially greater variability than ensembles (an obvious result) and, in this case, a rich array of variability and potential future climates. The approach used here lends itself to a relatively inexpensive version of a sensitivity analysis, in that ecologists can compare the effects on vegetation given the range of projections available. Recent research shows that using comprehensive sets of metrics to choose a set of "best models" does not necessarily result in better projections, but rather that including more models increases the likelihood of producing robust projections (Pierce et al. 2009)—when projections rely on estimates of mean quantities.

One prospect for future research is to use regional climate model simulations to downscale GCM projections and retain fine-scale dynamical processes (e.g., Diffenbaugh et al. 2005). In the meantime, the approach used here, bracketing the ensemble GCMs results with plausible, but different, individual projections is one that can be applied to robust decision-making approaches advocated by some decision scientists (e.g., CCSP 2009). The use of statistically downscaled GCM projections can provide a starting point for considering the range of vegetation conditions resource managers might face in the near future and by the end of the century.

ACKNOWLEDGMENTS

This work was funded by Department of Energy National Institute for Climate Change Research, project MPC35TV-02, "Regional Dynamic Vegetation Model for the Colorado Plateau: a Species-Specific Approach." The authors acknowledge Dr. Andrew Ellis, Arizona State University, for providing estimated potential evapotranspiration data for the Colorado Plateau region. The authors also acknowledge the communications and design staff of the University of Arizona Institute of the Environment for assistance with graphics.

REFERENCES

Adams, H. D., M. Guardiola-Claramonte, G. A. Barron-Gafford, J. Camilo Villegas, D. D. Breshears, C. B. Zou, P. A. Troch, and T. E. Huxman. 2009. Temperature sensitivity of drought-induced tree mortality portends increased regional die-off under global-change-type drought. Proceedings of the National Academy of Sciences 106(17):7063–66. doi: 10.1073/pnas.0901438106

Bachelet, D., R. P. Neilson, T. Hickler, R. J. Drapek, J. M. Lenihan, M. T. Sykes, B. Smith, S. Sitch, K. Thonicke. 2003. Simulating past and future dynamics of natural ecosystems in the United States. Global Biogeochemical Cycles 17(2):14-1–14-21.

Brekke, L. D., Dettinger, M. D., Maurer, E. P., and Anderson, M. 2008. Significance of model credibility in estimating climate projection distributions for regional hydroclimatological risk assessments. Climate Change 89, 371–94. doi: 10.1007/s10584-007-9388-3.

Breshears, D. D., N. S. Cobb, P. M. Rich, K. P. Price, C. D. Allen, R. G. Balice, W. H. Romme, J. H. Kastens, M. L. Floyd, J. Belnap, J. J. Anderson, O. B. Myers, and C. W. Meyer. 2005. Regional vegetation die-off in response to global-change-style drought. Proceedings of the National Academy of Sciences 102(42):15144–48.

Burkett, V. R., D. A. Wilcox, R. Stottlemyer, W. Barrow, D. Fagre, J. Baron, J. Price, J. L. Nielsen, C. D. Allen, D. L. Peterson, G. Ruggerone, and T. Doyle. 2005. Nonlinear dynamics in ecosystem response to climatic change: Case studies and policy implications. Ecological Complexity 2:357–94.

CCSP. 2009. Best Practice Approaches for Characterizing, Communicating, and Incorporating Scientific Uncertainty in Decisionmaking, edited by M. Granger Morgan, Hadi Dowlatabadi, Max Henrion, David Keith, Robert Lempert, Sandra McBride, Mitchell Small, and Thomas Wilbanks. A Report by the Climate Change Science Program and the Subcommittee on Global Change Research. National Oceanic and Atmospheric Administration, Washington, D.C.

CCSP. 2008. Climate Models: An Assessment of Strengths and Limitations. A Report by the U.S. Climate Change Science Program and the Subcommittee on Global Change Research, authored by D. C. Bader, C. Covey, W. J. Gutowski, I. M. Held, K. E. Kunkel, R. L. Miller, R. T. Tokmakian and M. H. Zhang. Department of Energy, Office of Biological and Environmental Research, Washington, D.C.

Christensen, J. H., B. Hewitson, A. Busuioc, A. Chen, X. Gao, I. Held, R. Jones, R. K. Kolli, W.-T. Kwon, R. Laprise, V. Magaña Rueda, L. Mearns, C. G. Menéndez, J. Räisänen, A. Rinke, A. Sarr, and P. Whetton. 2007. Regional Climate Projections. In Climate Change 2007: The Physical Science Basis. Contribution of Working Group I to the Fourth Assessment Report of the Intergovernmental Panel on Climate Change, edited by S. Solomon, D. Qin, M. Manning, Z. Chen, M. Marquis, K. B. Averyt, M. Tignor, and H. L. Miller. Cambridge University Press, Cambridge, United Kingdom, and New York, New York.

Christensen, N., and D. P. Lettenmaier. 2007. A multimodel ensemble approach to assessment of climate change impacts on the hydrology and water resources of the Colorado River basin. Hydrology and Earth System Sciences Discussions 11(1417–34).

Coquard, J., P. B. Duffy, K. E. Taylor, and J. P. Iorio. 2004. Present and future climate in the western USA as simulated by 15 global climate models. Climate Dynamics 23:455–72.

Covey, C., K. M. AchutaRao, U. Cubasch, P. Jones, S. J. Lambert, M. E. Mann, T. J. Phillips, and K. E. Taylor. 2003. An overview of results from the Coupled Model Intercomparison Project. Global and Planetary Change 37:103–33.

Daly, C., R. P. Neilson, and D. L. Phillips. 1994. A statistical-topographic model for mapping climatological precipitation over mountainous terrain. Journal of Applied Meteorology 33:140–58.

Daly, C., G. H. Taylor, W. P. Gibson, T. W. Parzybok, G. L. Johnson, P. Pasteris. 2001. High-quality spatial climate data sets for the United States and beyond. Transactions of the American Society of Agricultural Engineers 43:1957–62.

Daly, C., W. P. Gibson, G. H. Taylor, G. L. Johnson, P. Pasteris. 2002. A knowledge-based approach to the statistical mapping of climate. Climate Research 22:99–113.

Dessai, S., M. Hulme, R. Lempert and R. Pielke, Jr., 2009. Do We Need Better Predictions to Adapt to a Changing Climate? EOS: Transactions of the American Geophysical Union 90(13):111–12.

Dettinger, M. D., 2005. From climate change spaghetti to climate change distributions for 21st century California. San Francisco Estuary Watershed Science 3(1):1–14.

Diffenbaugh, N. S., J. S. Pal, R.J. Trapp, and F. Giorgi. 2005. Fine-scale processes regulate the response of extreme events to global climate change. Proceedings of the National Academy of Sciences 102(44):15774–78.

Duffy, P. B., B. Govindasamy, J. P. Iorio, J. Milovich, K. R. Sperber, K. E. Taylor, M. F. Wehner, S. L. Thompson. 2003. High-resolution simulations of global climate, part 1: Present climate. Climate Dynamics 21:371–90.

Eischeid J. K., Pasteris P. A., Diaz H. F., Plantico M. S., Lott N. J. 2000. Creating a serially complete, national daily time series of temperature and precipitation for the western United States. Journal of Applied Meteorology 39:1580–91.

Ellis, A. W., T. W. Hawkins, R. C. Balling, and P. Gober. 2008. Estimating future runoff levels for a semiarid fluvial system in central Arizona. Climate Research 35:227–39.

Frey, H. C., and S. R. Patil. 2002. Identification and review of sensitivity analysis methods. Risk Analysis 22(3):553–78.

Govindasamy, B., P. B. Duffy, and J. Coquard. 2003. High-resolution simulations of global climate, part 2: Effects of increased greenhouse gases. Climate Dynamics 21:391–404.

Hamlet, A. F., P. W. Mote, M. P. Clark, and D. P. Lettenmaier. 2007. Twentieth-century trends in runoff, evapotranspiration, and soil moisture in the western United States. Journal of Climate 20:1468–86.

Hamon W. R. 1961. Estimating potential evapotranspiration. Proceedings of the American Society of Civil Engineering 871:107–20.

Higgins, R. W., Y. Chen, and A. V. Douglas, 1999. Interannual variability of the North American warm season precipitation regime. Journal of Climate 12:653–80.

Hoerling, M., J. Eischeid, X. Quan, T. Xu. 2007. Explaining the record U.S. warmth of 2006. Geophysical Research Letters, 34, L17704. doi:10.1029/2007GL030643.

IPCC. 2007. Climate Change 2007: The Physical Science Basis. Contribution of Working Group I to the Fourth Assessment Report of the Intergovernmental Panel on Climate Change, edited by S. Solomon, D. Qin, M. Manning, Z. Chen, M. Marquis, K. B. Averyt, M. Tignor and H. L. Miller. Cambridge University Press, Cambridge, United Kingdom. and New York, New York.

Knowles, N., M. D. Dettinger, and D. R. Cayan. 2006. Trends in snowfall versus rainfall in the western United States. *Journal of Climate* 19(18):4545–59.

Kripalani, R. H., J. H. Oh, H. S. Chaudhari. 2007. Response of the East Asian summer monsoon to doubled atmospheric CO_2: Coupled climate model simulations and projections under IPCC AR4. Theoretical and Applied Climatology 87:1–28.

Lin, J.-L., B. E. Mapes, K. M. Weickmann, G. N. Kiladis, S. D. Schubert, M. J. Suarez, J. T. Bacmeister, and M.-I. Lee. 2008. North American monsoon and convectively coupled equatorial waves simulated by IPCC AR4 coupled GCMs. Journal of Climate 21(12): 2919–37. doi:10.1175/2007 JCLI1815.1.

Logan, J. A., J. Régnière, and J. A. Powell. 2003. Assessing the impacts of global climate change on forest pests. Frontiers in Ecology and the Environment 1:130–37.

Meehl G. A., C. Covey, T. Delworth, M. Latif, B. McAvaney, J. F. B. Mitchell, R. J. Stouffer, and K. E. Taylor. 2007. THE WCRP CMIP3 multimodel dataset: A new era in climate change research, BAMS, 88:1383–94. doi 10.1175/BAMS-88-9-1383.

Meehl, G. A., C. Tebaldi, and D. Nychka, 2004. Changes in frost days in simulations of twentyfirst century climate. Climate Dynamics 23:495–511.

Milly, P. C. D., K. A. Dunne, et al. 2005. Global pattern of trends in streamflow and water availability in a changing climate. Nature 438:347–50.

Moss, R., M. Babiker, S. Brinkman, E. Calvo, T. Carter, J. Edmonds, I. Elgizouli, S. Emori, L. Erda, K. Hibbard, R. Jones, M. Kainuma, J. Kelleher, J. Francois Lamarque, M. Manning, B. Matthews, J. Meehl, L. Meyer, J. Mitchell, N. Nakicenovic, B. O'Neill, R. Pichs, K. Riahi, S. Rose, P. Runci, R. Stouffer, D. van Vuuren, J. Weyant, T. Wilbanks, J. Pascal van

Ypersele, and M. Zurek. 2008. Towards New Scenarios for Analysis of Emissions, Climate Change, Impacts, and Response Strategies. Intergovernmental Panel on Climate Change. Geneva, Switzerland.

Mote, P. W., A. F. Hamlet, M. P. Clark, and D. P. Lettenmaier. 2005. Declining mountain snowpack in western North America. Bulletin of the American Meteorological Society 86:39–49.

Nakicenovic, N., J. Alcamo, G. Davis, B. de Vries, J. Fenhann, S. Gaffin, K. Gregory, A. Grübler, T. Y. Jung, T. Kram, E. L. La Rovere, L. Michaelis, S. Mori, T. Morita, W. Pepper, H. Pitcher, L. Price, K. Raihi, A. Roehrl, H.-H. Rogner, A. Sankovski, M. Schlesinger, P. Shukla, S. Smith, R. Swart, S. Van Rooijen, N. Victor, and Z. Dadi. 2000. Special Report on Emissions Scenarios: A Special Report of Working Group III of the Intergovernmental Panel on Climate Change. Cambridge University Press, Cambridge, United Kingdom. Available online at: *http://www.grida.no/climate/ipcc/emission/index.htm*.

Pierce, D. W., T. P. Barnett, B. D. Santer, and P. J. Gleckler. 2009. Selecting global climate models for regional climate change studies. Proceedings of the National Academy of Sciences 106(21):8441–46.

Rauscher, S. A., J. S. Pal, N. S. Diffenbaugh, and M. M. Benedetti. 2008. Future changes in snowmelt-driven runoff timing over the western U.S. Geophysical Research Letters, 35, L16703. doi:10.1029/2008GL034424.

Reichler, T., and J. Kim. 2008. How well do coupled models simulate today's climate? Bulletin of the American Meteorological Society 89:303–11.

Salathé, E. P. 2005. Downscaling simulations of future global climate with application to hydrologic modelling. International Journal of Climatology 25:419–36.

Savage, M., P. M. Brown, and J. Feddema. 1996. The role of climate in a pine forest regeneration pulse in the southwestern United States. Ecoscience 3:310–18.

Schrag, A. M., A. G. Bunn, and L. J. Graumlich. 2008. Influence of bioclimatic variables on tree-line conifer distribution in the Greater Yellowstone Ecosystem: Implications for species of conservation concern. Journal of Biogeography 35(4):698–710. doi: 10.1111/j.1365-2699.2007.01815.x

Seager, R., M . Ting, I . Held, Y . Kushnir, J . Lu, and G. Vecchi. 2007. Model projections of an imminent transition to a more arid climate in southwestern North America. Science 316(5828):1181–84.

Stephenson, N., D. Peterson, D. Fagre, C. Allen, D. McKenzie, and J. Baron. 2006. Response of western mountain ecosystems to climatic variability and change: The Western Mountain Initiative. Park Science. 34(1):24–29.

Stewart, I. T., D. R. Cayan, et al. 2004. Changes in snowmelt runoff timing in western North American under a 'business as usual' climate change scenario. Climatic Change 62:217–32.

Stewart, I. T., D. R. Cayan, et al. 2005. Changes toward earlier streamflow timing across western North America. Journal of Climate 18:1136–55.

Stewart, I. T. 2009. Changes in snowpack and snowmelt runoff for key mountain regions. Hydrological Processes 23(1):78–94. doi: 10.1002/hyp.7128

Swetnam, T.W., and J. L. Betancourt. 1998. Mesoscale disturbance and ecological response to decadal climatic variability in the American Southwest. Journal of Climate 11:3128–47.

Timbal, B., P. Hope, and S. Charles. 2008. Evaluating the consistency between statistically downscaled and global dynamical model climate change projections. Journal of Climate 21:6052–59.

USDA Forest Service Southwest Region. 2008. Locations of bark beetle activity: Arizona and New Mexico. Aerial survey results posted at *http://www.fs.fed.us/r3/resources/health/beetle/index.shtml.* Accessed on 9/21/2009.

van Mantgem P. J., N. L. Stephenson, J. C. Byrne, L. D. Daniels, J. F. Franklin, P. Z. Fulé, M. E. Harmon, A. J. Larson, J. M. Smith, A. H. Taylor, T. T. Veblen. 2009. Widespread increase of tree mortality rates in the western United States. Science 323:521–24. doi: 10.1126/science.1165000

Weiss, J. L., C. L. Castro, and J. T. Overpeck. In press. Distinguishing pronounced droughts in the southwestern U.S.A.: Seasonality and effects of warmer temperatures. Journal of Climate. Revised version sent 18 March 2009.

Widmann, M., C. S. Bretherton, and E. P. Salathé. 2003. Statistical precipitation downscaling over the northwestern United States using numerically simulated precipitation as a predictor. Journal of Climate 16:799–816.

INTEGRATING RESTORATION AND CONSERVATION OBJECTIVES AT THE LANDSCAPE SCALE: THE KANE AND TWO MILE RANCH PROJECT

Thomas D. Sisk, Christine Albano, Ethan Aumack, Eli J. Bernstein, Timothy E. Crews, Brett G. Dickson, Steve Fluck, Melissa McMaster, Andi S. Rogers, Steven S. Rosenstock, David Schlosberg, Ron Sieg, and Andrea Thode

ABSTRACT

Across the Colorado Plateau, conservation and restoration projects are progressing in a dizzying array of overlapping efforts. Increasing the efficiency and effectiveness of these efforts requires integration at scales relevant to multiple ecological processes and management practices. While watersheds provide natural and discrete units for organizing and coordinating efforts in many regions, the unique geomorphology and land tenure of the Colorado Plateau suggests a landscape-scale approach for organizing restoration and conservation efforts. In 2005, the Grand Canyon Trust and The Conservation Fund purchased the historic Kane and Two Mile Ranches and initiated an experiment in public-private partnership in science and land management spanning 340,000 ha of the Arizona Strip, including some of the West's most remote places and spectacular scenery. This effort provides an ongoing case study of how collaborative projects are promoting new understanding and strategic approaches to landscape planning and management efforts that cross ecosystem and jurisdictional boundaries. In this chapter, we present four ongoing projects that illustrate the interrelationships between research and management: 1) the influence of severe fire on plant communities and fuels, 2) native shrub restoration and habitat improvement efforts for mule deer (*Odocoileus hemionus*), 3) restoration of native cool-season grasses in the House Rock Valley, and 4) the development of spatial models to predict the occurrence of cheatgrass (*Bromus tectorum*). Together, these efforts illustrate the value of integrating research efforts to guide a landscape-level conservation and restoration program that is visionary, scientifically grounded, and practical.

INTRODUCTION

In 2005, the Grand Canyon Trust (GCT) and The Conservation Fund purchased the Kane and Two Mile Ranches, placing into conservation ownership livestock grazing permits extending across 340,000 ha of mostly public lands between the Grand Canyon and the Arizona-Utah state line (Figure 1). This vast expanse encompasses a 2,000 m elevational gradient and a broad range of ecosystems that support and, in turn, are sustained by a highly diverse biota. Despite the remoteness of the ranch lands and their national profile as destinations for tourism and recreation, industrial logging, heavy livestock grazing, and fire suppression have altered and, in some cases, degraded ecosystems, leaving them vulnerable to unnatural fire regimes, drought, and the stresses associated with climate change. These and other influences have reduced habitat quality for a number of ecologically valuable and/or imperiled species, while impacting ecosystem services such as the provision of water, maintenance of soil fertility, sequestration of carbon, and amelioration of fire behavior. While the landscape retains its open and natural character with most native species present, trends in environmental quality and ecosystem function are not encouraging, and the prospect of climate change creates even greater uncertainty regarding the future of this spectacularly beautiful region. This chapter presents a model for how a diversified but focused science program can provide new information, guidance, and ongoing monitoring of land and resource management activities, transforming management challenges into ongoing experiments that will guide conservation efforts and inform an adaptive approach across large areas of mixed public and private ownership. The unprecedented commitment of conservation resources to the Kane and Two Mile Ranches, combined with ongoing commitments of many public and private partners, poses numerous questions and new opportunities for conserving Western landscapes; meanwhile, the commitment to a strong and collaborative science program to guide future restoration and conservation efforts provides a new model for public-private partnerships in an era of declining federal and state budgets and unprecedented environmental change.

Ultimately, the success of both science and management must be judged in the context of clear goals and objectives. From these, desired outcomes can be articulated, and questions regarding appropriate actions, their effectiveness, and the efficiency of management approaches can be addressed. Only then can science be employed in a constructive way, such that it is responsive to key challenges that require a broadly informed perspective, site-specific knowledge, and a forward-looking approach that provides a predictive capability to guide management under changing conditions. In a concerted effort to ground ranch management in a conservation ethic, work on the Kane and Two Mile Ranches is motivated by the following management goals and ecological objectives:

Goals for Managing the Kane and Two Mile Ranches

1. Restore productive grassland, shrubland, woodland, forest, and riparian ecosystems.
2. Protect unique and sensitive natural resources, such as springs, ancient forests, and remnants of native grasslands.
3. Restore and maintain thriving, viable populations of the full range of native species.
4. Maintain ecologically and economically sustainable land uses to benefit local economies and support ongoing management activities.
5. Promote inclusive, conservation-based land management by engaging citizens and local, state, tribal, and federal government agencies.
6. Manage livestock grazing in a manner consistent with restoration and maintenance of ecological and scenic integrity.

Ecological Objectives

1. Ecosystems across the ranches support viable populations of all native species.
2. If native species are missing, they may be reintroduced or allowed to colonize by natural means.
3. Ecosystems consist of indigenous species to the greatest practicable extent.
4. Ecosystems across the ranches support characteristic assemblages of native species, to the greatest extent possible.
5. Ecosystem functions, including natural disturbance regimes, are entrained with

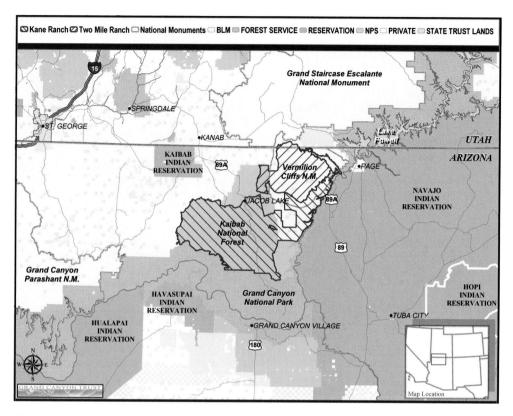

Figure 1. Location of the Kane and Two Mile Ranches, Coconino County, Arizona. The ranches's location, between Grand Canyon National Park and several National Monuments and National Recreation Areas, confers an important role for restoring and sustaining ecosystem function, habitat connectivity, and the viability of regional conservation plans.

and respond freely to current and future climate variability and change.
6. Ecosystems occur within their natural range of variability and are suitably integrated into a larger ecological matrix or landscape.
7. Potential threats to the health and integrity of all ecosystems from the surrounding landscape have been eliminated or reduced as much as possible.
8. Ecosystems are sufficiently resilient to endure the normal periodic stress events in the local environment that serve to maintain the integrity of the ecosystem.

BASELINE ECOLOGICAL ASSESSMENT

While these goals and objectives provide a formal articulation of GCT's vision for the Kane and Two Mile Ranches and a context for developing partnerships, they are by nature quite general and do little to direct the thousands of specific actions required to manage this expansive land base. To provide a context for management, and to establish a baseline for tracking changes and informing management, GCT launched an ambitious assessment effort shortly after purchasing the Kane and Two Mile Ranches. Guided by an independent science advisory council, and in consultation with partnering land management agencies, universities, and other conservation organizations, GCT

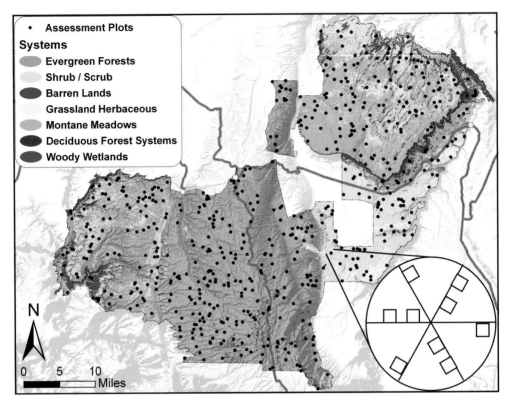

Figure 2. Sampling design and plot layout developed for the Grand Canyon Trust's "Baseline Assessment" project. This extensive field effort established the first cross-jurisdictional, science-driven data set to guide integrated land and resource management over this 340,000 ha of public lands. Inset sampling plot has a radius of 7.32 m.

designed and implemented an assessment plan that engaged several of the co-authors of this chapter.

We employed a randomized design, stratified on soil and vegetation type, to locate over 600 ground plots in a representative manner across the ranches (Figure 2). Over a six-month period in 2005, we combined an extensive field effort with analysis of remotely sensed data to guide a rapid ecological assessment of the ranches. Results from this effort allowed GCT to identify opportunities for linking restoration and livestock management strategies across the project area. Analysis of landscape-scale vegetation and soil characteristics has allowed us to prioritize appropriate locations for site-specific restoration efforts. We have initiated more than a dozen grassland, woodland, forest, and stream restoration projects whose results already are contributing, alone and in combination, to broader landscape-level restoration activities. Additional efforts are being planned to focus on key ecological gradients to account for and learn from the significant heterogeneity existing across the project area. We have used assessment results to inform the development of long-term, multiscaled monitoring and adaptive management efforts across the project area. By working adaptively and systematically across scales, we intend to develop and implement restoration approaches that are efficient, effective, and relevant to emerging public lands management challenges across the southwestern United States.

METHODS

We chose four case studies to illustrate the integration of research and monitoring to inform landscape-level restoration and conservation efforts across the Kane and Two Mile Ranches. These cases addressed (1) the influences of severe fire on plant communities and fuels, (2) native shrub restoration and habitat improvement efforts for mule deer (*Odocoileus hemionus*), (3) restoration of native cool-season grasses in the House Rock Valley, and (4) the development of spatial models to predict the occurrence of cheatgrass (*Bromus tectorum*). For each case, we provide a brief introduction to the purpose and nature of the study, a minimalist account of the methods employed in each study, and an integrated look at preliminary results and their interpretation. By examining the development of these case studies, each linking science with management, we demonstrate how a clear set of objectives, a knowledge of the landscape, and a capacity to bring rigorous science to bear on emerging challenges can help to focus research and monitoring efforts, such that they contribute to on-the-ground management efforts in a practical, timely, and scientifically rigorous manner.

CASE STUDY 1

Effects of the Warm Fire on understory vegetation, ponderosa pine crown mortality, and fuels: Implications for post-fire management

INTRODUCTION

On 8 June 2006, a lightning strike ignited the Warm Fire on the northeastern edge of the Kaibab Plateau. The fire was initially managed as a Wildland Fire Use (WFU) fire, which means that this naturally ignited fire was allowed to burn in order to help meet management objectives, but changing weather conditions led to its reclassification as a wildfire and, over a several-week period, it burned 24,000 ha of the Kaibab National Forest, despite fire-suppression efforts. The North Kaibab Ranger District had recently developed a WFU program, which was implemented with success for the first stages of the Warm Fire. Unusual weather conditions pushed the fire beyond the intended fire management area, where it burned intensely and with severe effects over large areas, forcing evacuations from Grand Canyon National Park, affecting local communities, and creating considerable public controversy about fire management practices on public lands.

Post-fire management often focuses on the regeneration of trees, neglecting the understory vegetation even though it includes the most plant biodiversity, regulates tree growth and regeneration, stabilizes soils, controls nutrient cycling, and provides habitat for wildlife (Kerns et al. 2006). In this collaborative study involving GCT, Northern Arizona University, and the Kaibab National Forest, we are initiating an assessment of fire impacts across two fire severity classes, and between seeded and unseeded areas. The objectives of this research are to: (1) characterize the understory vegetation response to fire severity, (2) determine the effects of seeding with exotic ryegrass (*Lolium perenne* var. *multiflorum*) on native plant communities, (3) qualitatively assess the differences in pre- and post-fire plant communities, (4) monitor changes in fuel loads, and (5) determine the factors that best predict post-fire ponderosa pine (*Pinus ponderosa*) mortality.

METHODS

This research is conducted on the northeastern portion of the Kaibab Plateau and focuses on fire effects in ponderosa pine ecosystems. We established 100 plots to sample understory plant composition across the landscape (Table 1). Our plot design combined elements of the GCT baseline assessment with recommendations of the North American Weed Management

Table 1. Distribution of plots by fire severity, seeding treatment, and year. Unburned plots were located within one km of the burn perimeter, outside the burn area.

	Fire Severity[1]				
	High, Seeded	High, Unseeded	Low	Unburned Controls	Total
Number of Baseline Assessment Plots	3	4	4	8	19
2007	10	13	12	10	45
2008–2010	25	25	25	25	100

[1] Severity is assessed using the Composite Burn Index (Key and Benson 2006)

Association (Stohlgren et al. 2003). Eleven burned and eight unburned plots were adopted from the baseline assessment; others were randomly selected within strata defined by fire severity, vegetation, soils, and elevation. (Figure 3; Table 1).

Wildfire areas with high burn severity (total extent = 1,360 ha) were seeded with 4.0kg/ha of ryegrass, a non-native biennial, to discourage establishment of less-desirable non-native plants, help stabilize soils and prevent erosion, and to improve aesthetics of the post-burn landscape. There is growing concern that aerial seeding with ryegrass may not be cost-effective and may cause more problems than it solves (Keeley et al. 2006, Barclay et al. 2004). We established plots in high severity areas to look at the community-level effects of fire severity and seeding. Understory response variables include density and cover of all species, herbaceous biomass, tree seedling recruitment, a complete species list, and shrub counts.

Within the Warm Fire burn area, various site-scale factors and management practices may exacerbate a continuing loss of trees that survived the initial burn. We are collecting overstory data and monitoring tree mortality for all live ponderosa pines in our plots over a four-year period. Initial measurements included: tree height; diameter at breast height; height to lowest green branch both pre and post-fire; maximum height and severity of bole scorch; minimum height and severity of bole scorch; presence of beetles; and percent consumption and scorch of the pre-fire crown. Overstory sampling techniques were modified slightly from McHugh and Kolb (2003). These data will allow the validation of existing logistic regression models (McHugh and Kolb 2003, Sieg et al. 2006) used to identify the best predictors of ponderosa pine mortality. Refining our ability to predict tree mortality will inform decisions related to reforestation, hazard tree removal, and salvage logging. Another goal of this aspect of the study is to provide information on natural tree seedling recruitment across the different fire severities, which can help in planning post-fire tree planting. We also are monitoring changes in post-fire fuels using Brown's

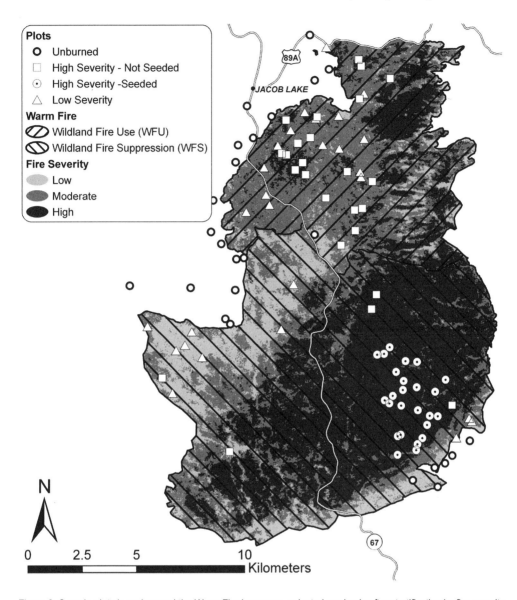

Figure 3. Sample plots in and around the Warm Fire burn area, selected randomly after stratification by fire severity, fire type, soils, and vegetation. Plots fall into three classes: low and high severity burn, and unburned controls.

(1974) transect methods for quantifying 1-, 10-, 100-, and 1,000-hour fuels. These data will help to inform managers of future re-burn potential and how it varies with fire severity and seeding treatment.

CURRENT STATUS AND EXPECTED RESULTS

We have completed two years of sampling on 100 plots and will continue to monitor understory vegetation responses, overstory mortality, and fuels over a four-year period. We are just beginning data analysis and, therefore, have only preliminary results indicating possible trends. One of the most pressing questions concerns the effects of seeding on vegetation composition and biomass. We expect a trend towards greater biomass in the high severity seeded areas, but with higher concentrations of non-native plants (our expectations serve as further hypotheses for testing). This trend could change if ryegrass abundance decreases after two years, as predicted; however, more subtle impacts on native plants and community composition are unknown. Preliminary data show a significant difference in plant community composition between seeded and unseeded sites, but long term effects are yet to be determined.

We expect tree seedling recruitment to be higher in areas of low burn severity, compared to high severity (seeded or unseeded) or unburned areas (Barclay et al. 2004). Initial data support this trend. We also expect that re-seeding of ryegrass could influence tree seedling recruitment; however, the magnitude of this influence is difficult to predict. Results from this study will provide useful information regarding the short- to mid-term effects of post-fire seeding of non-native understory species on plant community composition, seedling recruitment, and presence of other non-natives. Remaining fuels vary greatly by fire severity, but these data have not yet been analyzed.

The Warm Fire provides a compelling opportunity to quantify fire effects and apply this knowledge to recovery and restoration planning in a timely and rigorous manner. Ongoing research will provide a framework for identifying and prioritizing post-fire rehabilitation and management efforts. At broader scales, this study will assist land managers in designing ecologically appropriate landscape-level approaches to forest restoration and post-fire management practices. In addition to informing the rehabilitation efforts across the Warm Fire burn area, this research will help inform future planning, management, and monitoring practices and help avoid any unintended and undesirable effects associated with future post-fire management.

CASE STUDY 2
Improving mule deer habitat on the western Kaibab Plateau: Collaborating on adaptive management

INTRODUCTION

The Kaibab Plateau supports the highest concentration of mule deer in Arizona, with a 2004 population estimate of about 10,000 animals (Wakeling 2007). These large-bodied, large-antlered deer are a valuable resource for hunters and tourists, and an important prey species for predators and scavengers, including the California condor (*Gymnogyps californianus*), recently reintroduced to Arizona. The Kaibab mule deer herd is well known in wildlife management circles because of its history of population oscillations (Russo 1964, Gruell 1986). For the last decade, there has been controversy over the relationship between mule-deer population size and its influence on winter-range shrub species. The west side of the Kaibab Plateau provides important winter habitat and transitional range used during seasonal migration (Haywood et al. 1987, Watkins et al. 2007). A history of intense disturbances, including wildfires, livestock grazing, and establishment of undesirable exotic species (specifically cheatgrass), have reduced winter browse species and degraded mule deer habitat in this critical area.

Conditions on the west side are best described as burned or unburned pinyon pine (*Pinus edulis*) and juniper (*Juniperus* spp.) woodlands with diminishing or absent understory vegetation, and valley bottoms dominated by seeded non-native grasses, little to no native grasses, and few shrubs. Preliminary vegetation sampling (2005–2007) from the Arizona Game and Fish

Department's (AZGFD) Research Branch indicates that winter range quality is currently low to moderate with a decreasing trend, based on criteria in Davis et al. (2005), with shrub decadency at 27% (exceeds optimal range of <20%), perennial grass cover at 6% (below optimal range of 8–15%), perennial forb cover at 1.4% (below optimal range of >5%), and annual grass at 10% cover (exceeds optimal range of <5%). In addition, GCT's model for cheatgrass occurrence (see below) indicates that cheatgrass is either present or within the ninety-fifth percentile for likelihood of occurrence over much of this area. Due to the low quality winter range habitat for deer, AZGFD, along with the U. S. Forest Service (USFS), have begun the initial stages of an adaptive landscape-scale project (9,700 ha) involving pinyon-juniper removal, native seeding of winter shrub species, and follow-up cheatgrass treatments (Figure 4). Lastly, GCT, USFS Rocky Mountain Research Station, and Northern Arizona University are assisting with pre- and post-treatment monitoring.

METHODS

Treatment methods are meant to promote the appropriate mix and age structure of native browse species, which have been shown to be important in improving nutritional requirements for Southwestern mule deer (Heffelfinger et al. 2006, Watkins et al. 2007). The approach is based on an influential monograph on range restoration that focuses on seed-bed preparation and minimized soil disturbance (Monsen et al. 2004). In addition, the protocol for this project calls for treating weeds with Plateau® (Imazapic), an emergent and pre-emergent herbicide that is particularly effective on cheatgrass, to reduce the risk of exotic species establishment following disturbance associated with mechanical seeding of native browse species. Currently, two treatments are being applied on the west side of the Kaibab Plateau:

1. Re-treatment of circa 1950s and 1960s bulldozer pushes through mechanical grinding. These treatments were originally implemented to increase forage for cattle by removing pinyon-juniper woodlands. Instead, woodlands have reoccupied this area, followed by an increase in cheatgrass. Current treatments aim to enhance the existing browse plant component by removing competing pinyon and juniper trees, and providing mulch to create favorable microsites for native grasses and shrubs while minimizing soil disturbance.

2. Seeding locally collected native shrubs, including cliffrose (*Purshia mexicana* var. *stansburiana*), sagebrush (*Artemisia* spp.), fourwing saltbush (*Atriplex canescens*), and winterfat (*Krascheninnikovia lanata*) into burned areas. These shrubs are important, protein-rich browse for wintering deer. To achieve multiple seed depths for the various species, we used a rangeland drill with hydraulic discs.

During spring and summer 2007, 1,740 ha of pinyon-juniper bulldozer pushes were treated at seven sites, using a mechanical grinder. The treatments removed most pinyon and juniper trees less than 30 cm dbh (diameter at breast height), and treatment unit edges were feathered toward non-treated areas in order to make the treatments look more like natural openings. All sites were treated in a mosaic pattern, leaving cover in known deer movement corridors. The mechanical grinder was a tracked vehicle and soil disturbance was minimal. Chip depth was 8 cm or less to provide appropriate microsite characteristics.

During fall 2007, 200 ha were seeded with native shrub species in four areas that were either burned in the 1996 Bridger Knoll Fire or re-burned in the 2007 Slide Fire. Selection of shrub species to be seeded was based on soil characteristics and pre-fire field data. Based on the recommendations of Monsen et al. (2004), large-seeded species such as cliffrose and fourwing saltbush were

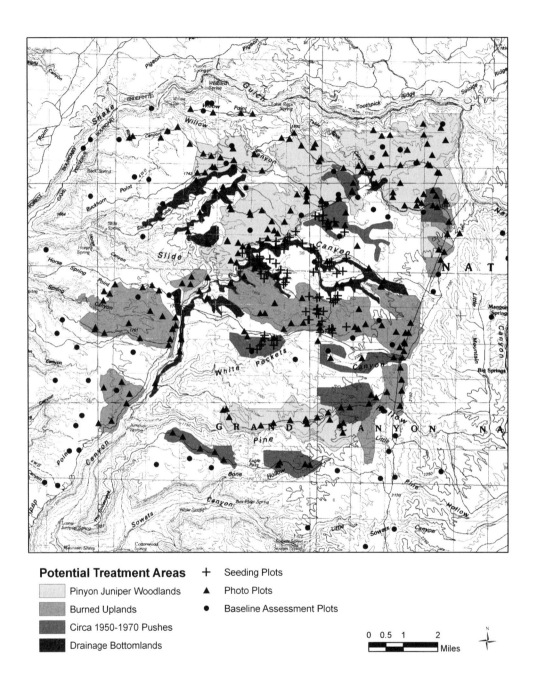

Figure 4. Study area of the Arizona Game and Fish Department's mule deer habitat-improvement project on the west side of the Kaibab Plateau. Implementation of this multiyear project began in 2007.

planted at roughly 1.3 to 2.6 cm deep, while small-seeded species such as sagebrush and winterfat were surface-planted over lightly scarified soil.

Within the seeding areas, 100 ha were sprayed with Plateau® herbicide about two weeks after seeding. We focused on areas that had extensive cheatgrass cover prior to seeding treatments. Herbicide was applied at a rate of 0.59 l/ha, with 1.2 l/ha of mentholated seed oil and 1.0 l/ha of dye.

CURRENT STATUS AND EXPECTED RESULTS

Habitat enhancement work, including additional seeding and weed treatments, will continue for several years to achieve desired landscape-level results. Because shrubs take several growing seasons to establish, it will take one to three years before seedling success can be fully evaluated. However, cheatgrass establishment could occur much more rapidly, necessitating a rigorous monitoring effort to track the effects of early treatments and use these to guide subsequent work. To maximize the area treated, initial funding for this restoration effort did not cover a monitoring plan for treated areas. AZGFD recognized that a robust monitoring plan was needed to enhance opportunities for learning from this effort and allow them to improve restoration techniques over the course of the project. To meet this need, AZGFD partnered with GCT and others to obtain additional funds to support monitoring and adaptive management.

Two monitoring efforts have been initiated to provide an adaptive management aspect to this ambitious habitat enhancement program. The first effort, supported through a grant from the Grazing Land Conservation Initiative, is monitoring plant-community dynamics in replicated 25 by 30 m plots established within seeded and unseeded areas, each with and without herbicide. Control plots were also established outside the treatment areas. A second effort is monitoring plots from the 2005 baseline assessment, augmented with randomly located vegetation plots, also stratified by seeding and herbicide treatments. This effort focuses on differences among treatment types with respect to cheatgrass cover and shrub seedling establishment. Lastly, several photo plots of treatment areas where pinyon-juniper woodland has been cleared were established to evaluate long-term effectiveness of tree removal.

AZGFD plans to continue mule deer habitat improvement treatments for several years; however, the cost for this project is quite high and effectiveness is difficult to assess over the short term. Monitoring projects provide an important expansion of this project at the implementation phase, with a potential for optimizing treatments and providing an early warning of any unintended consequences that might arise. Monitoring will also provide data for assessing success, an important step in sustaining support and developing collaborative approaches that will be necessary if AZGFD is to complete the ambitious treatment plan and meet landscape-scale objectives for the reestablishment of native shrub communities and enhancement of mule deer winter habitat.

CASE STUDY 3

Restoration of native cool-season grasses: Proceed with caution

INTRODUCTION

Roughly half of the western United States is public land, and 70% of this area is used to graze livestock. According to the U.S. Department of Agriculture, "over half of public rangelands are in unsatisfactory condition, and about two-thirds of these rangelands are not responding to current management practices" (Peters et al. 2006). Much of the degradation has resulted from livestock grazing practices during the twentieth century; however, removal of livestock, by itself, is unlikely to lead to rapid recovery (Allen 1995, Bekker et al. 1998, Loeser et al. 2007). Active restoration, including reseeding of native bunchgrasses, is often proposed for degraded rangelands, yet the slow growth of most native perennial grasses, coupled with low and increasingly erratic precipitation in arid regions, has resulted in highly variable responses to grassland restoration efforts (Westoby et al. 1989, Cortina et al. 2006). Broad-scale restoration of degraded arid rangelands represents an ongoing challenge that is ecologically uncertain, often expensive, and logistically difficult. This is compounded by the fact that most public rangeland restoration must be reconciled within the context of working ranches, because existing policy generally requires livestock allotments to be occupied, and extended rest or retirement of allotments is controversial and rarely successful (Hess and Holechek 1995, Reese 2005).

When GCT purchased the Kane and Two Mile Ranches, a critical challenge was to discover if and how landscape-level grassland restoration and conservation could be achieved in the context of a working cattle ranch. Native grass seed is expensive, and germination and recruitment of seedlings is typically episodic, with high interannual variability, making active reseeding efforts a financial and ecological gamble. To explore these issues and prepare for possible larger restoration efforts across the lower elevations of the Kane and Two Mile Ranches, we researched common practices for reseeding native cool-season perennial grasses, the component of the local plant community that has experienced greatest declines in recent decades. Perhaps the most degraded rangeland on the ranches is found in the House Rock Valley (HRV), a 49,200-ha expanse between the Vermilion Cliffs to the north and the Marble Canyon gorge to the south. Near the center of the HRV, on an in-holding of private land, we conducted reseeding trials and followed bunchgrass recruitment to address two primary research questions: (1) which seeding methods are most successful and (2) to what extent might precipitation determine success?

METHODS

Over the last few decades, efforts to improve degraded public rangelands have shifted from planting exotic species intended to enhance livestock forage to the reintroduction of native plant species, with the aim of moving beyond rangeland improvement and toward the restoration of ecological function and native biodiversity (Call and Roundy 1991, Allen 1995). Pursuing ecological restoration of arid rangelands through reseeding has been encouraged when guided by appropriate theory and methods (Aronson et al. 1993, Monsen et al. 2004). However, success of reseeding efforts is probably closely tied to seasonal precipitation (Bakker et al. 2003) and seedbed microsite characteristics (Von Winkel et al. 1991). These conditions can be difficult to manipulate in restoration treatments.

To investigate how managers might most effectively carry out restoration of native grasslands on a working cattle ranch, we conducted an experiment comparing three seeding techniques commonly used in rangeland restoration: mechanical drill-seeding; broadcast seeding; and broadcast seeding followed by cattle trampling. We also employed cattle trampling without broadcast seeding to isolate the effects of cattle and to test the germination response drawing on the soil seed bank alone. To simulate a wet winter, we supplied additional water to half the plots, while the other half experienced ambient conditions during a drier-than-average growing season. These treatments, along with watered and unwatered controls, resulted in 10 distinct experimental treatments. We used a replicated, block design to establish 100 experimental plots, each 3 square meters,

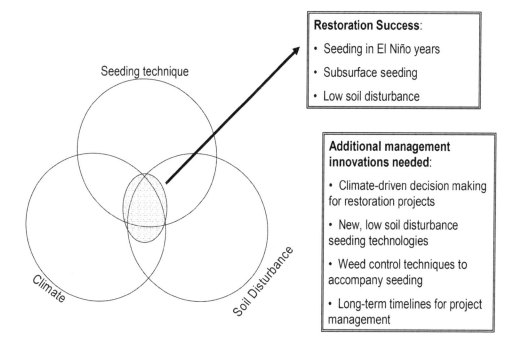

Restoration success depends on specific combinations of climate scenarios and seeding techniques and may necessitate additional management innovations.

Figure 5. In a grassland restoration experiment in the House Rock Valley, reseeding success depended on drill-seeding techniques during simulated wet winter conditions, which comes at the cost of soil disturbance that may increase erosion and the likelihood of weedy species germination, forcing managers to make difficult decisions involving complicated ecological, budgetary, and planning tradeoffs.

near the center of the HRV, where soil and vegetation conditions were representative of the valley as a whole. We randomly assigned one of 10 treatments to each plot, resulting in 10 replicates per treatment.

We selected four native, C_3 grasses for seeding: Indian ricegrass (*Achnatherum hymenoides*), needle and thread (*Hesperostipa comata*), Sandberg bluegrass (*Poa secunda*), and bottlebrush squirreltail (*Elymus elymoides*). These four species are well adapted to the site's arid climate and well-drained soils, but are present in the HRV in limited abundance, presumably due to historical grazing pressures during winter and spring, when cool-season grasses germinate and initiate growth. In addition to germination counts, we collected data on plant community composition and soil stability to further assess the impacts of the treatments on this arid rangeland. This replicated, controlled experiment, carried out in small plots over a limited area of one part of the Kane and Two Mile Ranches, focuses on the proximate factors affecting germination and recruitment of cool-season grasses, acknowledging that long-term restoration can only be achieved after successful recruitment and growth over large and heterogeneous areas.

CURRENT STATUS AND EXPECTED RESULTS

Following the initial winter and spring season, germination of seeded species was observed, at low densities, and only in the drill-seeded plots that received additional water. This result supports previous findings that high precipitation and favorable seedbed preparation strongly influence re-seeding success. Even in these plots, however, germination rates were low and many seedlings died during the growing season. Seedling recruitment, which will be measured over the coming two growing seasons, appears to be low initially, further calling into question the feasibility of active reseeding as a central technique for arid grassland restoration.

Despite the relative success of drill-seeding under experimental conditions that provided above-average precipitation, this treatment also had undesirable effects on the plant community. In the plots where the germination of native grasses was highest, we also observed a proliferation of exotic weeds, especially Russian thistle (*Salsola tragus*) and decreases in soil stability. These initial findings are the subject of ongoing research. However, it appears that while the mechanical drill-seeding was essential for native species germination, the disturbance it caused had the unintended consequence of increasing exotic plant germination and the risk of soil erosion, two key drivers of rangeland degradation of particular concern to GCT and its partners. One outcome of this research is an evolving management model of the components of restoration success under challenging conditions (Figure 5).

We have yet to complete this experiment, but one result is clear: there are tradeoffs to commonly used reseeding techniques and a careful approach is warranted. Our decision to invest two or three years in experimentation prior to broad implementation of a particular restoration treatment has already provided relevant information on the reasonable expectations regarding native grass establishment, the influence of precipitation, and potentially overwhelming unintended consequences of treatments that involve heavy soil surface disturbances. This phased, experimental approach to grassland restoration already has provided a context for developing informed strategies for lower-elevation projects and represents a first, necessary step in developing ecologically and economically viable approaches for re-establishing native plant species and restoring degraded arid rangelands.

The ecological tradeoffs implicit in our early results suggest that managers should be cautious about pushing ahead with reseeding treatments without first testing them under local conditions. Even the most successful seeding technique, applied at the wrong time, could favor invasive weeds over native grasses, because natives require favorable conditions, whereas weedy species that favor disturbance are often present in the soil seedbank.

Finding ways to reconcile active livestock grazing and restoration will not be an easy task in the HRV. Our preliminary results showed little support for the idea that germination can be improved by the seedbed disturbance cause by livestock trampling, and because grass seedlings are themselves vulnerable to trampling, livestock probably should be removed from pastures where germination of seeded species is underway.

The relative infrequency of high precipitation years that are associated with successful germination, as well as the low germination and recruitment rates of native bunchgrasses, even under favorable conditions, suggests that restoration of grasslands in the HRV will proceed very slowly. Nonetheless, continued experimentation remains an important objective. Given the difficulty of balancing the need for restoring native grasslands with the policy requiring livestock grazing, and acknowledging the high climatic variability and its unpredictable nature, a significant investment in learning is needed, such that the costly tradeoffs identified in this study can be lessened or at least managed. Arid grassland and shrub steppe communities account for 26% of the Colorado Plateau, and they have suffered an estimated 20% decline since 1850 (Tuhy 2002). Reversing these trends will not happen easily or through good intention alone. The integration of ranching and conservation provides opportunities to carry out much-needed research in a pragmatic manner, in collaboration with land management agencies. Intelligent sequencing of controlled experiments, field trials, and adaptively managed restoration projects at landscape scales can improve our understanding of ecological tradeoffs and increase the likelihood of success in the challenging practice of arid rangeland restoration.

CASE STUDY 4

Modeling cheatgrass occurrence to target restoration priorities

INTRODUCTION

In the Southwest, invasive plant species pose a substantial threat, especially in areas where the structure and composition of native vegetation has been affected by fire, fire suppression policies, or livestock grazing (Sieg et al. 2003). During the 2005 baseline ecosystem assessment, vegetation, ground cover, and soil characteristics were measured at 606 random plot locations. Of the 34 exotic and invasive plant species we detected, cheatgrass was the most widely distributed, occurring on 41% of the plots. Given recent disturbances, such as the 2006 Warm Fire, and degraded conditions associated with historical livestock grazing and fire suppression, there is high potential for further invasion on this landscape, and the identification of environmental conditions that influence cheatgrass occurrence is important for targeting restoration efforts and deterring further spread. To address these issues, we are developing statistical models of cheatgrass occurrence, using a novel model-selection framework to identify the model or models best supported by existing data. Our objective is to build a robust, continuous, probabilistic model of cheatgrass occurrence across the entirety of the Kane and Two Mile Ranches. While this effort is ongoing, we present preliminary model results as an example of how field inventory, expert opinion, and advanced spatial modeling techniques can be combined to help advance understanding of patterns of rapid cheatgrass establishment.

METHODS

We drew on information from 606 baseline assessment plots collected by GCT in summer 2005, multiple logistic regression models, and model selection and multi-model inferential methods to build a spatial data layer that predicts cheatgrass occurrence. Data collection methodologies were designed by the Forest Ecosystem Restoration Analysis (ForestERA) Project at Northern Arizona University, in consultation with GCT and its Science Advisory Council. Data were collected by a field crew supervised by GCT. First, 42 biophysical landcover types were identified using USGS Southwest ReGAP vegetation data and USFS and Natural Resources Conservation Service (NRCS) digital soil data layers. Next, assessment plot locations were randomly assigned to the 42 landcover types in proportion to the areal extent of each type. Within each 8 m diameter circular assessment plot, data on vegetation, presence or absence of cheatgrass, and soil stability characteristics were measured and aggregate soil samples were collected for analysis in the lab. During this 2005 sampling period, cheatgrass occurred on 41% of the 606 baseline assessment plots.

Information on landscape-scale variables was derived using a geographic information

Table 2. Cumulative AIC weights, model-averaged parameter estimates, and unconditional standard errors for all variables included in the global model of cheatgrass occurrence across the Kane and Two Mile Ranches in 2005.

Parameter	AIC Weight	Estimate	SE
Spring climate	1.000	1.276	0.153
Roughness	1.000	-1.263	0.249
Sandy-sandy loam	1.000	-1.272	0.262
Fire	1.000	3.154	0.787
Bedrock-rock outcrop	0.874	-1.104	0.666
Roads	0.871	0.263	0.155
Grazing	0.693	0.163	0.150
Rocky loam	0.679	0.367	0.350
Spatial term	0.632	0.166	0.178
Rocky sand	0.455	0.168	0.281
Silt-clay loam	0.446	0.235	0.385
Topographic heat load	0.414	0.109	0.199

system (GIS; ArcGIS V9.2, Environmental Systems Research Institute, Redlands, CA, USA) at a 30 m resolution. Variables were identified and classified using a standardized, expert- and literature-based information collection process. For the statistical modeling, we characterized climatic conditions by generating spatial input models (i.e., data layers) of spring (March–June) 2005 climate (average precipitation × average temperature) derived using 4 km resolution PRISM data. We classified soil texture into five classes using digital Terrestrial Ecosystem Survey data obtained from the USFS and Soil Survey geographic data downloaded from the NRCS. We used information from the USFS to identify large fire perimeters for the period 1996–2005. We estimated topographic roughness on the landscape using USGS National Elevation Data and the Spatial Analyst extension to ArcGIS. We calculated topographic heat load (McCune and Keon 2002) by incorporating slope, aspect, and latitude to approximate direct incident radiation and heat load. We derived a data layer to quantify distances from all road types on the study area using roads data obtained from the USFS and GCT. As a surrogate measure for grazing intensity, we used information on the location of known water sources across the study area to create a layer that quantified distance from water sources. Lastly, we derived a spatial trend term for the statistical model that was the mathematical product of the latitudinal and longitudinal coordinates for a given cell. All continuous variables were standardized prior to statistical analysis.

We used multiple logistic regression and Akaike's Information Criterion (AIC; Burnham and Anderson 2002) to predict cheatgrass occurrence as a function of the landscape-scale variables indicated above. Because we were interested in exploring the relationship between cheatgrass oc-

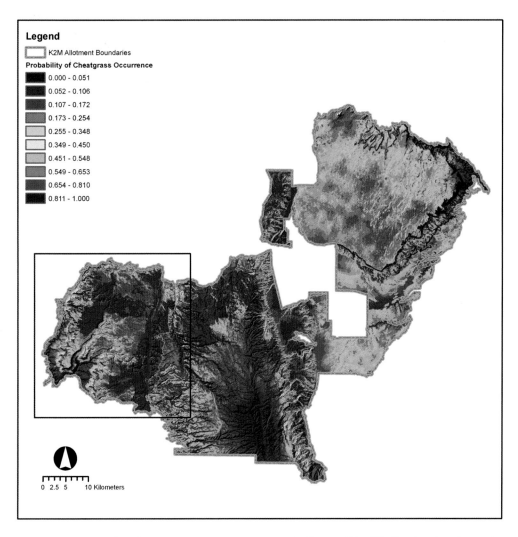

Figure 6. Draft model of predicted cheatgrass occurrence across the Kane and Two Mile Ranches. Inset box on the left is the area of Figure 7. (Note: These draft figures should not be used to guide management; contact C. Albano for further information.)

currence and the landscape variables, we focused our analysis on variable importance rather than on the specification of the best predictive model(s). Thus, we used a global (i.e., fully parameterized) model and multi-model inference to compute model-averaged regression coefficients and their unconditional standard errors from all possible combinations of variables (Burnham and Anderson 2002). We used cumulative AIC weights to rank the relative importance of the variables in the global model (Burnham and Anderson 2002), and considered a cumulative AIC weight of 0.50 to be strong evidence for cheatgrass response to a given variable (Barbieri and Berger 2004). We conducted all analyses in SAS (V9.1, SAS Institute, Cary, NC, USA) and used maximum-likelihood and PROC LOGISTIC for parameter estimation. We

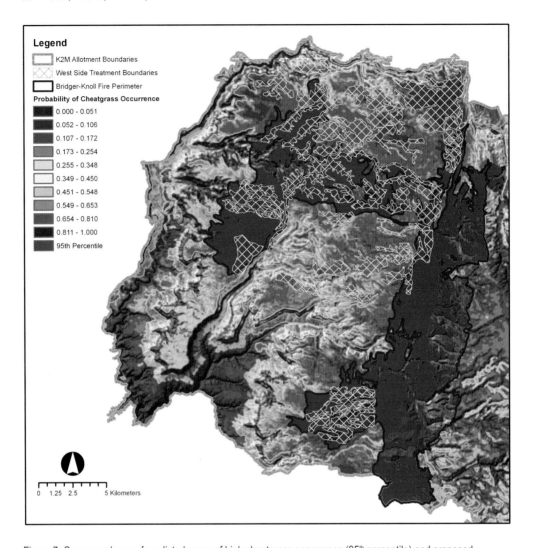

Figure 7. Correspondence of predicted areas of high cheatgrass occurrence (95th percentile) and proposed restoration projects designed to benefit winter habitat for mule deer. The possibility that mechanical treatment could lead to cheatgrass establishment has highlighted the need for rigorous monitoring of the herbaceous plant community following drill seeding of restoration project areas.

used an intercept-only model to evaluate how well our global model approximated the data.

Because we had a large sample size and AIC is asymptotically equivalent to common cross-validation and bootstrap techniques (Shtatland et al. 2004), we used AIC in all steps of model building and assessment. To evaluate model classification accuracy, we computed the area under the receiver operating characteristic (ROC) curve. The ROC value provides a likelihood-based measure of discrimination between predicted presences and absences. We considered ROC values of 0.70 to provide good discrimination.

CURRENT STATUS AND EXPECTED RESULTS

We developed a preliminary model of predicted cheatgrass occurrence across the Kane and Two Mile Ranches in 2005 (Figure 6). The predictive integrity of this model was high, about 217 AIC units better than the intercept-only (i.e., null) model, and the ROC value was 0.85, suggesting good discrimination. The maximum probability predicted by the final model was 0.997. Spring climate, topographic roughness, sandy to sandy loam soils, and fire-affected areas were the strongest predictors (i.e., AIC weights approximately 1.0) of cheatgrass occurrence (Table 2). Rocky sand soils and silt-clay loam soils were relatively weak predictors of occurrence.

The model presented here should be viewed as preliminary, subject to ongoing changes and refinement as this research continues (i.e., it should not be used in its current form to make management decisions, without prior communication with C. Albano or B. Dickson). We have, however, used this model to inform a preliminary exploration of cheatgrass management issues on the ranches. For example, we overlaid the model onto known areas of high cheatgrass occurrence and found strong concordance between predictions of high probability of cheatgrass occurrence and sites where this species is known to have previously established, such as on the west side of the Kaibab Plateau where cheatgrass invaded and established extensive, nearly monospecific stands following the Bridger Knoll complex of fires in 1996. Interestingly, some project areas proposed for native shrub restoration, for the purpose of mule deer habitat enhancement, also coincide with very high probability of cheatgrass occurrence (Figure 7). This finding generated a re-examination of the restoration program focusing on the west side of the Kaibab Plateau, a discussion of which follows.

DISCUSSION

Research and Monitoring to Guide Landscape-level Restoration, Conservation, and Management

The way we manage America's public lands is changing. The increasing demands of a rapidly expanding human population, combined with the poorly understood but increasingly apparent influence of climate change, indicate a desperate need for new science that is both rigorous and practical. Land and resource management cannot look to the past for answers, as global change is clearly showing that future environments will not necessarily resemble those from any period in the past. Instead, we should expect novel climates, unique plant communities, and ecological surprises (Williams and Jackson 2007). To prepare for this uncertain future, managers must be ready to integrate current scientific understanding with their experience and work in collaborative teams to identify the best way forward in protecting biodiversity, ecosystem function, and the services that are provided by healthy ecosystems, including soil fertility, water purification and delivery, and the appropriate occurrence of fire on the landscape.

This is a new and by no means easy task, but one that is critical to a more hopeful future. We believe that the science-informed management now occurring across the Kane and Two Mile Ranches provides a rough sketch of how such integrated, collaborative, and science-based efforts might move forward in the future. To illustrate just one example that emerged from the four ongoing case studies examined in this chapter, consider the issue of cheatgrass invasion.

Given that shrub restoration involves drill-seeding and the associated soil disturbance, and given the preliminary finding that this disturbance can exacerbate exotic species problems in other areas of the ranch, we met to discuss management plans in light of the emerging science. Subsequent discussions and planning efforts benefitted from the integration of preliminary results from each of the projects outlined in this paper, resulting in a new approach to ecological monitoring of the west-side mule deer habitat restoration effort. We concluded that there was sufficient reason for caution on the ambitious mechanical treatment plans, given the findings from the House Rock Valley grassland restoration experiment and the predictions of the preliminary cheatgrass model. Drawing on experiences from the baseline assessment and protocols for expanding monitoring within the Warm Fire burn area, we worked with partners to design the vegetation monitoring program for the west side project described as Case Study 2. This monitoring effort is designed to track changes in the understory plant community, specifically with respect to the establishment of cheatgrass and other exotic annuals. Monitoring data will provide an early warning of any unintended consequences, allowing improvement or suspension of the seeding effort, if necessary, and it may also identify particularly successful outcomes and guide their replication across the treatment area.

This integration of experimental research, spatial modeling, and monitoring is a clear example of the new way that science can work with management to increase our understanding of environmental change, while simultaneously informing and improving management practice. Implementing this new understanding requires strong collaborative relationships among landowners, land and natural resource management agencies, and the public. However, at the same time that public involvement increases through participation in influential collaborative teams and public interest groups such as GCT, federal budgets that support land and resource conservation continue to decline. This dynamic has resulted in an emerging "capacity vacuum" that is only now becoming apparent, and it is unclear how it will be filled. It appears highly unlikely that the budgets for federal land management agencies will increase to support the level of engagement that typified the twentieth century. A new model is therefore needed.

The Kane and Two Mile Ranches project is a response to this situation on the southern Colorado Plateau, where the changing economics of livestock ranching, combined with the increasing responsibilities of strained federal agencies, have created a need for new public-private partnerships to pursue collaborative management across 340,000 ha of stunningly beautiful public lands ringed by national parks and monuments. At the core of this novel experiment is a science program that integrates ecological assessment, targeted experimentation, and an adaptive management approach powered by efficient monitoring of environmental change and management effectiveness. The ranches have tremendous ecological importance, significant social value, and spectacular beauty. Conservation and restoration opportunities across the area are unsurpassed in the American Southwest.

ACKNOWLEDGEMENTS

The staff, board, and supporters of GCT made this research possible through their purchase of the Kane and Two Mile Ranches and their support for a science-based approach to ecological restoration and land and water management across this incredibly diverse and beautiful region. We also wish to thank the staff of the USFS, North Kaibab Ranger District of the Kaibab National Forest, and the Arizona Strip Office of the Bureau of Land Management for their willingness to partner with us and support these and other research efforts. Many participants in GCT's Volunteer Program contributed to the research described in this paper, and we thank volunteer coordinators Kate Watters, Travis Wiggins, and Lauren Berutich for coordinating this important work. The projects described in this paper are supported by grants from AZGFD, the Doris Duke Charitable Foundation, the Joint Fire Science Program, the National Forest Foundation, and the Wilburforce Foundation.

LITERATURE CITED

Allen, E. B. 1995. Restoration ecology: limits and possibilities in arid and semiarid lands. In Proceedings—Wildland shrub and arid land restoration symposium, Las Vegas, Nevada; 19–21 October 1993, edited by B. S. Roundy, E. D. McArthur, J. S. Haley, and D. K. Mann, pp. 7–15. U.S. Department of Agriculture, Forest Service General Technical Report INT-315. Intermountain Forest and Range Experiment Station, Ogden, Utah.

Aronson, J., C. Floret, E. Lefloc'h, C. Ovalle and R. Pontanier. 1993. Restoration and rehabilitation of degraded ecosystems in arid and semiarid lands: I. A view from the south. Restoration Ecology 1:8–17.

Bakker, J. D., S. D. Wilson, J. M. Christian, X. Li, L. G. Ambrose, and J. Waddington. 2003. Contingency of grassland restoration on year, site, and competition from introduced grasses. Ecological Applications 13:137–53.

Barbieri, M. M., and J. O. Berger. 2004. Optimal predictive model selection. Annals of Statistics 32:870–97.

Barclay, A. D., J. L. Betancourt, and C. D. Allen. 2004. Effects of seeding ryegrass (*Lolium multiflorum*) on vegetation recovery following fire in a ponderosa pine (*Pinus ponderosa*) forest. International Journal of Wildland Fire 13:183–94.

Bekker, R. M., J. P. Bakker, U. Grandin, R. Kalamees, P. Milberg, P. Poschlod, K. Thompson and J. H. Willems. 1998. Seed size, shape and vertical distribution in the soil: Indicators of seed longevity. Functional Ecology 12:834–42.

Brown, J. K. 1974. Handbook for inventorying downed woody material. U. S. Department of Agriculture, Forest Service General Technical Report INT-16. Intermountain Forest and Range Experiment Station, Ogden, Utah.

Burnham, K. P., and D. R Anderson. 2002. Model selection and multimodel inference: A practical information-theoretic approach. Springer-Verlag, New York.

Call, C. A., and B. A. Roundy. 1991. Perspectives and processes in revegetation of arid and semiarid rangelands. Journal of Range Management 44:543–49.

Cortina, J., F. T. Maestre, R. Vallejo, M. J. Baeza, A. Valdecantos, and M. Perez-Devesa. 2006. Ecosystem structure, function, and restoration success: Are they related? Journal for Nature Conservation 14:152–60.

Davis, J. N., D. S. Summers, D. B. Eddington, and D. B. Davis. 2005. Utah big game range trend studies, volume 1. State of Utah, Department of Natural Resources, Wildlife Resources. Publication No. 06–07. Department of Natural Resources, Salt Lake City, Utah.

Gruell, G. E. 1986. Post-1900 mule deer eruptions in the Intermountain West: Principle cause and influences. U. S. Department of Agriculture, Forest Service General Technical Report INT206. Intermountain Research Station. Ogden, Utah.

Haywood, D. D., R. L. Brown, R. H. Smith, and C. Y. McCulloch. 1987. Migration patterns and habitat utilization by Kaibab mule deer. Research Branch Final Report, Federal Aid in Wildlife Restoration, Project W-78-R. Arizona Game and Fish Department, Phoenix.

Heffelfinger, J. R., C. Brewer, C. H. Alcalá-Galván, B. Hale, D. L. Weybright, B. F. Wakeling, L. H. Carpenter, and N. L. Dodd. 2006. Habitat guidelines for mule deer: Southwest desert ecoregion. Mule Deer Working Group, Western Association of Fish and Wildlife Agencies. http://www.wildlife.state.nm. us/conservation/habitat_handbook/index.htm. Last accessed 6 August, 2009.

Hess, K. Jr., and J. L. Holechek. 1995. Policy roots of land degradation in the arid regions of the United States: An overview. Environmental Monitoring and Assessment 37:123–41.

Keeley, J. F., C. D. Allen, J. L. Betancourt, G. Chong, C. J. Fotheringham, and H. D. Safford. 2006. A 21st Century perspective on postfire seeding. Journal of Forestry 104:103–04.

Kerns, B. K., W. G. Thies, and C. G. Niwa. 2006. Season and severity of prescribed burn in ponderosa pine forests: Implications for understory native and exotic plants. Ecoscience 131:44–55.

Key, C. H., and N. C. Benson. 2006. Landscape assessment: Sampling and analysis methods. In FIREMON: Fire effects monitoring and inventory system, edited by D. C. Lutes, R. E. Keane, J. F. Caratti, C. H. Key, N. C. Benson, S. Sutherland, and L. J. Gangi, pp. LA1–LA55. U.S. Department of Agriculture, Forest Service General Technical Report RMRS-GTR-164-CD. Rocky Mountain Research Station, Fort Collins, Colorado.

Loeser, M. R, T. D. Sisk, and T. E. Crews. 2007. Impact of grazing intensity during drought in an Arizona grassland. Conservation Biology 21:87–97.

McCune, B., and D. Keon. 2002. Equations for potential annual direct incident radiation and heat load. Journal of Vegetation Science 13:603–06.

McHugh, C. W., and T. E. Kolb. 2003. Ponderosa pine mortality following fire in northern Arizona. International Journal of Wildland Fire 12:7–22.

Monsen, S. B., R. Stevens, and N. L. Shaw (editors). 2004. Restoring western ranges and wildlands. U.S. Department of Agriculture, Forest Service General Technical Report RMRS-GTR-136-vol-1. Rocky Mountain Research Station, Fort Collins, Colorado.

Peters, D., D. Anderson, K. Havstad, S. Tartowski, B. Bestelmeyer, M. Lucero, A. Rango, R. Estell, E. Frederickson, E. Frederickson, E. Herrick, and J. Herrick. 2006. Technologies for Management of Arid Rangelands, Agricultural Research Service project 406494, *http://www.ars.usda.gov/research/projects/projects.htm?ACCN_NO=406494&showpars=true&fy=2006.* Accessed on 23 May 2009.

Reese, A. 2005. The Big Buyout. High Country News Online, *http://www.hcn.org/servlets/hcn.Article?article_id=15398.* Published 04 April 2005; accessed on 21 December 2007.

Russo, J. P. 1964. The Kaibab deer herd: Its history, problems, and management. Wildlife Bulletin 7. Arizona Game and Fish Department, Phoenix, USA.

Shtatland, E. S., K. Kleinman, and E. M. Cain. 2004. A new strategy of model building in PROC LOGISTIC with automatic variable selection, validation, shrinkage and model averaging. SUGI '29 Proceeding, Paper 191-29, SAS Institute, Inc. Cary, North Carolina.

Sieg, C. H., B. G. Phillips, and L. P. Moser. 2003. Exotic invasive plants. In Ecological restoration of southwestern ponderosa pine forests, edited by P. Friederici, pp. 251–67. Island Press, Washington, D.C.

Sieg, C. H., J. D. McMillin, J. F. Fowler, K. K. Allen, J. F. Negron, L. L. Wadleigh, J. A. Anhold, and K. E. Gibson. 2006. Best predictors for post-fire mortality of ponderosa pine trees in the Intermountain West. Forest Science 52:718–28.

Stolhgren, T., Barnett D. T., and S. Simonson. 2003. Beyond North American Weed Management Association Standards. North American Weed Management Association. Meade, Kansas.

Tuhy, J. S., P. Comer, D. Dorfman, M. Lammert, J. Humke, B. Cholvin, G. Bell, B. Neely, S. Silbert, L. Whitham, and B. Baker. 2002. A conservation assessment of the Colorado Plateau ecoregion. The Nature Conservancy, Moab, Utah.

Von Winkel, K., B. A. Roundy and, J. R. Cox. 1991. Influence of seedbed microsite characteristics on grass seedling emergence. Journal of Range Management 44:210–14.

Wakeling, B. F. 2007. Wildlife management decisions and Type I and II Errors. In Proceedings of the Sixth Western States and Provinces Deer and Elk Workshop – 2005, edited by M. Cox, pp. 35–42. Nevada Department of Wildlife, Reno.

Watkins, B. E., C. J. Bishop, E. J. Bergman, A. Bronson, B. Hale, B. F. Wakeling, L. H. Carpenter, and D. W. Lutz. 2007. Habitat guidelines for mule deer: Colorado Plateau shrubland and forest ecoregion. Mule Deer Working Group, Western Association of Fish and Wildlife Agencies. *http://www.wildlife.state.nm.us/conservation/habitat_handbook/index.htm.* Accessed 6 August 2009.

Westoby, M., B. Walker, and I. Noy-Meir. 1989. Opportunistic management for rangelands not at equilibrium. Journal of Range Management 42:266–74.

Williams, J. W., and S. T. Jackson. 2007. Novel climates, no-analog communities, and ecological surprises. Frontiers in Ecology and the Environment 5:475–82.

Assessing Monitoring Frameworks and Systems

MAPPING ECOLOGICAL SITES FOR LONG-TERM MONITORING IN NATIONAL PARKS

Steven L. Garman, Dana Witwicki, and Aneth Wight

ABSTRACT

The Northern Colorado Plateau Inventory and Monitoring Network (NCPN) of the National Park Service is responsible for the design and initiation of long-term monitoring of upland ecosystems across 16 National Park units on the Colorado Plateau. Monitored indicators include soil properties, as well as vegetation structure and composition. Targeted populations for monitoring are defined by individual ecological site types, which have variable resistance to disturbances. Monitoring of individual ecological sites largely ensures sufficient samples for status and trend evaluations, and facilitates interpretation of observed changes in indicators on the basis of soil and other properties. The sampling and survey design approach used by the NCPN requires explicit mapping of ecological sites. Existing soil surveys provide maps of soil map units that can contain one to many ecological sites and provide estimates of site proportions within a map unit. We used two predictive mapping methods to enhance the mapping of ecological sites from existing soil maps. Methods included feature-extraction modeling, which employs machine-learning algorithms to generate feature extraction rules, and decision tree models. Feature-extraction methods were applied to selected ecological sites in Canyonlands National Park, and both methods were used to map selected ecological sites in Capitol Reef National Park. Models were parameterized and tested using field samples of ecological sites. Prediction accuracy of the feature-extraction model in Canyonlands National Park ranged from 64 to 71 percent for three of the five target ecological site types, but overall accuracy was only 43 percent. A hierarchical re-learning process with enhanced training samples improved overall accuracy to 75 percent, although accuracy of the re-learned models was based on model-development data. Decision-tree models for Capitol Reef National Park separated non-targeted from targeted site types, but predicted mixtures of targeted ecological sites. Prediction accuracy for these mixtures was 52 to 65 percent. Feature-extraction models developed to predict one of the ecological site mixtures in Capitol Reef National Park were as accurate as the decision tree models. To determine accuracy and efficiency gains with predictive mapping, Monte Carlo simulations were used to estimate potential "accuracy" in locating a targeted ecological site within soil map units using only the reported distributional percentages. Compared to simulations, both predictive mapping methods afforded greater accuracy in mapping ecological sites, which translates operationally to greater efficiency of sample selection. Small sample sizes and lack of independent data for testing re-learned feature-extraction models limited the rigor of model comparisons and assessments. Despite these limitations, results of this study indicate the potential for predictive mapping methods to enhance the spatial delineation of ecological sites. The advantage of enhanced mapping of ecological sites is increased efficiency of sampling designs for monitoring, and more exact delineation of the spatial extent of targeted populations and, thus, sampling inference.

INTRODUCTION

The Northern Colorado Plateau Inventory and Monitoring Network (NCPN) of the National Park Service is responsible for the design and initiation of long-term monitoring of upland ecosystems across 16 National Park units on the Colorado Plateau (O'Dell et al. 2005). The goal of upland monitoring is to provide early warning of system degradation to inform park-management decision making. Indicators of upland monitoring include vegetation structure and composition, and soil properties such as soil aggregate stability and biological soil crust. Focused discussion between NCPN and park resource-management staff identifies general vegetation types for monitoring, such as shrublands, grasslands, and woodlands. Landscape variability in soils, landform, and localized climatic conditions supporting the selected vegetation types is then considered in determining areas for monitoring. The use of sampling and survey designs is a key approach of the NCPN in aquiring monitoring observations, which are most effective when the variability in monitored observations due to biophysical factors is minimized. For instance, the response of shrubland communities on sandy soils to climatic change and other stressors is likely to differ from shrublands on deeper, sandy loam soils due to inherent differences in resistance and resilience of these soils. Targeting upland monitoring efforts to specific combinations of soil, landform, and localized climate provides more precise monitoring information, and enhances the ability to understand reasons for observed changes in monitored indicators.

Ecological site types (SRM 1989, 1995) are defined by biophysical properties that influence system processes, and are used by the NCPN as the basis for targeting upland monitoring efforts in grasslands and shrublands. The ecological site is an expansion of the rangeland-site concept and is attributed by soil properties (e.g., texture, depth), parent material, landform, localized climate, and potential vegetation. Additionally, processes leading to altered states and characterization of these states are described to provide a context for determining ecological degradation. Multiple ecological site types are associated with a physiognomic type, such as grasslands, but types differ in key soil properties that influence soil hydrologic processes, soil-nutrient cycling, and resistance to disturbances. Landform and localized climatic conditions (e.g., desert vs. semi-desert) also tend to be factors differentiating ecological site types. The use of separate ecological site types as distinct target populations and the development of separate sampling designs for each target population are important features of NCPN upland monitoring. Focused monitoring of individual ecological site types ensures that sufficient monitoring observations for status and trend evaluation of indicators are collected, and that data are characteristic of the same underlying processes (O'Dell et al. 2005). The latter especially facilitates understanding reasons for observed changes. Also, separate sampling designs for ecological site types are more efficient, in that sample sizes can be adjusted to account for site-type differences in the inherent variability of monitoring indicators.

Survey designs used by the NCPN to monitor upland systems require the explicit delineation of a finite, target population (Cochran 1977). Explicit mapping of the spatial extent of each ecological site selected for monitoring is thus a critical first step in designing a sampling scheme for monitoring. A map of an ecological site defines the inference domain and delineates the maximum number of possible sampling locations. These maps are the basis for the sampling frame from which sampling locations are selected, and are required input to sample-selection software (e.g., Theobald et al. 2007). Existing soil surveys (e.g., SSURGO – *http://soildatamart.nrcs.usda.gov*) describe the ecological site types for grasslands and shrublands in NCPN park units; they also provide soil maps (1:24,000). However, existing soil maps have limitations. The basic mapping unit, called a soil map unit, contains from one to many ecological site types and inclusions (e.g., rock outcrop). Furthermore, the reported proportions of ecological sites within a map unit are a generalization over multiple polygons of the map unit on a landscape. Thus, the number—and even types—of ecological sites within a map-unit polygon can differ considerably from the described average. Using the existing soil maps to define the spatial extent of targeted ecological sites can lead to over-estimation of the area of targeted ecological sites, and could lead to the selection of numerous monitoring samples in non-targeted ecological sites. Both of these factors can contribute to misrepresentation of the targeted population and inefficient sampling designs.

Predictive mapping methods have the potential to discern individual ecological site types within soil map units. Predictive mapping is commonly used to map ecological attributes over unsampled areas, or to improve existing maps. Effective mapping models render the spatial associations between biophysical and ecological properties to statistical relationships or rules (MacMillian et al. 2007). The spatial components of these relationships or rules originate from gradients (e.g., elevation gradient) or are contextual in nature (i.e., local neighborhood pattern). Classification (Decision) and Regression Tree models are popular for generating mapping rules because they can handle both continuous and discrete data, as well as non-additive and non-linear relationships in an interpretable manner (Brieman et al. 1984, De'ath and Fabricius 2000). Decision-tree models based on spatial gradients have been widely employed to map vegetation patterns (e.g., Franklin 1998, Brown de Colstoun et al. 2003, Poulos et al. 2007), to evaluate landscape heterogeneity of soils (McKenzie and Ryan 1999), and to map soil orders (Scull et al. 2005). Other rule-based and probability models to map soils have been developed by Lagacherie et al. (1995). Moran and Bui (2002) used contextual features in decision trees to enhance soil-mapping accuracy. More advanced mapping methods have been developed which incorporate contextual features with feature extraction (objective recognition), and employ fuzzy logic and various machine-learning algorithms. These methods have been used to map soils (Zhu et al. 2001), ecological landforms (MacMillan et al. 2007) and downed wood (Swain 2007).

Our goal was to assess methods to map individual ecological site-types using existing soil maps and other relevant biophysical data layers. We evaluated two predictive mapping methods. The first was object recognition using an existing commercial software package. This approach emphasized the use of contextual information to differentiate among site types and is referred to as feature-extraction modeling. The second method we evaluated was decision tree models. A feature-extraction model was developed and tested for Canyonlands National Park. Decision-tree and feature-extraction models were applied to Capitol Reef National Park.

STUDY AREA

Canyonlands National Park encompasses 136,610 ha and is located on the Colorado Plateau in southeast Utah (Figure 1). The park is characterized by deep canyons formed

Table 1. Ecological sites selected for predictive mapping in Canyonlands National Park, and number of field samples used in developing (D) and testing (T) the feature-extraction model. Semidesert site types are associated with annual precipitation levels of 23-30 cm. Annual precipitation of desert site types is 15-23 cm.

Ecological site Name [NRCS Identification Code code]	No.	Number of samples		General description
		D	T	
Semidesert Sandy Loam (4-wing saltbush) [035XY215UT]	215	39	14	deep alluvium and eolian deposits dominated by Indian ricegrass, needle and thread, 4-wing saltbush, Mormon tea
Desert Sandy Loam (4-wing saltbush) [035XY118UT]	118	35	34	deep alluvium and eolian deposits dominated by Indian ricegrass, galleta, dropseeds, 4-wing saltbush
Semidesert Shallow Sandy Loam (blackbrush) [035XY236UT]	236	46	37	shallow residuum and eolian deposits dominated by blackbrush, UT juniper, and pinyon pine
Desert Shallow Sandy Loam (blackbrush) [035XY133UT]	133	41	31	shallow residuum and eolian deposits dominated by blackbrush
Semidesert Sand (4-wing saltbush) [035XY212UT]	212	9	114	deep eolian sands on dunes dominated by Indian ricegrass, 4-wing saltbush, Mormon tea

Table 2. Ecological sites selected for predictive mapping in Capitol Reef National Park, and number of field samples used in developing the decision-tree and feature-extraction models. CR - Cathedral–Rock Springs allotments, H - Hartnet allotment. Test samples were only collected for a predicted ecological site complex (nos. 201/215), with 26 samples for the combined Cathedral-Rock Springs allotments, and 23 samples for the Hartnet allotment.

Ecological site Name [NRCS Identification Code]	No.	Number of samples		General description
		CR	H	
Semidesert Alkali Sandy Loam (alkali sacaton) [035XY201UT]	201	97	94	deep alluvium and eolian deposits dominated by alkali sacaton, greasewood, Indian ricegrass, shadscale
Semidesert Sandy Loam (4-wing saltbush) [035XY215UT]	215	145	39	deep eolian deposits dominated by Indian ricegrass, 4-wing saltbush, Mormon tea
Semidesert Stony Loam (shadscale) [035XY242UT]	242	0	44	deep alluvium and colluvium dominated by galleta, shadscale, Torrey Mormon tea, Bigelow sagebrush, winterfat, Indian ricegrass
Semidesert Very Steep Stony Loam (Salina wildrye) [035XY260UT]	260	0	37	deep soils from basalt glacial outwash dominated by Indian ricegrass, shadscale, Torrey Mormon tea, Bigelow sagebrush, Salina wildrye

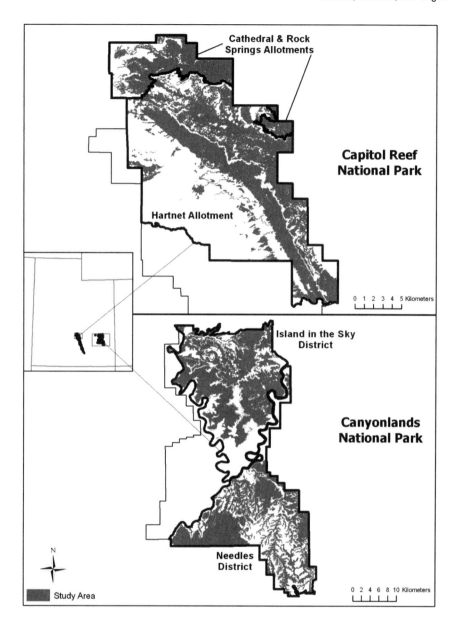

Figure 1. Study area for the evaluation of predictive mapping of ecological site types in Canyonlands National Park and Capitol Reef National Park, UT.

by the Green and Colorado Rivers. Pinyon pine (*Pinus edulis*), Utah juniper (*Juniperus osteosperma*), and blackbrush (*Coleogyne ramosissima*) occur along canyon slopes and on rocky upland soils; desert grasses and shrublands prevail on deeper soils on mesa tops and valley bottoms. Vegetative cover is generally sparse, with intervening areas of slick rock and biological soil crust. The NCPN upland monitoring goals are specific to each park and pertain to topical resource management issues. Concern about the impacts of social trailing on soil erosion and overall degradation of herbaceous and shrubland ecosystems resulted in the selection of dominate grassland (n=3) and shrubland (n=2) ecological site types for monitoring (Table 1) across the Island in the Sky and Needles districts of the park. Total area evaluated for these site types was 48,815 ha (Figure 1).

Capitol Reef National Park is located on the Plateau in south-central Utah, and is 97,895 ha in size (Figure 1). The northern portion of the park was initially selected for NCPN monitoring, and consists of two retired grazing allotments (Cathedral and Rock Springs), and an active allotment (Harnet). Large expanses of grasslands, and pinyon pine and Utah juniper dominate the retired allotments. Grazing has promoted the dominance of shrubs over native bunch grasses in the Hartnet section. Grassland dynamics and grazing-effect issues motivated the selection of four herbaceous ecological sites for monitoring in this park (Table 2). The study area in the combined retired allotments of Cathedral Rock Springs was 3,354 ha, and the Hartnet allotment (Figure 1) encompassed 11,009 ha.

METHODS

Predictive Mapping in Canyonlands National Park. Field reconnaissance and initial inspection of aerial photography (color Digital Orthophoto Quads – DOQQs) indicated noticeable differences in reflectance and contextual properties among the five targeted ecological site types. These properties are the basis of object-recognition mapping; thus, we selected this approach for predictive mapping. We used the Feature Analyst package (VLS 2002) to generate feature extractions rules for mapping ecological sites. Use of this specific package was largely dictated by software costs. Feature Analyst employs inductive machine-learning algorithms to develop pattern recognition rules from the spectral, spatial, and ancillary information of training sites classified to a feature class type. The minimum spatial structure of a feature class is specified by a pattern template (contextual classifier), which is used in evaluating the contextual signal of a feature. Contrasts in the pattern of spectral and related information among feature classes are the basis of the extraction rules. Once rules are developed, they are applied to an image to produce a map of predicted features. Options are available to resample input images to reduce redundancy and processing requirements and to set minimum feature patch-size (aggregate area) in a predicted map. The latter minimizes "slivers." An additional feature of this program is hierarchical learning. Using the original rules and indication of correct and incorrect predictions, it re-learns from the previous misclassifications and generates a refined set of extraction rules. Accuracy metrics of feature-extraction rules are not provided. Accuracy of predictions must be manually assessed using independent data.

We used the most recent soil survey for the park (SCS 1991) to guide sampling for training sites for initial model development. Training sites were obtained in 2005 by sampling 376 locations in soil map units reported or suspected to contain targeted ecological sites. Sample locations were randomly generated, but access and time constraints limited sampling to within 5 km of transportation networks. Additionally, 85 samples were contributed by a rangeland study conducted in the southern portion of the park (M. Miller, USGS). Field crews used GPS to navigate to the pre-determined coordinates of sampling locations. At each location, the dominant ecological site type around the point was assigned based on site-type descriptions in the soil survey, and aerial extent was estimated (up to 2 ha). Comprehensive assessment of an ecological site type requires inspection of subsurface soil properties; however, digging holes in Canyonlands National Park was

Table 3. Digital attributes used in the feature-extraction model for Canyonlands National Park.

Attribute	Resolution	Range	Source, description
Color Digital Orthophoto Quads (DOQQs) 3-Band, orthorectified	1 m	na	Horizons Inc., Salt Lake City, UT, Bureau of Reclamation, 1:12,000, aerial photographs, 2001
Elevation	10 m	1140-2189 m	USGS digital elevation models (DEMS) 1:24,000 quadrangles
Geology (bedrock)	1:62,500	19 classes	Digitized from Geologic Map of Canyonlands National Park and Vicinity, Utah (Huntoon, Peter W., Billingsley, George H., and Breed, William J., 1982, The Canyonlands Natural History Association, Moab, UT)
Soil map units	1:24,000	18 classes	NRCS Soil Survey Geographic (SSURGO) database, mmu = 2.04 ha

problematic due to park policy. For this reason, subsurface properties were not directly inspected. Although field crews were trained to recognize key features of ecological sites, the inability to directly assess soil depth and horizons may have resulted in incorrect assignment of ecological site types. We generally feel that the number of incorrect assignments was nominal, but recognize this source of error in our field assessment of ecological sites. Rock-outcrop inclusions were prominent for certain site types. A location was assigned to a site type if inclusions were <30% of the described area, but the occurrence of an inclusion was recorded. Where conditions differed from existing site descriptions or inclusions exceeded the tolerance limit, the location was assigned to other. Also, locations with undesirable terrain features (large drainage gullies, cliffs) were scored as "other." Samples representing small (<1.0 ha) patches and nontargeted ecological site types were dropped from further consideration. This resulted in 170 samples for the targeted ecological sites, with 35 to 46 samples for each of four site types (Table 1). Only nine samples were acquired for the semidesert sand ecological site (#212). Numerous samples were recorded but most were small patches. Large slick-rock patches occur throughout the park, but are too numerous to remove from input data layers with manual methods. We instead included slick rock as a feature to be modeled, with the intent of removing all predicted slick-rock areas from further consideration. Forty-one training sites of slick rock were digitized from color DOQQs (Table 3).

Feature-extraction models were developed using the ecological site type and biophysical properties (Table 3) of the training sites. Color DOQQs provided spectral information that noticeably differed among ecological site types. Spectral differences were due to differences in vegetation type and patterns, and to differences in soil properties among individual ecological sites. Initial experimentation with feature-extraction models explored the use of various terrain data layers in addition to bedrock geology. The ability of predictive models to correctly predict the training sites was used to judge the importance of individual and combined terrain data layers. We employed this non-independent assessment to provide a relatively objective basis for selecting the "final" data layers used in model development. Moreover, an independent data set for accuracy assessment was not available at the time of initial model development. Results of these initial tests indicated that elevation and geology were the best predicators. The use of elevation is logical since it correlates with climate regime. Bedrock geology is a surrogate for parent material and soil development, which similarly is a logical choice for differentiating ecological site types. We clipped all data layers to an accessibility map, which eliminated areas that were too steep for reasonable access (>50% slope) or that were isolated by steep slopes. The resulting data layers represented areas that could be monitored from a practical perspective. In addition, we clipped data layers to the extent of soil map units selected for prediction of ecological sites. We eliminated paved roads and developed areas from data layers. The minimum spatial resolution was set by the DOQQs (1 m), but we used a re-sampling factor of 2 to expedite processing. Reflectance properties were perceived to be a key distinguishing characteristic among ecological sites. Thus, we used the resolution of the DOQQs as the minimum to facilitate accurate separation of ecological site types. Differences in the resolution of data layers are acknowledged as being problematic. However, from a practical perspective, these data are typically available at different resolutions, and it is often impractical to independently acquire large-scale physical data models at similar resolutions. Furthermore, the resolutions of the data layers we used in model development typify what is commonly used in predictive mapping of natural resources. We used a 13-pixel Manhattan contextual classifier, which is an optimal pattern for extracting non-linear features (VSL 2002), and a minimum of 200 pixels for aggregate area. The output from the model was a GIS shape file of predicted ecological sites and slick rock. The predictive model generated in this initial procedure is referred to as the Original feature-extraction model.

Accuracy-assessment data were collected in 2006 and 2007 as part of a pilot monitoring effort. Sample locations were randomly selected from the predicted ecological site map. The number of locations was determined by pilot monitoring requirements, but was expanded where prediction accuracy was exceptionally low. Following procedures described above for acquiring training sites, each location was scored to ecological site type, and aerial extent of the site type was estimated. If the aerial extent was <1 ha, the site type of the sample was still noted but was scored as invalid due to small patch size. Prediction accuracy was assessed by an error matrix, which shows the number and percentage of assessment samples correctly and incorrectly predicted by ecological site type. Overall accuracy was derived, along with a measure of chance-corrected accuracy (kappa). Accuracy assessment of predicted slick rock was not considered.

After performing the accuracy assessment, we used the hierarchical re-learning process of Feature Analyst to improve predictions for ecological sites with low accuracy levels. Accuracy assessment sites for the semidesert shallow sandy loam (#236) and the semidesert sand (#212) ecological sites were labeled as correct or incorrect, and input to the re-learning process. Accuracy assessment results also indicated the need to refine the classification of training sites, as well as the need for additional training sites. Spectral properties noticeably differed between correct and incorrect semidesert shallow sandy loam (#236) samples, indicating obvious differences in surface-rock cover. The original training sites for this ecological site were viewed on DOQQs (Table 3), and we visually estimated rock cover. Where rock cover within the estimated extent of a training site exceeded 15 percent, the site was re-scored as invalid. This threshold level was one-half that originally used to accept a training site. Given the low number of training sites (n=9) and low accuracy for the semidesert sand (#212) ecological site, additional sites from large patches of this site type were digitized from color DOQQs. The modified training data set along with the re-coded accuracy assessment samples were input to the re-learning process. Since the re-learning process operates on a single feature class (i.e., ecological site type), generation of new feature-extraction rules for the semidesert sandy loam (#215) site type was necessary to "back fill" area omitted in the new mapping of the semidesert sand ecological site (#212). The re-learning process produced new extraction rules and predictive maps for ecological sites numbers 236, 212, and 215. The predictive models generated by the re-learning process are referred to as re-learned feature-extraction models. Additional independent data were not available for accuracy assessment of the re-learned predictive models. Reported accuracies of the re-learned models are based on the original assessment samples which were employed in the re-learning process, and thus are not independent accuracy measures. Jackknife and bootstrapping methods can be used for a more credible evaluation of accuracy in such situations (Manley 1997), but these procedures were not attempted in this study.

Predictive Mapping in Capitol Reef National Park. Initial field investigations indicated correspondence between ecological site types and combinations of categorical features (vegetation associations and soil map units). Also, contextual differences among at least similar site types were not readily discernable. Based on these observations, we selected decision tree models as the appropriate predictive mapping method.

Decision trees accommodate multiple categorical classes of a dependent variable (e.g., ecological site class), and continuous and categorical independent variables (e.g., biophysical attributes) (Breiman et al. 1984, Ripley 1996). This method recursively partitions a training data set into increasingly homogenous groups, producing a dendritic pattern of terminal nodes (Scull et al. 2005). Starting at the root node (the entire training data set), each branching of the tree is determined by statistics that measure the decrease in heterogeneity within the resulting subsets. Branching ceases whenever tolerance limits are exceeded. Variables and associated values that best predict a branching are derived. Nodes contain any number of training samples, and can be pure (a single dependent class type) or mixed (multiple classes). Each node is numbered and labeled according to the majority class. The decision rules of a tree collectively provide a predictive model. Decision rules can be applied to mapped independent variables to predict node numbers over a study area. The interpretation of a predicted node depends on its purity. A pure node maps a single class. Mixed nodes can be interpreted by the majority class, or by probabilities of class type based on the class proportions contained in the node.

We used a vegetation-association map (NCPN-CARE Vegetation Mapping, in prep.) and soil maps (USDA NRCS 2004) to guide the selection of training sites for decision-tree-model development. In initial field assessments of the soil map, we found vegetation classes closely corresponded to ecological site type. Additionally, we found discrepancies in the soil mapping that could be corrected using the vegetation information. For instance, certain vegetation classes were characteristic of the semidesert stony loam ecological site, yet these classes tended to overlap soil-map units not reported to contain this ecological site type. Based on these observations, we considered the

Table 4. Digital attributes used in the decision tree and feature-extraction models for Capitol Reef National Park (NP). Note - Color Orthophoto Quads were only used in the feature-extraction models.

Attribute	Resolution	Range	Source, description
Color Digital Orthophoto Quads (DOQQs) - 3-Band, orthorectified	1 m	na	Horizons Inc., Salt Lake City, UT, Bureau of Reclamation, 1:12,000, aerial photographs, 2001
Elevation	10 m	1633-2281m	USGS digital elevation models (DEMS) 1:24,000 quadrangles
Percent slope	10 m	0-50 %	derived from elevation
Slope position	10 m	0-100	derived from elevation using USDA Forest Service SLOPEPOSITION Function (http://www.fs.fed.us/digitalvisions/toods). Indicates relative slope position with 0 = valley floor, 100 = ridge top
Geology (bedrock)	1:62,500	14 classes	Digitized from Geologic Map of Capitol Reef National Park and Vicinity, Utah (Billingsley, G. H., P. W. Huntoon, and W. J. Breed, 1987. Canyonlands Natural History Association, Moab, UT)
Soil map units	1:24,000	15 classes	NRCS Soil Survey Geographic (SSURGO) database, mmu = 2.04 ha
Vegetation associations	1:12,000 maximum	11 classes	NCPN Capitol Reef NP vegetation map (in prep.) (derived using DOQQ's), mmu = 0.5 ha

soil and vegetation maps to be primary determinants of ecological site types. We selected soil map units reported to contain target ecological site types and that contained vegetation classes we identified to typify targeted ecological sites. The vegetation map was clipped to the spatial extent of selected soil map units. Maps were then clipped to accessible area, and rasterized to a 10 m grain size. The two retired allotments were combined and treated separately from the active allotment. Training sites were acquired in 2007 by randomly selecting from soil-vegetation combinations. Sites were scored to ecological site type following methods described above for Canyonlands National Park and also tagged with soil map unit and vegetation-class type, which carried over from the site selection procedure. We collected 360 samples across the Cathedral-Rock Springs allotments and 438 samples in the Hartnet allotment. After removing training sites of insufficient spatial extent (<1 ha), 39 to 145 samples were available for targeted ecological sites (Table 2). Also, 89 training samples representing 8 non-targeted ecological site types in Cathedral-Rock Springs, and 224 training samples representing 11 non-targeted site types in Hartnet were used in model development.

Additional physical attributes were added to the training sites layers through GIS operations. Training sites were overlaid on rasterized layers (10 m grain size) of elevation, percent slope, slope position, and bedrock geology (Table 4). These data were selected for their potential to aid in discriminating among site types. Using a square 1 ha window as the size of a training site, averages of numeric data and the majority class of categorical data were generated. Training sites for model generation thus consisted of an ecological site type, and means and majority class of biophysical features.

We used the Statistica (StatSoft, Inc. 2004) software package to generate decision trees for the combined Cathedral-Rock Springs allotments and the Harnet allotment, and imposed a cross-validation cost of one-standard error (Breiman et al. 1984) as the tolerance limit to branching. The resulting decision tree models provided branching rules based on training site attributes and indicated the ecological site types comprising the numerically coded end nodes of the tree.

We applied the rules of a decision tree model to the maps used to derive the training samples. The Cathedral-Rock Springs model was applied to the Cathedral-Rock Springs study area, and the Harnet model was applied to the Harnet study area. We used a 1 ha moving window with a centroid-assignment scheme to score each 10 m landscape cell to a node number. Within each window, data averages and majority classes of the data layers used in the decision rules were generated; the decision rules were applied using these data; and the centroid was assigned to the predicted node number. This resulted in a predicted map of decision-tree node numbers, which was interpreted based on the purity of nodes.

We collected accuracy assessment data in 2007 as part of a pilot monitoring effort (Table 2). Samples were randomly selected from the predicted maps of decision-tree nodes which we had interpreted to ecological sites. Due to programmatic constraints, we only collected samples for nodes predicted as semidesert alkali sandy loam (#201) and semidesert sandy loam (#215) ecological sites. The lack of assessment samples across the full spectrum of predicted ecological sites limits the completeness of our accuracy assessments. However, this limited assessment provided a useful, initial performance evaluation of decision trees for mapping ecological sites.

For comparison with decision tree models, we developed feature-extraction models for targeted ecological sites for the combined Cathedral-Rock Springs allotments, and for the Harnet allotment. Procedures used to derive the original feature-extraction model for Canyonlands National Park were followed. However, the terrain data layers used in the decision tree models in addition to color DOQQs (Table 4) were used in the feature-extraction models.

Simulation of "Accuracy." In addition to increased spatial accuracy of target populations, a goal of ecological site mapping is to increase field efficiency. Accurate maps translate to lower rejection rates of non-target sites in the field. Use of the NRCS soil map units as the basis for selecting monitoring locations was perceived to lead to high rates of sample rejection, due to inexact mapping of ecological sites, and thus result in low field efficiency. To evaluate efficiency gains with our predictive mapping methods, we used Monte Carlo simulations to sample NRCS map units solely on the basis of the reported proportions of an ecological site type within a map unit. This approach essentially determines the ability to locate a targeted ecological site type without refined mapping. The soil map units used for predictive mapping in Canyonlands and Capitol Reef National Parks were inputs to the simulation. Information for each map unit included the total area of the soil map unit across the study area, and the proportion of a targeted ecological site type reported to be included in the map unit. The number of simulated locations for each targeted ecological site type (n) was equal to the number of independent accuracy assessment samples used in model testing. Simulations first selected a soil map unit using an area-weighting approach. The percentage of the targeted ecological site in the map unit served as the probability of locating a sample on the targeted site type. This probability was compared to a uniform random number (u.r.v.), and if it was less than or equal to the u.r.v. the simulated location was scored as a correct sample for the targeted ecological site. Otherwise, it was scored as incorrect. One-thousand replications of n samples were simulated. For each replication, the percentage of correct samples was recorded. Graphical methods were used to compare the simulated distribution of percent-correct predictions with accuracy assessment values of predictive mapping methods.

RESULTS

Canyonlands National Park. The feature-extraction model predicted 36,445 ha to contain targeted ecological sites. Prediction accuracy was >64% for three of the five ecological sites (see Producer accuracy in Table 5), but overall accuracy was only 43% (k=0.35). Low accuracy was primarily due to prediction errors of the semidesert sand type (#212). This type is eolian deposits, characterized as sand dunes with sparse vegetation. It occurs as localized, large patches on the landscape, but also is interspersed with grasslands on semidesert sandy loam (#215). More than half of the predicted semidesert sand samples were the semidesert sandy loam ecological site type (Table 5). This error was attributed to inadequate representation of the contextual properties of semidesert sand sites in the original training data set. Prediction accuracy of semidesert shallow sandy loam (#236) was marginal (Table 5). This site is dominated by sparse cover of pinyon pine, Utah juniper, and blackbrush, with intervening patches of biological soil crust, bare soil, and varying levels of rock. Thirty-six of the 37 assessment samples actually contained this ecological site. However, rock inclusions dominated 16 of the samples, and were scored as other.

Results with the re-learned feature extraction models suggest improvement in overall classification (from 43% to 74.8%) (Table 6). This was due to improved predictions of the semidesert shallow sandy loam (#236), semidesert sand (#212), and semidesert sandy loam (#215) ecological sites, all of which were subjected to the re-learning process. The reported accuracies, however, are not based on independent data since accuracy assessment samples were employed in the re-learning process as well as used to derive accuracy values of the re-learned models. Thus, accuracies shown for the re-learned models are only suggestive of improvement in predictive capability.

Capitol Reef National Park. The decision tree for the combined Cathedral-Rock Springs allotments is shown in Figure 2. Soil map unit and vegetation class were the dominant components of decision rules. Relatively pure (77-100%) nodes for non-targeted ecological sites were generated (nodes 3, 9 in Figure 2). Other terminal nodes were dominated by mixtures of the two targeted ecological sites (nodes 5, 11, 12-14, 16, 17 in Figure 2), but with small numbers of non-target sites. The ratio of the two targeted ecological sites varied among these nodes, but was as high as 2:1, and the dominant site type varied among nodes. The decision tree for the Hartnet allotment was similar, where soil map unit and vegetation classes dominated the decision rules, the tree contained relatively pure nodes of non-targeted site types, and targeted site types co-occurred in nodes. However, similar target site types tended to be co-dominate, with the semidesert sandy loam types (#201, 215) occurring in the same nodes and the two stony loam types (#242, 260) occurring in the same nodes. Although mapping of individual

Table 5. Error matrix for target ecological sites with the original feature-extraction model for Canyonlands National Park. Overall accuracy was 43% (k=0.35).

Observed Ecological Site (number)	Predicted Ecological Site (number)						Row total	User accuracy (%)
	215	118	236	133	212	other		
215	10	0	0	0	60	0	70	14.3
118	0	22	0	0	0	0	22	100.0
236	2	0	20	0	14	0	36	55.5
133	0	5	0	21	0	0	26	80.7
212	0	0	0	0	27	0	27	100.0
other	2	7	17	10	13	0	49	0.0
Column total	14	34	37	31	114	0		
Producer accuracy (%)	71.4	64.7	54.1	67.7	23.7			

Table 6. Error matrix for target ecological sites with the three re-learned feature-extraction models for Canyonlands National Park. Overall accuracy was 74.8% (k=0.68); however, accuracies of the re-learned models (ecological site nos. 236, 212, 215) were derived using model-development data (see text for explanation).

Observed Ecological Site (number)	Predicted Ecological Site (number)						Row total	User accuracy (%)
	215	118	236	133	212	other		
215	68	0	0	0	2	0	70	97.1
118	0	22	0	0	0	0	22	100.0
236	6	0	22	0	5	3	36	61.6
133	0	5	0	21	0	0	26	80.7
212	3	0	0	0	24	0	27	88.9
other	8	7	4	10	5	15	49	30.6
Column total	85	34	26	31	36	18		
Producer accuracy (%)	80.0	64.7	84.6	67.7	66.7	83.3		

site types was compromised by the lack of node purity, the degree of separation of non-targeted from targeted ecological sites was deemed valuable. The predicted maps for all allotments consisted of non-targeted and a targeted site complexes (nos. 201-215 in Cathedral-Rock Springs, nos. 201-215 and nos. 242-260 in Hartnet). About 2,249 ha were predicted as targeted ecological sites in the Cathedral-Rock Springs allotments, and 3,993 ha in the Hartnet allotment.

Accuracy assessment data were only collected for the one target-site complex (nos. 201-215) in Cathedral-Rock Springs and Harnet allotments. Accuracy was based on a sample matching either one of the target sites. Of the 26 samples predicted as a targeted ecological site in Cathedral-Rock Springs, 65% were correctly predicted. Three sites were non-targeted ecological sites and six sites contained a targeted site type but more than 30% of the 1 ha area of a sample location also included undesirable terrain features (deep stream drainages).

Of the 23 samples in the Hartnet allotment, four were non-targeted ecological sites, and seven corresponded to a targeted site type but with terrain infractions, resulting in only 52% accuracy.

Feature-extraction models were developed to predict only the one targeted-site complex (nos. 201-215). Model accuracies for the combined Cathedral-Rock Springs allotments and the Hartnet allotment were identical to the accuracies acquired with corresponding decision tree models.

Comparisons with Simulations. Simulated accuracy for most ecological site types tended to be lower than predictive map accuracy (Figures 3, 4). Accuracy for the semidesert sand (#212) ecological site with the original feature-extraction model for Canyonlands National Park was even lower than the accuracy range from simulation (Figure 3). This changed with the re-learned feature extraction model for this ecological site where prediction accuracy exceeded the ninety-fifth percentile of simulated results. However, the non-independent accuracy

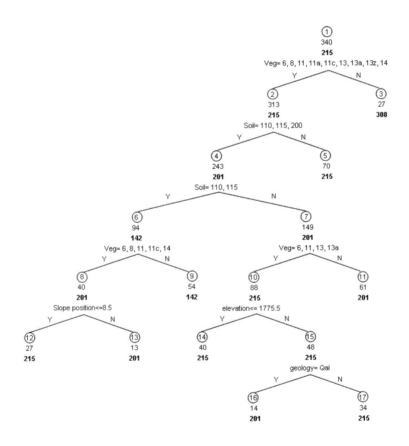

Figure 2. Decision tree for ecological sites in the Cathedral-Rock Springs allotment, Capitol Reef National Park. At each node, the encircled number is node number, the second line is the number of samples, the third line is the majority ecological site class (see Table 2), and for non-terminal nodes, the fourth line is the splitting criterion. The left branch marked 'Y', contains observations from the parent node for which the splitting criterion is true. Criterion codes: Soil = soil map unit number (USDA NRCS 2004), Veg = vegetation associations (6 - saltbush shrublands, 8 - mixed grasses, 11 - desert wash/shrubland mosaic, 11a – greasewood, 11c – sagebrush, 13 – mixed desert shrublands, 13a – Ephedra shrublands, 13a – rabbitbrush, 14 – woodland/grassland mosaic), geology (Qal – alluvial deposits).

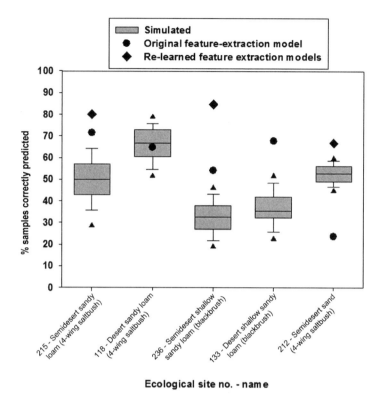

Figure 3. Comparison of predictive mapping (feature-extraction models) and simulated accuracy for target ecological sites in Canyonlands National Park. Numbers of accuracy assessment samples are shown in Table 1. Simulated results are based on 1000 replications (see text), and are shown as Box-Whisker plots. Bottom and top of the box is the 25th and 50th percentiles, respectively; bottom and top whiskers are 10th and 90th percentiles, respectively, and bottom and top triangles are the 5th and 95th percentiles, respectively. The line within the box is the median. Means were very similar or identical to medians, and are not shown.

measure of re-learned models should be viewed with caution. Accuracy of the desert sandy loam ecological site mapping with the original Canyonlands National Park feature-extraction model was slightly lower than the mean of simulations, but both were relatively high. Accuracy of predictive maps for the ecological site complexes in Capitol Reef National Park exceeded the simulated 95th percentile (Figure 4). The predictive map for the Hartnet allotment included three soil map units we found to include targeted ecological sites, but which were not reported to contain these ecological sites. For comparative purposes, these map units were included in the simulation, but had zero probability of containing a targeted ecological site. This in part accounts for the noticeably low simulated accuracy for this allotment.

DISCUSSION

Accuracy levels of our predictive maps were slightly lower than results achieved in related studies. Prediction accuracy of 65-88% for soil mapping is considered reasonable (e.g., Moran and Bui 2002; Scull et al. 2005). The low overall accuracy of 43% of the original feature-extraction model for Canyonlands National Park was due primarily to low prediction accuracy of one of the five ecological site types. The re-learning process resulted in an overall accuracy of 75%, but this result is confounded by the use of model-development data to determine accuracy for three of the five ecological sites. Accuracy of the decision tree models for the one ecological-site complex was marginal (65%) in the Cathedral-Rock Springs allotments, and low (52%) in the Hartnet allotment.

Despite the marginal accuracy of the predicted maps, comparisons with simulation results suggest that predictive mapping provided greater accuracy in the delineation of ecological sites and greater efficiency in selecting samples for targeted ecological sites than relying only on the currently available soil maps (Figures 3, 4). Accuracy of predictive maps generally exceeded the range of simulated accuracy. Accuracy of the feature-extraction model for desert sandy loam (#118) ecological site was lower than the mean of the simulations (Figure 3). The same was true for the semidesert sand (#212) ecological site in the original feature extraction model, but accuracy of the re-learned model exceeded the range of simulated accuracy. These results alone indicate the advantage of using predictive mapping to delineate the spatial extent of ecological sites. Relative to relying solely on the soil maps, predictive mapping provides a better representation of the spatial extent of targeted ecological sites, and reduces inclusion of non-targeted ecological sites.

An additional advantage of the predictive mapping process is the discovery and rectification of errors in existing maps. For example, sampling for training sites in Capitol Reef National Park revealed the occurrence of our target ecological site types in soil map units not reported to contain these types. Information acquired in the predictive mapping process led to a more comprehensive mapping of target ecological sites.

Two major sources of error were evident in the feature-extraction modeling effort.

In the Canyonlands National Park model, it was apparent that contextual features of the original training sites for the semidesert sand type were too similar to one of the grassland site types, resulting in a high level of misclassification. This over-prediction was greatly reduced using semidesert sand training sites from within large patches, providing a more distinctive context. Use of marginal training samples for the semidesert shallow sandy loam ecological site impacted predictive performance. The original samples included a gradient of rock outcrop within prescribed, acceptable levels (<30% total ground cover); however this resulted in extraction rules that were too liberal. Filtering out training sites with excessive rock cover appeared to enhanced prediction accuracy, although independent testing of our re-learned models is necessary to substantiate this conclusion. When using contextual information in predictive modeling, it is critical that training sites are representative of the feature to be extracted (i.e., ecological site) and provide sufficient context to distinguish among target and non-target features.

Insufficient data preparation was a source of error in our predictive mapping in Capitol Reef National Park. Twenty-eight percent of the accuracy assessment samples in Capitol Reef National Park included the predicted ecological site type within a 1 ha area centered on the sample point, but also included narrow but relatively steep-sided and deep stream drainages that were <10 m wide. We scored these samples as incorrect predictions since the stream drainages would not be acceptable areas for monitoring. This error impacts the utility of predictive maps but derives from insufficient filtering of areas deemed inappropriate for locating monitoring plots. We delineated steep slopes and relatively wide stream drainages as inaccessible or unacceptable areas for monitoring using a 10 m DEM. The more narrow drainages were undetectable at this

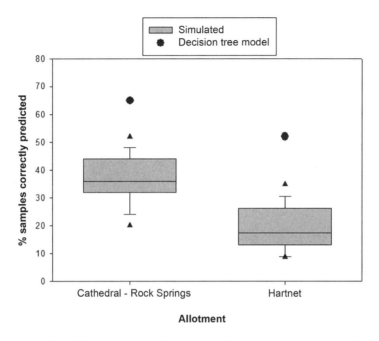

Figure 4. Comparison of predictive mapping (decision tree models) and simulated accuracy for the ecological site mixture (nos. 201/215) in Capitol Reef National Park. See Figure 3 for explanation of Box-Whisker plots.

resolution. On landscapes with fine-scale terrain features that are unacceptable for monitoring, high-resolution terrain data layers or air photos should be used to more effectively delineate the targeted area for predictive mapping.

Our decision tree models provided reasonable predictions of ecological site complexes rather than reliable predictions of individual ecological sites. Although this seems unfavorable given our intent of mapping individual ecological sites, models at least were able to separate targeted from non-targeted ecological sites. Excluding non-targeted areas from a sampling frame is important and contributes to operational efficiency of a sampling design. The use of maps of ecological site complexes, however, requires special consideration when crafting a sampling design. Predictive maps derived decision tree models indicate the location and extent of predicted nodes. Using the proportions of ecological site types in a node, which is a model output, sample sizes can be proportionally distributed across predicted node types to achieve desired numbers for each ecological site. We implemented this strategy in our pilot monitoring effort in Cathedral-Rock Springs allotments and found good agreement between ecological site proportions (nos. 201, 215) in the field and modeled proportions for individual nodes. Proportional sampling across node types provided desired sample sizes of individual ecological sites.

Despite the various limitations of our study, our results indicate that both of the predictive mapping methods we assessed can provide more accurate spatial representation of ecological sites than what is possible using currently available soil maps. Our results, however, do not illustrate one method to be superior to the other. Both predictive mapping methods provided similar levels of mapping accuracy across the two park units. In our limited direct comparison of the two methods in Capitol Reef National Park, feature-extraction and decision tree models provided similar levels of accuracy for the one ecological site complex evaluated. We were unable to rigorously evaluate and determine the conditions under which one method would be more reliable or effective than the other. However, it is reasonable to assume that feature-extraction is the method of choice where spectral and contextual features are largely different among targeted ecological sites. Conversely, decision tree models may be optimal where these features are less distinguishable among ecological sites and where the combination of soil maps and other biophysical data provide separation of targeted from non-targeted ecological sites.

In conclusion, we feel that the advantages of predictive mapping of ecological site types outweigh the costs of model-development efforts. Survey designs are preferred for natural resource monitoring since they have few assumptions and provide reliable and legally defensible parameter estimates (O'Dell et al. 2005). A fundamental requirement of survey designs is the delineation of a finite, target population (Cochran 1977). Ecological sites differentiate among key soil and other properties that influence above and below ground processes, and thus provide a sound basis for delineating the target population(s) of upland monitoring efforts. Accurate mapping of ecological sites is thus critical to comprehensive delineation of the targeted population on a landscape, to bound the domain of statistical inference of monitoring observations, and to identify the maximum number of samples. These three factors are essential components of effective and efficient survey designs for monitoring. Given the planning horizon of monitoring programs such as the NCPN I&M program (80+ yrs), developing an effective survey design has long-term benefits. Additionally, in terms of operational efficiency, accurate mapping of ecological sites decreases the number of samples rejected in the field due to non-targeted conditions. In large national parks where travel time to sampling locations can constitute the bulk of monitoring costs, this increased efficiency can be a substantial benefit in the initiation of operational, long-term monitoring.

ACKNOWLEDGMENTS

A portion of the training sites for predictive mapping of ecological sites in Canyonlands National Park was acquired from a rangeland study conducted by M. Miller and T. Troxler of the U.S. Geological Survey Southwest Biological Science Center, and funded by the USGS: Biological Resource Discipline Status and Trends of Biological Resources Program.

LITERATURE CITED

Breiman, L., J. H. Friedman, R. A. Olshen, and C. G. Stone. 1984. Classification and regression trees. Wadsworth International Group, Belmont, California.

Brown de Colstoun, E. C., M. H. Story, C. Thompson, K. Commisso, T. G. Smith, and J. R. Irons. 2003. National Park vegetation mapping using multitemporal Landsat 7 data and a decision tree classifier. Remote Sensing of Environment 85:316–27.

Cochran, W. G. 1977. Sampling techniques. John Wiley & Sons, New York.

De'Ath, G., and K. E. Fabricius. 2000. Classification and regression trees: A powerful yet simple technique for ecological data analysis. Ecology 8:3178–92.

Franklin, J. 1998. Predicting the distribution of shrub species in southern California from climate and terrain-derived variables. Journal of Vegetation Science 9:733–48.

Lagacherie, P., J. P. Legros, and P. A. Burrough. 1995. A soil survey procedure using the knowledge of soil pattern established on a previously mapped reference area. Geoderma 65:283–301.

MacMillan, R. A., D. E. Moon, and R. A. Coupe'. 2007. Automated predictive ecological mapping in a forest region of B.C., Canada, 2001–2005. Geoderma 140:353–73.

Manlcy, B. F. J. 1997. Randomization, bootstrapping and Monte Carlo methods in biology. Chapman & Hall, London, United Kingdom.

McKenzie, N. J., and P. J. Ryan. 1999. Spatial prediction of soil properties using environmental correlation. Geoderma 89:67–94.

Moran, C. J., and E. N. Bui. 2002. Spatial data

mining for enhanced soil map modelling. International Journal of Geographical Information Science 16:533–49.

O'Dell, T., S. Garman, A. Evenden, M. Beer, E. Nance, D. Perry, R. DenBleyker, et al. 2005. Northern Colorado Plateau Inventory and Monitoring Network, Vital Signs Monitoring Plan, National Park Service, Inventory and Monitoring Network, Moab, Utah.

Poulos, H. M., A. E. Camp, R. G. Gatewood, and L. Loomis. 2007. A hierarchical approach for scaling forest inventory and fuels data from local to landscape scales in the Davis Mountains, Texas. Forest Ecology and Management 244:1015.

Ripley, B. D. 1996. Pattern recognition and neural networks. Cambridge University Press, Cambridge, United Kingdom.

SCS (USDA Soil Conservation Service). 1991. Soil survey of Canyonlands area, parts of Grand and San Juan Counties, Utah. U.S. Department of Agriculture, Natural Resources Conservation Service, Salt Lake City, Utah.

Scull, P., J. Franklin, and O. A. Chadwick. 2005. The application of classification tree analysis to soil type prediction in a desert landscape. Ecological Modelling 181:1–15.

SRM (Society for Range Management). 1989. Glossary of terms used in range management, 3rd edition. Society for Range Management, Denver, Colorado.

SRM (Society for Range Management). 1995. New concepts for assessment of rangeland condition. Journal of Range Management 48:271–82.

StatSoft, Inc. 2004. STATISTICA, version 6. www.statsoft.com.

Swain, J. 2007. The effect of spatial resolution on an object-oriented classification of downed timber. M.S. thesis, North Carolina State University, North Carolina.

Theobald, D. M., D. L. Stevens Jr., D. White, S. Urquhart, A. R. Olsen, and J. B. Norman. 2007. Using GIS to generate spatially balanced random survey designs for natural resource applications. Environmental Management 40:134–46.

USDA NRCS. 2004. Soil survey report for Capitol Reef National Park, Utah. Unpublished interim report. USDA NRCS, Richfield, Utah.

VSL (Visual Learning Systems). 2002. User manual, feature analyst extension for ArcView 3.2. Visual Learning Systems, Inc., Missoula, Montana.

Zhu, A-Xing, B. Hudson, J. Burt, K. Lubich, and D. Simonson. 2001. Soil mapping using GIS, expert knowledge, and fuzzy logic. Soil Science Society of America Journal 65:1463–72.

QUANTIFYING SAMPLE BIAS IN LONG-TERM MONITORING PROGRAMS: A CASE STUDY AT BANDELIER NATIONAL MONUMENT, NEW MEXICO

Chris L. Lauver, Jodi Norris, Lisa Thomas, and Jim DeCoster

ABSTRACT

The Southern Colorado Plateau Network (SCPN) of the National Park Service (NPS) is developing a long-term program to monitor the condition of natural resources at 19 NPS units on the Colorado Plateau. We present a case study of our soil and vegetation monitoring at Bandelier National Monument and quantify bias associated with our sampling design. To stratify natural variation in soils and vegetation for sampling, we selected soil map units from a recent soil survey to represent the initial spatial sampling frame. In modifying the sampling frame to select monitoring sites that are logistically feasible, a large portion of the frame was found to contain undesirable elements, including steep slopes and inaccessible areas. Removing these areas resulted in a final sampling frame that contained only 37.2% of the initial frame area. We tested for a bias in the distributions of elevation, slopes, and soils between the initial and final sampling frames, and for a site selection bias of these same features by comparing potential sampling sites to the final sampling frame. Mean percent slope for the initial frame was significantly higher than that of the final frame, and mean elevation for the initial frame was significantly lower than that of the final frame. Reasons for these differences include user-specified constraints on slope and accessibility. No bias was found in soil features between the initial and final sampling frames, and no site selection bias was found for slope, elevation, or soils. The analysis provided explicit data on some of the consequences of creating a practical sampling design.

INTRODUCTION

The Southern Colorado Plateau Network (SCPN) of the National Park Service (NPS) is charged with developing a statistically valid program to monitor the condition of natural resources at 19 NPS units on the Colorado Plateau. Long-term monitoring data on status and trends of natural resources will be used by resource managers to maintain or restore ecological integrity and to minimize ecological threats to park ecosystems. Considering these ambitious goals, the extensive areas involved, and the limited resources available, it is vital to ensure and implement an efficient and well-planned monitoring design (Silsbee and Peterson 1991).

Developing a sound monitoring plan requires several steps, including clearly stating goals and objectives, developing conceptual models to show system functioning and responses to stressors, providing a rationale for selected indicators, and establishing a sampling design and methods of measurement, as well as procedures for connecting the monitoring results to the decision process (Noon 2003). The Southern Colorado Plateau Network has addressed these steps and completed a monitoring plan

(Thomas et al. 2006) that documents the planning process and the resulting decisions. The network is currently developing monitoring protocols and implementing sampling designs for several suites of ecological indicators. As part of this process, we are working closely with local experts and resource managers at individual parks to determine specific target populations for monitoring, and to derive spatially explicit sampling frames for site selection.

In this paper, we present our strategy to monitor upland soils and vegetation at many NPS units within the network, focusing on Bandelier National Monument as a case study. The monitoring objectives are to determine long-term status and trends in three interrelated ecological indicators: vegetation composition and structure, upland hydrologic function, and soil stability. These indicators are key attributes of dryland ecosystems in the Colorado Plateau, and justification for their selection is contained in the SCPN monitoring plan (Thomas et al. 2006) and supporting work by several federal agencies (Miller 2005, Pellant et al. 2005). The foundation of our monitoring design is based on survey sampling, which includes defining a finite, target population and using probability-based sampling (Cochran 1977). The target population, or the ecological resource for which estimates are desired, is generally confined to a predominant vegetation type within a park.

Through collaborations with staff and resource experts at Bandelier National Monument, the pinyon-juniper ecosystem was selected as a target population for upland soils and vegetation monitoring. While developing a physical representation of this target population, or sampling frame, that is sensible and logistically feasible, some issues of concern related to the natural topography at Bandelier National Monument have surfaced. These concerns include the presence of steep slopes and the effort required to access certain areas of the target, both of which influence selection and establishment of sampling sites. It is undesirable to sample steep slopes for several reasons, including concerns about the safety of field crews and the potential to cause resource damage and increase erosion through sampling efforts. Additionally, a logistical problem is encountered in trying to sample across the entire extent of the pinyon-juniper ecosystem because of limited access by existing roads and trails.

Areas that are determined to be off-limits or too costly for sampling must then be eliminated from the sampling frame. However, these sub-populations may compose a significant portion of the target population, and removing them from the sampling frame raises some questions concerning the true ranges of natural features portrayed in the final sampling frame. The purpose of this paper is to address the following questions: Was a bias introduced in the distributions of elevation, slopes, or soils by reducing the initial coverage of a sampling frame to a more feasible sampling frame? And was there a site selection bias (Olsen et al. 1999) of these same abiotic features when comparing attributes of potential sampling sites to attributes of the final sampling frame?

METHODS

The lands within Bandelier National Monument that currently and potentially support pinyon-juniper woodlands represent the study area (Figure 1). The dominant overstory species are *Pinus edulis* (pinyon pine) and *Juniperus monosperma* (one-seed juniper). This vegetation type covers about one third of the Monument (Kleintjes et al. 2004) and dominates the uplands in elevations ranging from 1700 to 2249 m, mostly occurring on mesa tops and associated slopes. Annual precipitation is variable around a mean of 42 cm, with most rain occurring during the summer months (Allen and Breshears 1998, Kleintjes et al. 2004).

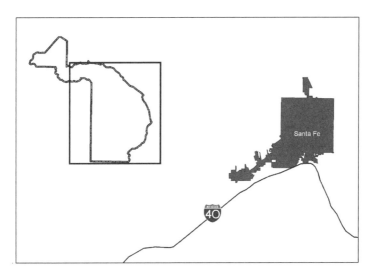
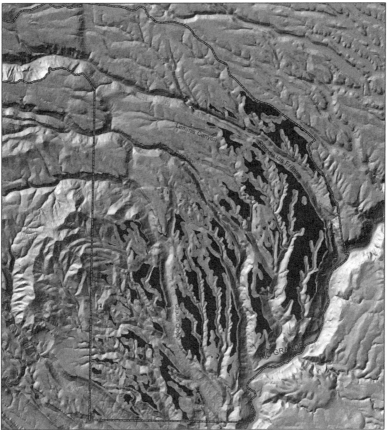

Figure 1. The location of Bandelier National Monument in New Mexico (top). The dark shaded areas in the inset (bottom) represent areas with pinyon-juniper woodlands and the initial sampling frame for soils and vegetation monitoring.

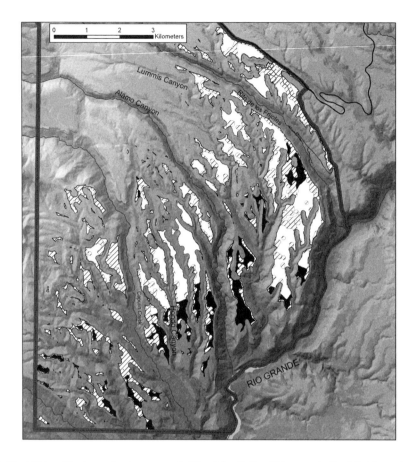

Figure 2. The initial and final sampling frames used at Bandelier National Monument. The initial frame (see also Figure 1) is shown by the collection of the white, black, and hatched polygons. The white polygons represent the final sampling frame. The hatched polygons are areas determined to have greater than 20% slopes, and were eliminated from the initial frame. The black polygons represent areas that were determined to be inaccessible, and were eliminated from the initial frame.

Soils are mostly shallow and were derived from volcanic ash and pumice (Natural Resources Conservation Service 2000).

To stratify natural variation in soils and vegetation for sampling, we used the concept of ecological sites. An ecological site is a landscape division with specific physical characteristics that differs from other landscape divisions in its ability to produce distinctive types and amounts of vegetation and in its response to management (Society for Range Management Task Group on Unity in Concepts and Terminology 1995). Ecological sites have characteristic soils, hydrology, plant communities, and disturbance regimes and responses (Natural Resources Conservation Service 2003). Ecological site descriptions are developed by the Natural Resources Conservation Service in conjunction with soil surveys. Ecological sites are associated with soil mapping units, are not expected to change and, thus, are appropriate for use as strata in long-term monitoring programs.

Ecological sites have not yet been formally defined for Bandelier National

Table 1. The 6 map units (Natural Resources Conservation Service 2000) that form the Mesa Pinyon-Juniper ecological site, and the percentages of each unit contained within initial and final sampling frames, and for the 50 potential sampling sites.

Map Unit Number	Map Unit Name	% of Initial Sampling Frame	% of Final Sampling Frame	% of Potential Sampling Sites
410	Palatka-Canuela-Rock Outcrop complex, dry, 2 to 20 percent slopes	33.8	29.7	22.0
400	Palatka-Canuela-Rock Outcrop complex, 2 to 20 percent slopes	30.4	6.8	38.0
413	Armenta very para-gravelly ashy coarse sand, 3 to 20 percent slopes	14.6	13.8	16.0
408	Adornado very para-gravelly ashy coarse sandy loam, 8 to 15 percent slopes	12.6	16.4	16.0
412	Canuela-Hackroy complex, 1 to 8 percent slopes	5.1	7.1	6.0
409	Hackroy-Nyjack complex, 2 to 12 percent slopes	3.5	6.2	2.0

Monument, so we constructed one for the target population in collaboration with local experts using a recent soil survey (Natural Resources Conservation Service 2000). The target ecological site for Bandelier National Monument, Mesa Pinyon-Juniper, is composed of six soil-mapping units (Table 1). We used spatial data that accompanied the soil survey of these six mapping units to form the initial sampling frame (Figures 1 and 2).

We adjusted the spatial coverage of the initial sampling frame to contain only the desired elements of the target population by conducting several GIS (geographic information system) modifications using ArcGIS software. These steps included: removing areas within a 100 m buffer along established roads; removing areas with slopes exceeding 20%, as identified by a 10 m digital elevation model (DEM); removing areas that had been disturbed by recent wildland fires; removing areas planned for vegetation management treatments that would significantly reduce overstory cover; removing small areas near the park boundary (which are subject to periodic treatments) by applying a 50 m buffer along the boundary; and removal of areas with accessibility issues. Accessibility is a concern for two logistical reasons: there is limited access provided by existing roads and trails, and field crews need approximately four hours to collect monitoring data from a site. Thus, using a cost-distance analysis involving slope, trail, and road data, we removed areas estimated to take more than two hours of hiking time from the main park trailhead, or more than two and a half hours hiking time from other roads or trailheads. The spatial coverage that remained following these GIS steps represented the final sampling frame.

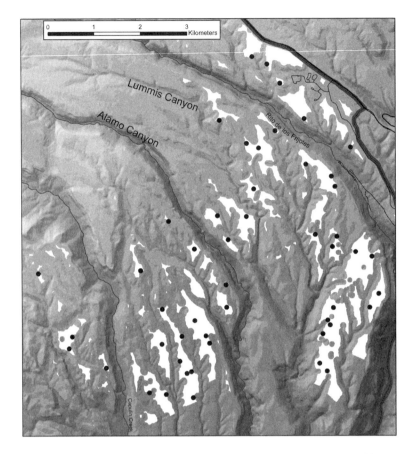

Figure 3. Fifty spatially-balanced points, representing potential sampling sites, within the final sampling frame. The points were produced using the Generalized Random-Tessellation Stratified design (Stevens and Olsen 2004).

To select sampling sites from the final sampling frame, we used the Generalized Random-Tessellation Stratified (GRTS) design (Stevens and Olsen 2004) and R statistical computing software to produce a spatially balanced sample of 50 points or sites (Figure 3). We estimate 40 sites will eventually be established and monitored for this target population. The distribution of the first 50 sites was used to represent the likely set of potential sampling sites, given that some sites may be rejected because they are found to be non-target. Strictly speaking, the analyses presented here addressing site selection bias (the second question posed in the purpose statement above) should be conducted again once final sites are established; however, given that the final set of sampling sites may not be fully implemented for several years, our analyses represent an approximation of possible site selection bias for this monitoring effort.

To compare the distributions of slope and elevation from the initial sampling frame to the final, we extracted data from both frames using a 10 m digital elevation model (DEM) and conducted an independent samples t-test. Testing for site selection bias in slope and elevation between the DEM data extracted from the final sampling frames and

Table 2. The coverage (in ha) of the initial and final sampling frames, and the resulting reductions imposed by the GIS (geographical information system) tasks conducted (see methods) to obtain the final sampling frame.

GIS tasks	Hectares	Reduction (ha)	% Remaining
Initial Sampling Frame	1647	-	100.0
Roads	1584	63	96.2
Slope	890	694	54.0
Capulin fire	887	3	53.8
Dome fire	887	0	53.8
Treatment area	857	30	52.0
50m boundary buffer	852	5	51.7
Accessibility / Final Frame	612	240	37.2

the potential sampling sites was also done using an independent samples t-test. To evaluate the distributions of soil features, we determined percentages of individual soil map units for the 2 sampling frames and the potential sampling sites (Table 1), and used Chi-square analysis to compare these percentages between the initial and final sampling frames, as well as to evaluate site selection bias between the potential sampling sites and the final sampling frame.

RESULTS

The GIS modifications to eliminate undesirable elements of the target population, such as steep slopes and inaccessible areas, resulted in a final sampling frame that contained only 37.2% of the area of the initial frame (Table 2 and Figure 2). The largest reductions in the frame were caused by eliminating slopes above 20% (reducing the frame by 694 ha, or 42%) and by applying the accessibility criterion (reducing the frame by 240 ha, or 14.6%) (Table 2).

The modifications to confine slopes in the final sampling frame to the range of zero to 20% resulted in a different distribution of slopes compared to the initial frame (Figure 4 and Table 3). Mean percent slope for the initial sampling frame (14.2%) was significantly higher ($t = 331.5$, $P < 0.001$) than that for the final sampling frame (7.2%) (Table 3). This result can be attributed to the greater range of slope values and higher median slope for the initial frame as compared to the final frame (Table 3).

No site selection bias was found ($t = 0.29$, $P > 0.975$) in comparing the distribution of slope values between the potential sampling sites and the final sampling frame (Figure 4). The data sets had the same mean (7.2%), median (7%), and similar standard deviations (3.22 and 3.28, respectively).

The large reduction in the final sampling frame compared to the initial frame did result in a different distribution of elevations (Figure 5). Mean elevation for the initial sampling frame (1946 m) was significantly lower ($t = 40.1$, $P < 0.001$) compared to that for the final sampling frame (1957m) (Table 2). This is partly a result of the difference in the ranges of elevation between the two frames, with the initial frame covering a wider range of lower elevations (Table 3 and Figure 5).

No site selection bias was found ($t = 0.10$, $P > 0.915$) in comparing distributions of elevation between the potential sampling sites and the final sampling frame (Figure 5). The data sets had nearly identical means

Table 3. Mean, standard deviation (SD), range, and other summary statistics of percent slope and elevation (m) for the initial and final sampling frames. A 10m digital elevation model (DEM) provided the base data source.

Feature	Sampling Frame	n	Mean	Median	SD	Min	Max	Range
Slope (%)	Initial	379,949	14.2	11.0	11.8	0	280	280
	Final	70,518	7.2	7.0	3.3	0	20	20
Elevation	Initial	379,949	1946	1939	78.0	1696	2249	553
	Final	70,518	1957	1943	65.1	1849	2218	369

and standard deviations (1958 ± 65.72 m and 1957 ± 65.18 m, respectively) and similar ranges (1861 to 2148 m; and 1849 to 2218 m, respectively).

In assessing bias of soil features, there was no difference ($\chi^2 = 4.98$, $p > 0.245$) between the percentages of individual soil map units (Table 1) of the initial sampling frame compared to those of the final frame. No site selection bias was found ($\chi^2 = 10.05$, $p > 0.050$) in comparing the percentages of soil map units of the potential sampling sites to those of the final sampling frame.

DISCUSSION

The biases found in slope and elevation in reducing the coverage of a sampling frame by nearly two thirds were anticipated, given the complex terrain of the study area. The magnitude of the reduction in the frame caused by the slope factor was unexpected because the soil map units that composed the frame were described as having slopes $\leq 20\%$ (Table 1). We assume these topographical errors arise from several potential sources of error, including the relatively large minimum mapping unit and commission errors in the soil survey geospatial data, and a potential difference between the high resolution (10 m) DEM that we used to model slope compared to the resolution that was used during the development of the soil survey data. Our decision to confine slopes to $\leq 20\%$ in the final sampling frame does limit our ability to detect trends in soil and vegetation characteristics to pinyon-juniper woodlands on level and moderately steep slopes. Lacking data from woodlands on steep slopes means that our interpretations of the observed trends will not include the full range of existing pinyon-juniper conditions. In our view, this is an unavoidable and acceptable consequence of creating a design where sites can be sampled safely and without causing undue disturbances to the sampled population.

The observed bias in elevation also has some ecological implications in that the monitoring effort may lack sample data from the drier end of the pinyon-juniper ecosystem. In light of recent drought-induced shifts involving this vegetation type and the related ponderosa pine forest (Allen and Breshears 1998), these lower elevation areas may represent ecotones that may respond more quickly to drought-related climate change than mid-elevation areas. The main reason lower elevation areas were eliminated from the sampling frame was because of the accessibility criterion. Access and its limitations on sampling are common issues (e.g., El-Ghani and Amer 2003, Reddy and Davalos 2003), particularly in efforts involving private land or complex terrain. Again, we believe that this observed bias is an acceptable consequence of the need to

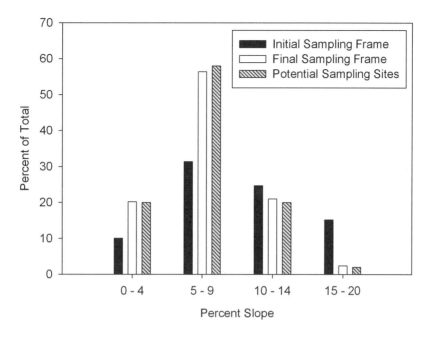

Figure 4. Percent slope distributions for the initial and final sampling frames, and for the 50 potential sampling sites (values obtained using a 10m DEM).

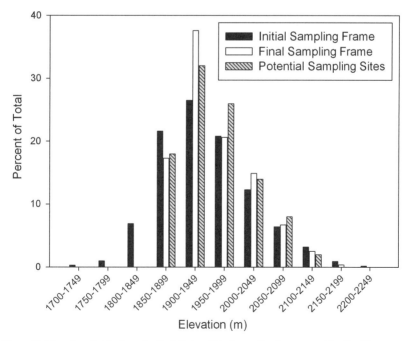

Figure 5. Elevation distributions for the initial and final sampling frames, and for the 50 potential sampling sites (values obtained using a 10m DEM).

create a sampling frame with sites that are logistically feasible to sample.

The finding of no site selection bias for the natural features tested here is an encouraging result regarding the GRTS spatial allocation method. Based on our preliminary results (final results will be obtained once final sampling sites have been established), this method of site selection performs well in obtaining a balanced sample that reflects the attributes of the population. Stevens and Olsen (1999, 2004) provide examples demonstrating the utility of this method to produce spatially balanced samples and provide for lower variance estimates (Stevens and Olsen 2003).

Collectively, our results indicate the utility of investigating bias in selecting monitoring sites not only for what the populations may be potentially missing, but also for what they contain. It is critical to know the composition of the population in the final sampling frame in order to ensure a match with the desired target population and the sampling objectives. When large reductions in the initial sampling frame are made for reasons of practicality, such as we demonstrate here, a review of the decisions that affect the frame coverage and their consequences is warranted. In light of lessons learned here, we intend to perform similar investigations of our sampling designs at NPS units across the southern Colorado Plateau prior to full establishment of long-term monitoring sites.

ACKNOWLEDGEMENTS

We thank B. Jacobs, C. Allen, and S. Fettig for their expertise and collaborations in determining monitoring targets and sampling sites at Bandelier National Monument. We also thank M. Swan for her expertise in the field in evaluating soil map units, and M. Miller for helpful discussions on applying the ecological site concept to long-term monitoring efforts.

LITERATURE CITED

Allen, C. D., and D. D. Breshears. 1998. Drought-induced shift of a forest-woodland ecotone: Rapid landscape response to climate variation. Proceedings of the National Academy of Sciences, USA 95:14839–14842.

Cochran, W. G. 1977. Sampling techniques. John Wiley & Sons, New York, NY.

El-Ghani, M. M. A., and W. M. Amer. 2003. Soil-vegetation relationships in a coastal desert plain of southern Sinai, Egypt. Journal of Arid Environments 55:607–628.

Kleintjes, P. K., B. F. Jacobs, and S. M. Fettig. 2004. Initial response of butterflies to an overstory reduction and slash mulching treatment of a degraded pinyon-juniper woodland. Restoration Ecology 12:231–238.

Miller, M. E. 2005. The structure and functioning of dryland ecosystems: Conceptual models to inform long-term ecological monitoring. U.S. Geological Survey Scientific Investigations Report 2005–5197.

Natural Resources Conservation Service. 2000. Special project soil survey of Bandelier National Monument. Interim report.

Natural Resources Conservation Service. 2003. National range and pasture handbook. USDA, NRCS, Grazing Lands Technology Institute.

Noon, B. R. 2003. Conceptual issues in monitoring ecological systems. In Monitoring ecosystems: Interdisciplinary approaches for evaluating ecoregional initiatives. edited by D. E. Busch and J. C. Trexler, pp. 27–71. Island Press, Washington, D.C.

Olsen, A. R., J. Sedransk, D. Edwards, C. A. Gotway, W. Liggett, S. Rathbun, K. H. Reckhow, and L. J. Young. 1999. Statistical issues for monitoring ecological and natural resources in the United States. Environmental Monitoring and Assessment 54:1–45.

Pellant, M., P. Shaver, D. A. Pyke, and J. E. Herrick. 2005. Interpreting indicators of rangeland health, version 4. Technical Reference 1734–6. U.S. Department of Interior, Bureau of Land Management, National Science and Technology Center, Denver, Colorado.

Reddy, S., and L. M. Davalos. 2003. Geographical sampling bias and its implications for conservation priorities in Africa. Journal of Biogeography 30:1719–1727.

Silsbee, D. G., and D. L. Peterson. 1991. Designing and implementing comprehensive long-term inventory and monitoring programs for National Park System lands. National Park Service, Natural Resources Report NPS/NRUW/NRR-91/04.

Society for Range Management Task Group on Unity in Concepts and Terminology. 1995. New concepts for assessment of rangeland condition. Journal of Range Management 48:271–282.

Stevens, D. L., and A. R. Olsen. 1999. Spatially restricted surveys over time for aquatic resources. Journal of Agricultural, Biological, and Environmental Statistics 4:415–428.

Stevens, D. L., and A. R. Olsen. 2003. Variance estimation for spatially balanced samples of environmental resources. Environmetrics 14:593–610.

Stevens, D. L., and A. R. Olsen. 2004. Spatially balanced sampling of natural resources. Journal of the American Statistical Association 99:262–278.

Thomas, L., M. Hendrie, C. Lauver, S. Monroe, N. Tancreto, S. Garman, and M. Miller. 2006. Vital signs monitoring plan for the Southern Colorado Plateau network. Natural Resource Technical Report NPS/SCPN/NRR-2006/002.

WISARDNET FIELD-TO-DESKTOP: BUILDING A WIRELESS CYBERINFRASTRUCTURE FOR ENVIRONMENTAL MONITORING

Kenji Yamamoto, Yuxin He, Paul Heinrich, Alex Orange, Bill Ruggeri, Holland Wilberger, and Paul Flikkema

ABSTRACT

The technology of wireless sensor networks has enabled new levels of spatial coverage and density in the monitoring of variables important in numerous applications, including ecological research and environmental management. These networks are composed of small, energy-efficient devices that wirelessly collaborate to gather data on temperature, light, soil moisture, sap flux, and other variables over space and time. However, a complete monitoring solution requires the conversion of data into useful information for a user, who may be anywhere in the world. To address this challenge, we have designed a complete Field-to-Desktop system for the acquisition, storage, and visualization of environmental data from a wireless sensor network using industry-standard networking, database, and web-development tools. This paper describes the architecture and capabilities of the system and lessons learned from a test implementation designed for public outreach at The Arboretum at Flagstaff.

INTRODUCTION

Ecologists and environmental scientists need many types of field data in order to answer complex ecological questions. In many research studies, this entails long-term monitoring of coupled environmental processes. These processes include meteorological variables, such as air and soil temperature and photosynthetically active radiation (PAR), as well as other environmental quantities and biotic responses, including soil moisture, sap flux, and tree trunk diameter. Measurement of most of these variables has been automated for years with the use of dataloggers of various types. These measurements are typically performed in conjunction with human field work (e.g., sampling seed traps and using portable soil-moisture measurement devices).

The dataloggers used have traditionally been of two types, wired and standalone devices. Wired arrays are relatively complex instruments with wired interfaces to multiple external high-precision transducers with a simple programming interface (e.g., dataloggers from Campbell Scientific). In addition to the complexity of installation, wired arrays are invasive and subject to rising costs of the required copper-wire interconnections. Standalone dataloggers (e.g., Onset Computer Hobo or Maxim iButton loggers) can provide similar data collection at greater spatial coverage since they are not constrained by the requirement for interconnections. However, each standalone unit must be periodically located and then connected to a computer or other smart device to upload the data samples. In either case, significant labor is involved in deploying the sensing infrastructure and/or gathering the data.

Applications often require sensing with high spatial resolution over a large coverage area in order to understand spatial variations across scales from meters to kilometers. Since this implies dozens, hundreds, or even thousands of sampling points, wired arrays and standalone dataloggers are inadequate solutions. In the last decade, research in wireless sensor networks has grown tremendously in response to the needs of numerous distributed sensing applications (Pottie and Kaiser 2000, Estrin et al. 2001). Wireless sensors can be viewed as sensors with the ability to communicate wirelessly. This capability allows greater density and spatial coverage area while minimizing the labor of data collection and the cost of wired networks. With appropriate algorithms implemented in the wireless sensor's software, these wireless sensors form a connected network to collaborate in sending the data to a central, terminal node, referred to as the network hub, where it can then be transmitted to the World-Wide-Web (Web) via Ethernet, telephone or satellite links.

Wireless sensor network technology is available or under development for a broad range of applications, including manufacturing process and inventory monitoring, control of built environments, monitoring of structural safety (e.g., bridges), public safety, and security. The authors have developed a wireless sensor network called Wireless Sensing And Relay Device Network (WiSARDNet) that specifically targets environmental and ecosystem sensing applications. Previous papers have reported on the technology of the WiSARDNET *in situ* wireless sensor network (West et al. 2001, Flikkema and West 2002, Flikkema et al. 2002, Flikkema and West 2003, Yang et al. 2005). This paper provides an overview of the design and capability of the complete WiSARD-Net Field-to-Desktop cyberinfrustructure, which integrates the *in situ* wireless sensor network and data-center hardware and software to provide a complete solution for sampling the environment, transferring the information to a networked server, and displaying the information via the Web anywhere in the world. Other wireless environmental-sensor-network experiments are reported in Mainwaring et al. (2002), and Tolle et al. (2005).

REQUIREMENTS

The goal of environmental and ecosystems sensing is to deliver science-quality data to the user; in the case of scientific research, the data should be delivered to a server accessible at the investigator's laboratory. To accomplish this, we built WiSARDNet using numerous wireless devices called WiSARDs. A WiSARD that collects data and forwards it along the network is referred to as a WiSARD sensor node. A WISARD that serves as the central, terminating point in the wireless network is called a WiSARD hub. WiSARD sensor nodes also serve as relays for other sensor nodes farther from the hub, since limited energy and power resources prevent every node from having a reliable wireless link to the hub. The WiSARD hub has the additional capability to transmit the gathered data over a wired, cellular, or satellite link. This link terminates at a database server, where the information collected by the WiSARDNet is processed and stored. An application specific website then presents the user with the information in the form of easy-to-interpret graphs and plots.

A number of requirements drove the WiSARDNet design. First, a deployed WiSARDNet needs to be minimally invasive to prevent disturbance of the environment, theft, and vandalism. This implies that the sensors should be as small and unobtrusive as possible. Because a wireless sensor node is in simple terms a datalogger with a radio, it consists of three main functional components: transducer interface electronics to gather the data, a radio for communication, and a tiny computer to plan and execute the sensor node's activities. Each wireless sensor node should interface with a broad spectrum of science-quality transducers, including temperature, PAR, and soil moisture, and also use a rugged, weather-resistant package. All of these characteristics must be achieved at a low cost, monetarily and in the effort required for installation and maintenance.

Perhaps the most challenging requirement for a wireless sensor network is long battery life, since battery replenishment is a labor-intensive, hence expensive, exercise. While this requirement is relatively easy to achieve in standalone dataloggers, the addition of a radio to support wireless networking demands a significant amount of energy. The problem is compounded when small sensor node size is desired (driven by the need to minimize invasiveness), limiting the size and capacity of batteries. In addition, many environments, such as forests with dense canopies, limit the potential of solar power.

When the network is deployed and the WiSARD sensor nodes are powered up, the network configures itself into a data-gathering configuration so that all data flows to the WiSARD hub. To achieve the full potential of wireless capability, the design supports transparent incremental deployment; in other words, the automatic integration of new nodes that are deployed to add coverage or to replace failed WiSARD sensor nodes. This allows the end user to deploy and maintain the WiSARD-Net without having to individually configure each unit. Even within the domain of environmental/ecosystems sensing applications, deployments differ significantly. For example, if a particular variable of interest has a low spatial rate of fluctuation (bandwidth), then the relevant data could be captured more efficiently with a deployment of lower spatial density and spatial extent. This implies that the wireless networking technology should be scalable in network coverage, size, and density.

These disparate (and often conflicting) requirements have driven the engineering design of the WiSARDNet system, which encompasses the disciplines of analog and digital electronics, radio-frequency electronics, signal processing, communication and networking, embedded computing system design, and mechanical packaging. We developed the WiSARDNet Field-to-Desktop system to enable the seamless flow of data from sensors to a user's desktop computer anywhere in the world. It is a complete end-to-end system, where the internal workings of the network and data processing are performed automatically to present the user with graphical data displays that are updated in real time (up to networking delays). To achieve these real-time displays, additional hardware and software were integrated into the WiSARDNet system.

DESIGN

The WiSARDNet sensor nodes employ a modular hardware/software architecture in order to meet the heterogeneous requirements of sensing, data wireless reception and transmission, and information processing and presentation. To support science-quality sensing, WISARD sensor nodes use a dual-computer board design, with the labor divided between a brain board that provides communication and networking services and a transducer data acquisition board that handles the details of the data acquisition. In the WiSARD hub nodes, the data acquisition board used in WiSARD sensor nodes is replaced by a board that provides communication interfaces, global time acquisition, and non-volatile memory for data archival. Both WiSARD sensor nodes and hubs use a third radio board designed around a wireless transmitter/receiver chip operating in the 900-928 MHz Industrial, Scientific, and Medical (ISM) radio frequency band. These boards, along with batteries, waterproof strain-relief fittings for transducers, and an antenna port,

Figure 1. WiSARDNET sensor node deployed on a tree trunk. The node is attached to the trunk using easily-obtainable UV-resistant cord. Note the antenna (pointing downward) and the wires from transducers entering the WiSARD enclosure.

are integrated into a rugged, weatherproof, polycarbonate enclosure (Figure 1).

Each WiSARD sensor node has a rich array of interfaces to high-accuracy external transducers for measurement of multiple channels of PAR (e.g., LI-COR photodiode-based transducers), temperature (type T thermocouples), and soil moisture (capacitance-based Decagon ECH_2O Probes). The nodes also include interfaces to smart peripherals such as the Vaisala WXT-520, which measures rainfall and hail, wind velocity and direction, and relative humidity. We have also developed and implemented the capability to measure sap flux using the Granier method with two type T thermocouples connected in series to measure the difference between heated and unheated probes. The data collected by the WiSARDNet can be archived on the WiSARD hub, waiting for the user to visit the deployment site and transfer the data to a laptop computer. Thus, raw data can be transferred and processed using a computer program to generate processed data files for easy viewing and analysis. The process requires the user to manage a large number of data streams, one for every transducer on every sensor node. While this is useful from an engineering standpoint, allowing easy access to raw data for debugging and testing purposes, it is not convenient for long-term field studies.

The components of the WiSARDNet Field-to-Desktop system are depicted in Figure 2. The first component is software, called NetBridge, which links a WiSARDNet to the Internet and the Web and runs on any standard desktop computer. This computer, referred to as the WiSARDNet server, is connected to the WiSARD hub via a standard computer data cable. Every data packet from the WiSARD sensor nodes received by the WiSARD hub is sent to the WiSARD server over this cable. The NetBridge software running on the WiSARD server listens for these data packets and records them. Instead of waiting for the user to run the data processing, the software processes the data on the fly, storing the information on the WiSARD server in a relational database. The processing includes formatting and conversion operations and allows for the storage of raw and processed information (e.g., the raw thermocouple reading vs. the calculated temperature). This data can be accessed locally, via standard keyboard and monitor, or it can be accessed remotely through a variety of methods. If the WiSARD server has access to the Internet, the data can be accessed from anywhere in the world. The NetBridge software is highly configurable, including an option to write data to multiple locations, in case the WiSARD server has limited storage space, the remote connection to WiSARD server has limited data throughput, or the data needs to be streamed to multiple destinations.

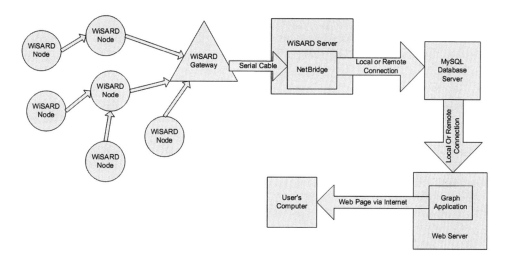

Figure 2. WiSARDNET Field-to-Desktop flow diagram. The MySQL database server can reside on same computer hosting the WiSARD Server, the computer hosting the web server, or another any other machine on the Internet. There can also be multiple database and web servers.

To facilitate data analysis, a web application was created to extract the data from the database and display it graphically. This web application consists of easy-to-use web pages that display the data based on user inputs. The user can select which transducers from a WiSARD sensor node to plot; multiple selections are allowed for easy comparison between sensors and nodes. This web application is easily configurable for any database. The application can be run on the WiSARD server at the deployment location or it can be run on a remote server better suited to handle web traffic and multiple users working in the application at the same time. Combined with the configurability of NetBridge, this gives the user a high degree of flexibility for any particular deployment, depending on the computer resources available. With the processing and the graphical display software, the user can view graphical plots of the data from any computer with Internet access. These plots can be updated in real time, due to the on-the-fly processing of the incoming data. No special software is needed on the user's computer as all of the processing and display is done on the WiSARD server and web server.

NetBridge is a combined communication and data processing system. The software currently runs on Linux operating systems; work is currently being done on a version of the software with the same functionality that will run on Microsoft Windows operating systems. The WiSARD hub connects to the WiSARD server via a standard nine-pin serial port. NetBridge listens for packets incoming on the serial port. When one arrives, the data processing is performed. The information is extracted and inserted into a database. A configuration file specifies the information needed to insert the data into any database accessible via the Internet. The requisite information includes the location of the database, local or remote, the user name and password needed to access the database, and the table in the database where the data will reside. By default, NetBridge always writes a backup copy of the data to the local database server.

If the deployment is located in an area with a local network or reliable Internet connection, then whenever NetBridge receives new data it can simply open up a connection to the remote database and write the data. NetBridge is capable of dealing with unreliable connections to the Internet, which can occur at remote sites. If the WiSARD server fails to connect to the remote database, then the data is stored in a queue. If the connection is successful the next time, NetBridge sends the data and clears the queue. All data remains in the queue until it is successfully transmitted to the remote database(s).

In some deployments the remote connection is intermittent. For example, satellite or cellular mobile phone connections to the Internet are expensive and may be charged on a "talk-time" basis. For these situations, NetBridge can be configured to temporarily store all sensor data in the queue and connect to the remote database at scheduled times, eliminating the need for a constant connection. This configurability was designed so the system is flexible enough to fit the widest possible range of deployment scenarios.

Any standard computer can be configured as a WiSARD server. The current version of NetBridge software is built to run on the Linux operating system due to the ease of adapting open-source industry-standard software tools to the required needs of the system. The web server framework that provides the capability to integrate the web server and database is known as a Linux Apache MySQL PHP (LAMP). A LAMP server consists of an Apache web server, a MySQL database server, and a PHP interpreter all resident on a Linux operating system. These three server applications integrate with each other extremely well, facilitating the development of the web application in PHP. MySQL is the database system used to store the processed data from the WiSARDNet processed by the NetBridge software. MySQL was chosen for its ability to be integrated into many different application platforms, including programs written in most modern computer languages and PHP web applications. The Apache server hosts the web page that the user interfaces with to display the graphs generated by the PHP application.

The web application and display page are compatible with any web server capable of supporting PHP scripts and viewable with any web browser. This allows the web site to be hosted on a different machine from the WiSARD server (and if needed, in a remote location), in case the WiSARD server isn't powerful enough to handle the web traffic and the processing software and displaying the web page for multiple users. Additionally, the database server can also be hosted remotely if the WiSARD server cannot handle the database size or high incoming traffic. In summary, the three server applications, Apache, MySQL, and NetBridge, can be run at three different locations, at the same location, or any combination. However, due to the number of server applications running and communicating with each other, it is important to verify that necessary security precautions are in place to prevent unauthorized use of the machine or corruption of the stored data. A security script is provided with the NetBridge software that can be run to secure the servers.

At remote sites, power is often provided by a battery-backed solar array, so the available power to run the WiSARD server and satellite/cellular modem is limited due to cost constraints. Typical personal computers can consume 80 W or more, requiring a large solar array. To control the cost associated with this or other types of remote power sources, we are porting NetBridge to a small ("single-board") Linux computer that is fanless and uses only solid-state (flash memory) disk for minimum power consumption.

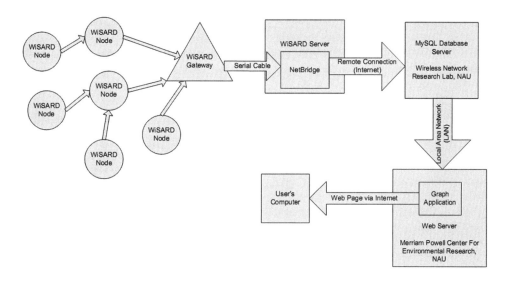

Figure 3. Instantiation of the WiSARDNET Field-to-Desktop system for the deployment at The Arboretum at Flagstaff. To minimize data traffic flow from the Arboretum, the database and web servers are located on the NAU campus.

RESULTS

The *in situ* WiSARDNET technology has been deployed to enable a new degree of data quality in ecological and environmental field studies, as well as other environmental monitoring applications. We currently have three separate WiSARDNet deployments operating across the United States. The first consists of three WiSARDNet networks in the Duke Forest of North Carolina that are supporting ecologists' efforts to characterize the role of the role of fine-scale environmental phenomena in the maintenance of ecosystem diversity in Eastern deciduous forests. The second deployment is mapping the effects of rainfall and soils in the coastal redwoods of California. And in the third deployment, we will determine the effects of scale on eddy covariance measurements of ecosystem energy balance in northern Arizona. We expect that the results of these studies will have global implications for biological diversity and ecosystem function.

The first WiSARDNet Field-to-Desktop system was deployed at The Arboretum at Flagstaff as part of a public outreach program in a joint venture with the Arboretum and Northern Arizona University's Merriam Powell Center for Environmental Research. Figure 3 shows how the Field-to-Desktop architecture is adapted to this particular deployment. The deployment consists of five WiSARD sensor nodes collecting data on microclimates at the Arboretum site. These five sensor nodes connect to a WiSARD hub that streams data to the WiSARD server. The Internet connection at The Arboretum at Flagstaff is reasonably fast, capable of sustained download rates of approximately 400 kilobytes per second. However, the connection is not capable of hosting a web page or database server. Thus, in this deployment, the WiSARD server writes the data to a database server in the Wireless Networks Research Lab located on the Northern Arizona University campus. The University's connection is capable of handling large amounts of incoming and outgoing traffic, making it idea to host the database.

The web application for the Arboretum is hosted on a server operated by the Merriam Powell Center and is accessible to Arboretum visitors at a touch screen kiosk. The web application also provides information on microclimates, specifically in northern Arizona, and how wireless sensor networks can help us understand how microclimates effect plant growth and ecosystem dynamics. Additionally, the application is made publically accessible so that any computer with an Internet connection can go to the website and view the same displays and information presented on the kiosk.

DISCUSSION

One of the most compelling applications of wireless sensor network technology is dense spatio-temporal sensing of environments in order to enable better understanding of environmental and ecosystem processes across multiple scales. In the WiSARDNet project, we have completed and explained the development of a prototype wireless environmental sensor network technology and how we have applied it in several field studies. In this paper, we have outlined the WiSARDNet Field-to-Desktop cyber-infrastructure that is adaptable to diverse applications and deployments, and have shown how the architecture is implemented in a deployment at The Arboretum and Flagstaff.

In ongoing work to improve the technology, we are currently developing new algorithms for data-gathering networks to further enhance WiSARD sensor node battery lifetime and to improve the network's performance in inferring data and models for ecosystem processes (Flikkema et al. 2006).

ACKNOWLEDGEMENTS

This work was partially supported by the National Science Foundation under grants BDEI-0131691 and EF-0308498, Northern Arizona University, and Microchip Technology, Inc. We would also like to thank our colleagues, J. Clark, B. Hungate, George Koch, and S. Sillett, as well as the NAU Merriam-Powell Center for Environmental Research, for their on-going collaboration and assistance on this project.

LITERATURE CITED

Estrin, D., L. Girod, G. Pottie, and M. Srivastava. 2001. Instrumenting the world with wireless sensor networks. In Proceedings of the International Conference of Acoustics, Speech and Signal Processing, 2001.

Flikkema, P., and B. West. 2002. Wireless sensor networks: From the laboratory to the field. National Conference for Digital Government Research, May 2002.

Flikkema, P., B. West, and B. Ruggeri. 2002. Network-aware lossless source coding of spatio-temporally correlated sensor data. Proceedings of the IEEE CAS Workshop on Wireless Communication and Networking, Pasadena, California, September 2002.

Flikkema, P., and B. West. 2003. Clique-based randomized multiple access for energy-efficient wireless ad hoc networks. IEEE 2003 Wireless Communication and Networking Conference, New Orleans, Louisiana, March 2003.

Flikkema, P., et al. 2006. Model-driven dynamic control of embedded wireless sensor networks: Workshop on dynamic data driven application systems. International Conference on Computational Science, Reading, United Kingdom, May 2006.

Mainwaring, A., J. Polastre, R. Szewczyk, D. Culler, and J. Anderson. 2002. Wireless sensor networks for habitat monitoring. Proceedings of the First ACM International Workshop on Wireless Sensor Networks and Applications, 2002.

Pottie, G., and W. Kaiser. 2000. Wireless integrated network sensors. Communications of the ACM 43:51–58.

Tolle, G., et al. 2005. A macroscope in the redwoods. Third ACM Conference on Embedded Networked Sensor Systems. San Diego, California, November 2005.

West, B., P. Flikkema, T. Sisk, and G. Koch. 2001. Wireless sensor networks for dense spatio-temporal monitoring of the environment: A case for integrated circuit, system, and network design. Proceedings of the 2001 IEEE CAS Workshop on Wireless Communications and Networking. Notre Dame, Indiana, August 2001.

Yang, Z., et al. 2005. WiSARDNET: A system solution for high performance. In Situ Environmental Monitoring. Second International Workshop on Networked Sensor Systems. San Diego, California 2005.

FINDING GAPS IN THE PROTECTED AREA NETWORK IN THE COLORADO PLATEAU: A CASE STUDY USING VASCULAR PLANT TAXA IN UTAH

Walter Fertig

ABSTRACT

About 19 percent of the Colorado Plateau region of Utah is currently under some form of permanent protective status. Most of these protected areas, however, were established for their scenic, cultural, or recreation value rather than for conservation of biological diversity. I used gap analysis methods to determine how well vascular plant species are represented in the existing protective network and to identify types of species, habitats, and geographic areas that are unprotected. At present 1948 of the 2859 species in the study area (68.1%) occur in permanent and highly protected areas (equivalent to GAP Status 1 or 2 lands), while 911 taxa (31.9%) are unprotected. Among the species on protected lands, 438 (22.5%) are Colorado Plateau endemics and 419 (21.5%) are listed as rare by the Utah Natural Heritage Program, while 1510 (77.5%) are widespread native or non-native species and 1529 (78.5%) are considered common. By comparison, plateau endemics (288 taxa) comprise 31.6 percent of the unprotected flora, and rare species (345 taxa) comprise 37.9 percent. Unlike distributions in most Gap studies, species from high elevation habitats in the Colorado Plateau are less likely to be protected than those from desert shrub, sagebrush, and pinyon-juniper communities. Almost 70 percent of the unprotected taxa occur in just 12 plant-diversity hotspots that are not currently part of the state's protected area network.

INTRODUCTION

Developing networks of nature reserves that contain the full array of native biological diversity and natural processes is one cornerstone of contemporary conservation planning (Groves et al. 2002, Margules and Pressey 2000). This strategy is especially important in regions with high endemism or species richness. One such area is the Colorado Plateau of southern Utah, southwestern Colorado, northern Arizona, and northwestern New Mexico; this area has the highest number of endemic vascular plant species of any ecoregion in North America and is among the top seven in overall plant species richness (Kartesz and Farstad 1999).

The Colorado Plateau also has the densest concentration of national parks in the United States (Newhouse 1992). In the Utah segment of the plateau alone, 12 national parks, monuments, or recreation areas and 6 designated Wilderness Areas have been established in the past century. These highly protected sites cover more than 1.7 million hectares, or nearly 19 percent of the Colorado Plateau ecoregion within Utah. Unfortunately, the state's protected-area network was created in an ad hoc fashion, with scenic, recreational, and historic values driving the selection process rather than biodiversity preservation. Due to such selection biases, a large number of species may still be missing from protected area networks (Groves et al. 2002, Scott et al. 2001).

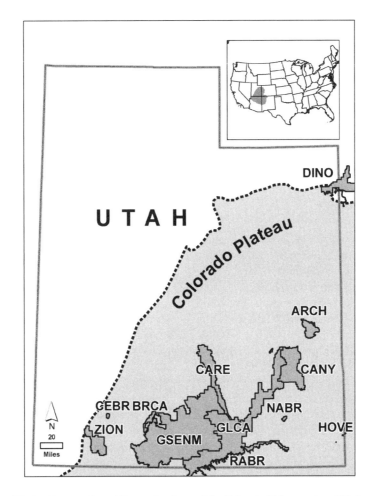

Figure 1. National Parks, Monuments, and Recreation Areas of the Colorado Plateau region of Utah scored as GAP status 1 or 2 lands for this study. Legend: ARCH (Arches National Park), BRCA (Bryce Canyon National Park), CANY (Canyonlands National Monument), CARE (Capitol Reef National Park), CEBR (Cedar Breaks National Monument), DINO (Dinosaur National Monument), GLCA (Glen Canyon National Recreation Area), GSENM (Grand Staircase-Escalante National Monument), HOVE (Hovenweep National Monument), NABR (Natural Bridges National Monument), RABR (Rainbow Bridge National Monument), ZION (Zion National Park). Map created by Matt Betenson, GIS specialist, Grand Staircase-Escalante National Monument.

Gap analysis is a research method for quantifying the biological diversity that is present (and absent) within a reserve network (Kiester et al. 1996). Typically this is accomplished by overlaying distribution data of target elements onto a map depicting land ownership and management status at a state or regional scale. Land areas are assigned a management status using a four-level rating scheme that ranges from permanently protected with an emphasis on biodiversity (status 1) to unprotected (status 4) (Scott et al. 1993). By intersecting distribution data over maps of GAP status 1 and 2 protected areas, researchers can begin to identify species, habitats, or geographic regions that are absent or poorly represented (i.e. "gaps") in the existing reserve network.

Traditionally, gap analyses have been applied mostly to terrestrial vertebrate species or major vegetation types using GIS-based habitat models and range maps, but gap methods can also be applied to other taxonomic groups (including vascular plants) using location data from sources such as herbarium records or species checklists (Fertig and Thurston 2001).

The purpose of this paper is to employ gap techniques to assess the degree of protection for vascular plant species in national parks, monuments, and recreation areas in the Colorado Plateau region of Utah. I will also compare the degree of protection afforded plants depending on their rarity, global distribution (i.e. endemics versus widespread taxa), and habitat affinities. Lastly, I will show how gap methods can help identify specific geographic areas and habitat types that are currently missing or under-represented in the protected area network.

METHODS

In this study I treated the Colorado Plateau portion of Utah in a broad sense to include the Colorado River drainage (the Colorado Plateau ecoregion proper) as well as the adjacent Utah High Plateaus ecoregion and the Uinta Basin segment of the Wyoming Basins ecoregion (ecoregion classification based on The Nature Conservancy system of Stein et al. 2000). This allowed me to include Bryce Canyon National Park, Cedar Breaks National Monument, and Dinosaur National Park, which technically fall outside the core area of the Colorado Plateau but share obvious floristic affinities (Figure 1). In all, I analyzed ten National Parks and National Monuments scored as GAP status-1 or -2 lands by the Utah Gap and Southwest ReGap programs (Edwards et al. 1995, Prior-Magee et al. 2007). In a departure from previous Gap studies, however, I included Glen Canyon National Recreation Area as a protected area (equivalent to GAP status 2, but scored as status 3 by others) based on its management by the National Park Service for natural values as well as recreation.

I developed a master vascular plant species list for the Colorado Plateau study area in Microsoft© Excel based on distributional data from Albee et al. (1988), Welsh et al. (2003), and Shultz et al. (2006), and on herbarium label data for selected taxa from the Brigham Young University Herbarium. Within this master list I incorporated species presence/absence data for eleven protected areas based on a review of pertinent literature (see Table 1 for citations of species checklists) and examination of herbarium collections and online specimen databases from each park and other regional repositories (Brigham Young University, Utah State University, and New York Botanical Garden). For parks that crossed state lines (Dinosaur and Hovenweep National Monuments and Glen Canyon National Recreation Area), I excluded species that had not been recorded in the Utah portion of the park. I also eliminated species known for a park only from cultivation (applicable primarily to Capitol Reef and Zion National Parks).

Unfortunately, data on vascular plant species were not available for several other highly protected areas recognized as GAP status-1 or -2 lands by Edwards et al. (1995) and Prior-Magee et al. (2007). These included Rainbow Bridge National Monument, six designated Wilderness Areas (Ashdown Gorge, Black Ridge Canyons, Box-Death Hollow, Dark Canyon, Paria Canyon-Vermilion Cliffs, and Pine Valley Mountain), Bureau of Land Management Areas of Critical Environmental Concern (ACECs), U.S. Forest Service Research Natural Areas (RNAs), National Wildlife Refuges, and private nature reserves managed by The Nature Conservancy. None of these areas were included in my analysis.

The final master species list was annotated with supplemental information on geographic distribution, rarity, and general habitat types. For geographic distribution, I determined

Table 1. Number of taxa and data sources for sampled GAP Status 1 and 2 lands in the Colorado Plateau area of Utah. The number of taxa per park does not include cultivated species.

Management Area	Number of Vascular Plant Taxa Confirmed or Reported	Source
Arches National Park	522	Fertig et al. (2009a), Harrison et al. (1964), Schelz et al. (2006)
Bryce Canyon National Park	587	Fertig and Topp (2009), Spence and Buchanan (1993)
Canyonlands National Park	594	Fertig et al. (2009b), Schelz et al. (2006)
Capitol Reef National Park	887	Fertig (2009a), Heil et al. (1993)
Cedar Breaks National Monument	345	Fertig (2009b), Roberts and Jean (1989)
Dinosaur National Monument	484*	Fertig (2009c), Naumann (2002)
Glen Canyon National Recreation Area	868*	Hill (2005), Spence and Zimmerman (1996), Welsh (1984)
Grand Staircase-Escalante National Monument	1002	Fertig (2005), Welsh and Atwood (2002)
Hovenweep National Monument	218*	Fertig (2009d), Schelz and Moran (2003)
Natural Bridges National Monument	429	Fertig (2009e), Schelz et al. (2006), Welsh and Moore (1968)
Zion National Park	991	Fertig and Alexander (2009), Welsh (1995)

*For parks that cross state lines, the number of taxa is based only on records from the Utah portion of the park.

whether each species was a local or regional endemic (found only within the Colorado Plateau ecoregion), widespread over western North America, or non-native through a literature review (Albee et al. 1988, Kartesz 2003, Welsh et al. 2003). Information on rarity was obtained from the Utah Natural Heritage Program (Utah Division of Wildlife Resources 1998). Habitat types were derived from the vegetation classification of Albee et al. (1988) which is analogous to the "class" level of the National Vegetation Classification hierarchy used in the Southwest ReGap project (Prior-Magee et al. 2007). Information on the presence of vascular plant species across habitat types was derived from Welsh et al. (2003). The more detailed vegetation classifications of Edwards et al. (1995) and Prior-Magee et al. (2007), which recognized 36–80 different land cover types for Utah were not used, due to a lack of adequate presence/absence information in the available literature for all vascular plant species.

Once the master data set was finalized, I calculated the number and percentage of species present in highly protected lands within the study area through simple data queries. Calculations were performed on the entire data set and on subsets sorted by geographic distribution (Colorado Plateau endemics versus widespread and non-native species), rarity (common versus rare species), and general habitat types. I also calculated the number and percentage of species occurring in 12 unprotected plant diversity hotspots within the Colorado Plateau region (Figure 2).

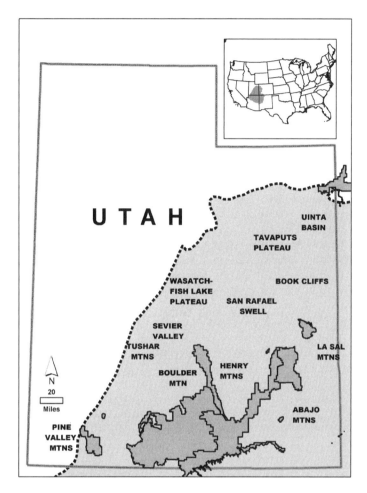

Figure 2. Unprotected vascular plant "hotspots" of the Colorado Plateau region of Utah. These physiographic regions have a high concentration of local and regionally endemic plant species that are largely absent from the existing network of highly protected national parks, monuments and recreation areas (shown in dark gray, see Figure 1 for labels). Precise boundaries are not shown as potential reserve boundaries have not been determined. The Pine Valley Mountains, located to the west of Zion National Park and sometimes considered part of the Great Basin physiographic region, are included within the Colorado Plateau in this study. Base map created by Matt Betenson.

RESULTS

As defined in this study, the Colorado Plateau region of Utah contains at least 2,859 vascular plant taxa (Table 2), which represents 74 percent of the native and naturalized flora of the state (Welsh et al. 2003). About 25 percent of the Colorado Plateau flora in Utah (726 taxa) is comprised of native species that are endemic to the plateau. Of the remaining species, 325 (11.4 percent) are not native to Utah or western North America and 1808 (63.2 percent) are native and occur widely across the state and region (Table 2). Rare species tracked by the Utah Natural Heritage Program (Utah Division of Wildlife Resources 1998) represent 26.7 percent of the flora of the Utah section of the Colorado Plateau (764

Table 2. Protection status of the flora of the Colorado Plateau region of Utah.

		Total Number of Taxa	Number of Taxa Present in GAP Status 1 or 2 Lands	Number of Taxa Not Present in GAP Status 1 or 2 Lands
All Vascular Plant Taxa in Colorado Plateau region of Utah		2859	1948	911
Geographic Distribution	Widespread Native Species	1808	1303	505
	Non-native Species	325	207	118
	Colorado Plateau Endemics	726	438	288
Rarity	Rare Plants (cited by Utah Division of Wildlife Resources 1998)	764	419	345
	Common Species (not cited by Utah Division of Wildlife Resources 1998)	2095	1529	566
Habitat	Desert Shrub	886	741	145
	Sagebrush	1193	935	258
	Pinyon-juniper Woodlands	1192	973	219
	Oak-ponderosa pine Montane Forest	1100	846	254
	Spruce-fir/Aspen Forest	782	502	280
	Alpine	217	92	125
	Riparian	777	563	214
	Disturbed Areas	323	225	98

species). Among the eight major vegetation classes in the Colorado Plateau region of Utah, sagebrush and pinyon-juniper have the greatest richness of vascular plant taxa (more than 1,190 species), while alpine vegetation has the lowest richness (217 species).

At least 1,948 (68.1 percent) of the plant species in the Utah portion of the Colorado Plateau are found in at least one of the 11 GAP Status 1 or 2 protected areas examined in this study (Table 2). The number of species occurring in more than one protected area exhibits a precipitous decline (Figure 3). Only 66 percent of protected plant taxa are found in two or more GAP status 1 or 2 areas, 23 percent occur in at least six protected areas, and just 1 percent are found in all 11 sites.

Among the protected species, 66.9 percent are widespread natives (1,303 taxa) and 78.5 percent are considered common within the state of Utah (1529 taxa) (Table 2). At least 207 protected taxa are not native to the state (10.6 percent). About 60 percent of the endemic plants of the Colorado Plateau (438 taxa) are protected at least once in GAP status-1 or -2 areas in Utah. Likewise, nearly 55 percent of the rare plants tracked by the Utah Natural Heritage Program (419 taxa) are found in protected areas. Over 75 percent of the plant species from desert shrub, sagebrush, pinyon-juniper, and oak-ponderosa pine habitats are under some protection (Table 2).

Presently at least 911 species (31.9 percent) from the Colorado Plateau flora are not found in GAP status-1 or -2 lands in the study area (Table 2). This group includes 118 non-native and 793 native

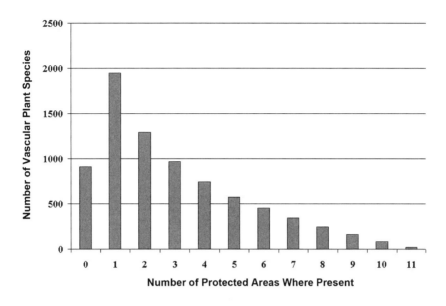

Figure 3. Representation of vascular plant species across 11 protected areas of GAP Status 1 and 2 lands in the Colorado Plateau region of southern Utah.

species. Among the natives, 505 (63.7 percent) are widespread species and 288 are Colorado Plateau endemics (36.3 percent). Rare species tracked by the Utah Natural Heritage Program account for 37.8 percent of the unprotected taxa (345 species). High elevation alpine and spruce-fir/aspen forest habitats have the highest proportions of unprotected species (Table 2).

Of the 911 unprotected species identified in this study, at least 631 (69.3 percent) occur in just 12 hotspots of high alpha diversity (Figure 2, Table 3) that are not currently part of the state's protected area network. Protecting representative examples of these 12 areas would capture 209 of the missing 288 Colorado Plateau endemics (72.6 percent) and 234 of 345 missing rare species tracked by the state natural heritage program (67.8 percent).

DISCUSSION

Three decades ago, Myers (1979) estimated that only one percent of the Earth's land surface was being protected for the sake of biological diversity. Based on island biogeographic theory, Myers suggested that 10–20 percent of the Earth's land area would need protection to capture the full array of species diversity. During the 1980s and 1990s, many conservation organizations and government agencies arrived at a consensus that protecting 10–12 percent of the area of a given region might be sufficient to preserve native biodiversity (Noss 1996). More recently, the *a priori* target of 10–12 percent has come under significant criticism for inadequately addressing complementarity, overemphasizing high-elevation "rock and ice" environments, and being too small for organisms with large area requirements (Margules and Sarkar 2007, Scott et al. 2001, Soule and Sanjayan 1998, Svancara et al. 2005). Ironically, this criticism comes at a time when the estimated area under protection worldwide is approaching 12 percent for the first time (Brooks et al. 2004).

Highly protected (GAP status 1 or 2) lands encompass nearly 19 percent of the Colorado Plateau region of Utah, making the

Table 3. Unprotected plant taxa in Colorado Plateau region of Utah

Category	Total Number of Taxa	Number of Colorado Plateau Endemics	Number of Rare Species
Number of Unprotected Taxa	911	288	345
Number of Unprotected Native Taxa	793	288	345
Number of Unprotected Taxa from Unprotected "Hotspots"	631	209	234
Number of Unprotected Native Taxa from Unprotected "Hotspots"	583	209	234

area a good test of the adequacy of the 10–20 percent protection goals advocated in the past literature. Based just on vascular plant species and data from 11 highly protected areas, the Colorado Plateau reserve network in Utah fails to include nearly one third of the region's flora. The degree of redundancy within the reserve network is also relatively low, with just two thirds of protected plant species occurring in at least two sites and less than one quarter found in as many as 6 (Figure 3). These findings corroborate the results of the Utah Gap and Southwest ReGap projects which found that 25–65 percent of terrestrial vertebrate species and 25–36 percent of major vegetation types were absent or inadequately (with < 10 percent of their predicted range) represented in the protected area network of Utah (Edwards et al. 1995, Prior-Magee et al. 2007).

Lack of protection for vascular plant species in the Colorado Plateau does not occur randomly. Unprotected plants are one and a half times more likely to be rare species and plateau endemics than common or widespread species. Rare plants and plateau endemics typically have small geographic ranges or occur in specialized environments that frequently do not overlap with the distribution of highly protected lands. Edwards et al. (1995) and Prior-Magee et al. (2007) reported similar results for underprotected terrestrial vertebrates and uncommon vegetation types. But in contrast to other Gap studies (Fertig and Thurston 2001, Pressey 1994, Prior-Magee et al. 2007, Scott et al. 2001), vascular plant taxa from alpine and high-elevation montane habitats in the Colorado Plateau were less likely to be protected than species from lower-elevation desert shrub, sagebrush, or pinyon-juniper habitats. This discrepancy results from the lack of parks and designated wilderness areas at high elevation sites in the Utah section of the Colorado Plateau, where such lands have historically been more profitable for agriculture (and thus remain in private ownership or are managed for multiple use) or other uses than low-elevation arid lands.

Geographically, the unprotected vascular plant species of the Colorado Plateau in Utah are also not found in random distribution. Unprotected species frequently occur in clusters or "hotspots" of alpha diversity, most of which also lack GAP status-1 or -2 protection. Just 12 such hotspots in

Utah (Figure 2) contain nearly 70 percent of the unprotected vascular plant taxa of the Colorado Plateau section of the state. Several hotspots—such as the La Sals, Abajos, Henrys, Boulder Mountain/Aquarius Plateau, Tushars, and Fishlake/Wasatch Plateau—are isolated mountain ranges with large numbers of local endemics. Other low-elevation hotspots (Uinta Basin, Book Cliffs, and Sevier Valley) also have significant rates of endemism and are highly threatened by ongoing or proposed mineral development. Only the Pine Valley Mountains along the boundary of the Colorado Plateau west of Zion National Park presently enjoy any protection afforded through wilderness designation, although better data are needed to ascertain which species are protected.

Species-level vascular plant data have been used infrequently in Gap analyses, but offer many advantages to complement findings from terrestrial vertebrates and major vegetation types. In the Colorado Plateau of Utah, the number of vascular plant species is more than six times greater than that of resident, migrant, or wintering terrestrial vertebrates (2,859 taxa compared to 453) (Dunn and Alderfer 2006, Reid 2006, Stebbins 2003). Likewise, the number of Colorado plateau endemics (726 taxa) and rare species (764) is significantly greater for vascular plants than for vertebrates (17 endemics and 81 rare species) (Bosworth 2003, Utah Division of Wildlife Resources 1998). Thus, using vascular plant species greatly increases the sample size available for Gap studies. In addition, large distributional datasets are available for vascular plants from herbarium records and species checklists, which are often lacking for poorly studied or cryptic vertebrates, including many reptiles, amphibians, and small mammal species.

Plant species lists based on verified herbarium vouchers and extensive sampling, such as those employed in this study, can be a useful alternative data source to distributional models used in conventional Gap analyses.

The primary advantage of species lists is that presence within a protected area can be verified, whereas models may mistakenly predict that a species is either present (false positives or commission error) or absent (false negatives or omission error). The Utah Gap program compared its vertebrate distribution models to species lists from eight National Park Service units and found error rates of 10 to 30 percent, with the majority of discrepancies being false positives (Edwards et al. 1995). While it is possible that these species were simply overlooked within the parks when the species lists were generated, commission error can be costly if species are presumed to already be represented in the protected area network when in fact they are not. Decisions to establish new protected areas based on such misinformation might fail to address unprotected species.

There are, however, some disadvantages to using species lists in Gap studies. Data are lacking for many protected areas, especially Bureau of Land Management ACECs, U.S. Forest Service RNAs, Wilderness Areas, and privately managed preserves. It is likely that the proportion of protected species in Utah's preserve network will be 5 to 10 percent higher than presently recognized once all components of the system are thoroughly inventoried. Species lists are also not static and new species are almost always being discovered that were either previously overlooked or are newly invading—even in well-studied areas. More importantly, species checklists often do not indicate the abundance of a species within an area, the number of discrete biological populations, or which populations are most likely to perpetuate themselves. Mere presence/absence data from species lists may be insufficient in determining whether necessary ecological processes or conditions are present in a given area, or if management policies are adequate or being enforced (Margules and Pressey 2000).

Although the foundation of a *systematic*

preserve system to protect the full array of biological diversity is in place in the Utah portion of the Colorado Plateau, many holes remain. Gap methods are a powerful tool for identifying species, habitats, and geographic areas that are missing or underrepresented in the preserve system. Filling the gaps will require identification, protection, and adequate management of areas of both alpha (species richness hotspots) and beta diversity (areas important because of complementarity). Hopefully, use of vascular plant species distribution and species checklists will gain wider use as a complement to more traditional gap studies of terrestrial vertebrates and major vegetation types. These results should prove useful to conservation planners in academia, the private sector, and at multiple levels of government for setting priorities for land acquisition or protection.

LITERATURE CITED

Albee, B. J., L. M. Shultz, and S. Goodrich. 1988. Atlas of the vascular plants of Utah. Occasional Publication 7, Utah Museum of Natural History, Salt Lake City, Utah.

Bosworth, W. R., III. 2003. Vertebrate information compiled by the Utah Natural Heritage Program: A progress report. Publication number 03-45. Utah Division of Wildlife Resources, Salt Lake City, Utah.

Brooks, T. M., M. I. Bakaar, T. Boucher, G. A. B. Da Fonseca, C. Hilton-Taylor, J. M. Hoekstra, T. Moritz, S. Olivieri, J. Parrish, R. L. Pressey, A. S. L. Rodrigues, W. Sechrest, A. Stattersfield, W. Straham, and S. N. Stuart. 2004. Coverage provided by the global protected-area system: Is it enough? Bioscience 54(12):1081–1091.

Dunn, J. L., and J. Alderfer. 2006. National Geographic field guide to the birds of North America. National Geographic Society, Washington, D.C.

Edwards, T. C., Jr., C. H. Homer, S. D. Bassett, A. Falconer, R. D. Ramsey, and D. W. Wight. 1995. Utah gap analysis: An environmental information system. Final Project Report 95-1, Utah Cooperative Fish and Wildlife Research Unit, Utah State University, Logan, Utah.

Fertig, W. 2005. Annotated checklist of the flora of Grand Staircase-Escalante National Monument. Moenave Botanical Consulting, Kanab, Utah.

Fertig, W. 2009a. Annotated checklist of vascular flora: Capitol Reef National Park. Natural Resource Technical Report, National Park Service, Fort Collins, Colorado.

Fertig, W. 2009b. Annotated checklist of vascular flora: Cedar Breaks National Monument. Natural Resource Technical Report, National Park Service, Fort Collins, Colorado.

Fertig, W. 2009c. Annotated checklist of vascular flora: Dinosaur National Monument. Natural Resource Technical Report, National Park Service, Fort Collins, Colorado.

Fertig, W. 2009d. Annotated checklist of vascular flora: Hovenweep National Monument. Natural Resource Technical Report, National Park Service, Fort Collins, Colorado.

Fertig, W. 2009e. Annotated checklist of vascular flora: Natural Bridges National Monument. Natural Resource Technical Report, National Park Service, Fort Collins, Colorado.

Fertig, W., and J. Alexander. 2009. Annotated checklist of vascular flora: Zion National Park. Natural Resource Technical Report, National Park Service, Fort Collins, Colorado.

Fertig, W., and R. Thurston. 2001. Gap analysis of the flora of Wyoming. Gap Analysis Bulletin 10:3–6.

Fertig, W., and S. Topp. 2009. Annotated checklist of vascular flora: Bryce Canyon National Park. Natural Resource Technical Report, National Park Service, Fort Collins, Colorado.

Fertig, W., S. Topp, and M. Moran. 2009a. Annotated checklist of vascular flora: Arches National Park. Natural Resource Technical Report, National Park Service, Fort Collins, Colorado.

Fertig, W., S. Topp, and M. Moran. 2009b. Annotated checklist of vascular flora: Canyonlands National Park. Natural Resource Technical Report, National Park Service, Fort Collins, Colorado.

Groves, C. R., D. B. Jensen, L. L. Valutis, K. H. Redford, M. L. Shaffer, J. M. Scott, J. V. Baumgartner, J. V. Higgins, M. W. Beck, and M. G. Anderson. 2002. Planning for biodiversity conservation: Putting conservation science into practice. Bioscience 52(6):499–512.

Harrison, B. F., S. L. Welsh, and G. Moore. 1964. Plants of Arches National Monument. Brigham Young University Science Bulletin 5(1):1–23.

Heil, K. D., J. M. Porter, R. Fleming, and W. H. Romme. 1993. Vascular flora and vegetation of Capitol Reef National Park, Utah. Technical Report NPS/NAUCARE/NRTR-93/01.

Hill, M. 2005. Vascular flora of Glen Canyon National Recreation Area, Utah and Arizona. Master's thesis, Northern Arizona University, Flagstaff, Arizona.

Kartesz, J. T. 2003. A synonymized checklist and atlas with biological attributes for the vascular flora of the United States, Canada, and Greenland, 2nd ed. In Synthesis of the North American Flora, Version 2.0 by J. T. Kartesz.

Kartesz, J., and A. Farstad. 1999. Multiscale analysis of endemism of vascular plant species. In Terrestrial Ecoregions of North America:

A conservation assessment by T. H. Ricketts, E. Dinerstein, D. M. Olson, C. J. Loucks, W. Eichbaum, D. DellaSala, K. Kavanagh, P. Hedao, P. T. Hurley, K. M. Caney, R. Abell, and S. Walters. 1999. Island Press, Washington, D.C.

Kiester, A. R., J. M. Scott, B. Csuti, R. F. Noss, B. Butterfield, K. Sahr, and D. White. 1996. Conservation prioritization using Gap data. Conservation Biology 10:1332–1342.

Margules, C. R., and R. L. Pressey. 2000. Systematic conservation planning. Nature 405:243–253.

Margules, C. R., and S. Sarkar. 2007. Systematic conservation planning. Cambridge University Press, Cambridge, United Kingdom.

Myers, N. 1979. The sinking ark: A new look at the problem of disappearing species. Pergamon Press, Oxford, United Kingdom.

Naumann, T. 2002. Dinosaur National Monument plant list. Dinosaur National Monument, Dinosaur, Colorado.

Newhouse, E. L., ed. 1992. National Geographic's guide to national parks of the United States. National Geographic Society, Washington, D.C.

Noss, R. F. 1996. Protected areas: How much is enough? In National Parks and Protected Areas: Their role in environmental protection, edited by R. G. Wright, pp. 91–120. Blackwell Science, Cambridge, Massachusetts.

Pressey, R. L. 1994. Ad hoc reservations: Forward or backward steps in developing representative reserve systems? Conservation Biology 8:662–668.

Prior-Magee, J. S., K. G. Boykin, D. F. Bradford, W. G. Kepner, J. H. Lowry, D. L. Schrupp, K. A. Thomas, and B. C. Thompson, eds. 2007. Southwest Regional Gap Analysis Project Final Report. U.S. Geological Survey, Gap Analysis Program, Moscow, Idaho.

Reid, F. A. 2006. A field guide to mammals of North America North of Mexico, 4th ed. Houghton Mifflin Co., Boston, Massachusetts.

Roberts, D. W., and C. Jean. 1989. Plant community and rare and exotic species distribution and dynamics in Cedar Breaks National Monument. Department of Forest Resources and Ecology Center, Utah State University, Logan, Utah.

Schelz, C. and M. Moran. 2003. Plants, Hovenweep National Monument. National Park Service, Southeast Utah Group, Moab, Utah.

Schelz, C., M. Moran, and A. Lafever. 2006. Southeast Utah group plants. National Park Service Southeast Utah Group (Arches NP, Canyonlands NP, Hovenweep NM, and Natural Bridges NM), Moab, Utah.

Scott, J. M., F. Davis, B. Csuti, R. Noss, B. Butterfield, C. Groves, H. Anderson, S. Caicco, F. D'Erchia, T. C. Edwards, J. Ulliman, and G. Wright. 1993. Gap analysis: A geographic approach to protection of biological diversity. Wildlife Monographs 123.

Scott, J. M., F. W. Davis, R. G. McGhie, R. G. Wright, C. Groves, and J. Estes. 2001. Nature reserves: Do they capture the full range of America's biological diversity? Ecological Applications 11(4):999–1007.

Shultz, L. M., R. D. Ramsey, and W. Lindquist. 2006. Revised atlas of Utah plants. College of Natural Resources, Utah State University, Logan, Utah. <http://earth.gis.usu.edu/plants>.

Soulé, M. E,. and M. A. Sanjayan. 1998. Conservation targets: Do they help? Science 179:2060–2061.

Spence, J. R., and H. Buchanan. 1993. 1993 update, checklist of the vascular plants of Bryce Canyon National Park, Utah. Great Basin Naturalist 53(3):207–221.

Spence, J. R., and J. A. C. Zimmerman. 1996. Preliminary flora of Glen Canyon National Recreation Area. Glen Canyon National Recreation Area, Page, Arizona.

Stebbins, R. C. 2003. A field guide to western reptiles and amphibians, 3rd ed. Houghton Mifflin Co., Boston, Massachusetts.

Stein, B. A., L. S. Kutner, and J. S. Adams. 2000. Precious heritage: The status of biodiversity in the United States. The Nature Conservancy and Association for Biodiversity Information, Oxford University Press, New York.

Svancara, L. K., R. Brannon, J. M. Scott, C. R. Groves, R. F. Noss, and R. L. Pressey. 2005. Policy-driven versus evidence-based conservation: A review of political targets and biological needs. Bioscience 55(11): 989–995.

Utah Division of Wildlife Resources. 1998. Inventory of sensitive species and ecosystems in Utah. Endemic and rare plants of Utah: An overview of their distribution and status. Report prepared for the Utah Reclamation Mitigation and Conservation Commission and U.S. Department of the Interior.

Welsh, S. L. 1984. Flora of Glen Canyon National Recreation Area. Report prepared for Glen Canyon National Recreation Area.

Welsh, S. L. 1995. Zion National Park annual report 1994–95: Rare plant survey of shuttle system and vascular plant scientific and common name list. Brigham Young University, Provo, Utah.

Welsh, S. L., and N. D. Atwood. 2002. Flora of Bureau of Land Management Grand Staircase-Escalante National Monument and Kane County, Utah. Brigham Young University, Provo, Utah.

Welsh, S. L., N. D. Atwood, S. Goodrich, and L. C. Higgins. 2003. A Utah flora, 3rd ed., revised. Brigham Young University, Provo, Utah.

Welsh, S. L., and G. Moore. 1968. Plants of Natural Bridges National Monument. Proceedings Utah Academy of Science, Arts, and Letters 45:220–248.

FIRE AND FIRE SURROGATE TREATMENT IMPACTS ON SOIL MOISTURE CONDITION IN SOUTHWESTERN PONDEROSA PINE FORESTS

Boris Poff, Daniel G. Neary, and Aregai Tecle

ABSTRACT

An experimental Fire and Fire Surrogate study (FFS) demonstrates use of silvicultural practices to replace and/or augment fuel reduction strategies in the semi-arid Southwest. The FFS study incorporates soil-moisture measurements into the existing experimental design. The data gathered in this study provide information on the amount of soil moisture available to plants at different rooting depths. Permanent plots have been established for the application of four treatments (control, burn only, cut only & cut and burn). Each treatment consists of 36 permanent plots. The soil-moisture-availability study was conducted near A1 Mountain in northern Arizona. There, ten permanent plots from each treatment were selected for installing semi-permanent soil-moisture probes. Of the ten plots, four were equipped with 15 cm probes, another four plots with 30 cm probes, while the remaining two plots have both 15 and 30 cm probes. Data were collected on a monthly basis from April 2006 to March 2007. An extensive statistical analysis of the soil moisture data was conducted, and the results of this study indicate no statistical significant difference between aspects, treatments, or stand density (basal area). However, some soil moisture trends are noticeable. Either the cut and burn or burn only units had the highest soil moisture content throughout the year at either depth.

INTRODUCTION

Forests in the western United States are more dense and have more downed fuels now than under historic conditions, mostly due to anthropogenic influences such as grazing, fire suppression, and logging. Past overgrazing practices have led to the deterioration of grass cover and allowed tree species, such as ponderosa pine, to become established in former grasslands (Fox et al. 1991). While some wildlife species—such as deer and elk (Patton 1991)—benefit from the forage and browse available in less dense stands, some of the same species—as well as others, such as wild turkey, elk and deer—have benefited from dense vegetation. These species require management practices that maintain this density to provide nesting, bedding, hiding, and escape cover (Wagner et al. 2000). However, this same dense vegetation increases the risk of wildfires and the spread of tree diseases and pathogens. Managers have recognized this problem and have acted to reduce stem density and fuels by thinning, burning, and/or fuel treatments. Although silvicultural treatments can mimic the effects of fire on structural patterns of woody vegetation, virtually no comparative data exist on how these treatments mimic ecological functions of fire. A nationwide Fire and Fire Surrogate study program was designed to conduct research into this issue.

The FFS Project was funded in 2000 by the Joint Fire Sciences Program. The

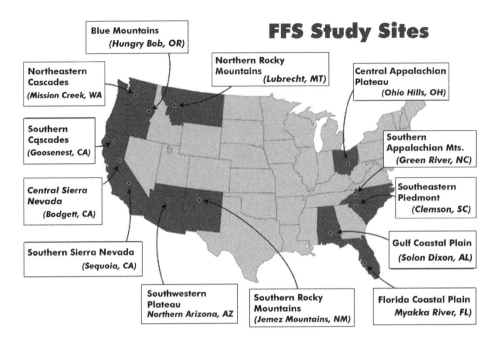

Figure 1. Location of Fire and Fire Surrogates (FFS) study sites in the United States.

objective of the project was to quantify the initial effects (first five years) of wildfire mitigation treatments on a number of specific ecosystem components (Weatherspoon 2000, Weatherspoon and McIver 2000). Twelve sites were established across the United States (Figure 1). Specific ecosystem variables within the following general groupings have been measured: (a) vegetation, (b) fuel and fire behavior, (c) soils and forest floor (including relation to local hydrology), (d) wildlife, (e) entomology, (f) pathology, and (g) treatment costs and utilization economics. Individual treatment plots were about ha. The small size of the FFS plots precluded making measurements of ecosystem responses to the treatments at a watershed scale in terms of the treatment effects on run-off, water-yield changes in sedimentation, and erosion.

Forested watersheds are some of the most important sources of water supply in the world. Maintenance of good hydrologic condition is crucial to protecting the quantity and quality of stream flow on these important lands. Watershed condition, or the ability of a catchment system to receive and process precipitation without ecosystem degradation, is a good predictor of the potential impacts of fire on water and other resources (e.g. roads, recreation facilities, riparian vegetation, etc.). The surface cover of a watershed consists of the organic forest floor, vegetation, bare soil, and rock. Disruption of the organic surface cover and alteration of the mineral soil by wildfire can produce changes in the hydrology of a watershed well beyond the range of historic variability (DeBano et al. 1998, Neary et al. 2005). Low-severity fires typical of prescribed fires rarely produce adverse effects on watershed condition, but high-severity fires usually do produce adverse effects. Successful management of watersheds in a post-fire environment requires an understanding of the changes in watershed condition and hydrologic responses induced by fire. Flood flows are the largest hydrologic response and most damaging to many resources (Neary 1995).

However, fire-induced changes in a soil-moisture regime are difficult to predict because of the different responses of the hydrological process affected by fire and/or surrogate treatments (DeBano et al. 1998). For example, the loss of understory, loss of overstory canopy, and reduced O-horizon density result in reduced precipitation interception. A reduction in leaf area reduces evapotranspiration, but usually increases ground surface evaporation due to greater insolation at the soil surface. Overall, these hydrological changes would be expected to lead to greater soil moisture following fire, as has been shown by Klock and Hevey (1976), Haase (1986), and Sayok et al. (1993). The reduction of canopy, O-horizon interception, and subsequent formation of water-repellent soils, on the other hand, can also result in reduced water infiltration to the soil, because of increases in water loss via run-off and, therefore, lead to lower soil moisture (Campbell et al. 1977, Milne 1979). As a result of the way these competing hydrologic processes are altered by fire or silvicultural treatments at a given site, no change in soil moisture was detected by Wells et al. (1979), Ower (1985), Ryan and Covington (1986), Macadam and Trowbridge (1988), Oakley (2004), and Moody et al. (2007).

It is important to understand the impacts of fire or fire-surrogate treatments on soil moisture, because soil moisture is crucial to soil microorganisms (Hart et al. 2005). These microorganisms (bacteria, fungi, etc.) serve as sources and sinks of key nutrients and as catalysts of nutrient transformations, influencing and maintaining soil structure (Paul and Clark 1996). Therefore, changes in soil moisture following a fire or surrogate treatment should result in changes in activity levels of these microorganisms (Paul and Clark 1996). Chen et al. (2006) found that other organisms, such as various insect species also respond both unfavorably and favorably to changes in soil-moisture levels following fire or fire-surrogate treatments in ponderosa pine forests. Higher levels of soil moisture in post-treatment ponderosa pine forest favored bark beetle outbreaks (Morehouse et al. 2008), while lower levels in the summer tended to favor the non-native and invasive plant species, such as diffuse knapweed (Wolfson et al. 2005).

In the context of a widespread drought in the western United States, questions are also being raised about the potential of fuels treatments in the West to increase supply of water to expanding urban areas in this mostly semi-arid region. Hence, the objectives of this study are to: (1) determine the small-scale hydrologic responses of ponderosa pine forest soils in Arizona to operational fuels treatments, and (2) gather information on the soil moisture that will be available to plants in different rooting depths as well as on that available for soil microbiology processes within the same National Forest and on similar soils.

STUDY SITE

Currently, there are three FFS sites on the Mogollon Rim of Arizona—two in the Coconino National Forest and one on the Kaibab National Forest. Both Forests are located in northern Arizona in the southwestern ponderosa pine forest ecosystem, which is considered to be the largest continuous stand of its kind in the United States. It stretches from southern Utah through northern and central Arizona to south-central New Mexico. This area constitutes about half of the 6.5 million ha of ponderosa pine forest ecosystem in the Rocky Mountains (USFS 1989). More than 65 percent of this forest ecosystem is under National Forest ownership (Conner et al. 1990), while the rest is in private, tribal, state, and National Park ownership. The current distribution of ponderosa pine is mostly affected by climatic factors, such as precipitation and temperature gradients, as well as fire regimes. However, anthropogenic

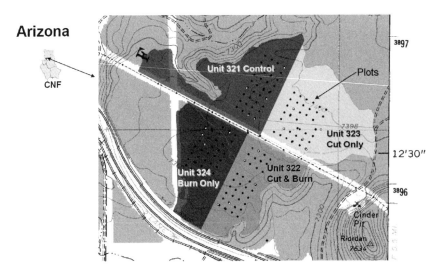

Figure 2. Location of Powerline FFS study site (32) in Northern Arizona.

influences such as grazing, timber-harvesting practices, and fire suppression also have some influence on its distribution and density (Covington et al. 1997).

The pilot study described in this paper was conducted at the Powerline site (FFS site number 32) on the Coconino National Forest, just west of Flagstaff, Arizona (see Figure 2). The site is broken down into four units. Unit 321 is the "control" unit, which received no silvicultural treatments. Unit 322 received a thinning treatment as well as a prescribed burn. Unit 323 only had a thinning treatment, and Unit 324 only had a prescribed burn conducted on it. Each unit has 36 permanent study plots.

METHODS

The instrument used to measure the soil moisture in this study was a TRASE System I, which is a self-contained, portable unit designed for field use; it provides fast and accurate measurements of moisture in various materials (including soil) using Time Domain Reflectometry (TDR). Measurements were taken monthly from April 2006 to March 2007, at the same time each month. One hundred and sixty pairs of semi-permanent probes were installed in the plots in November 2005. Eighty pairs of 15 cm and 80 pairs of 30 cm probes were distributed among the different treatment units (321, Control; 322, Cut & Burn; 323, Cut Only; 324, Burn Only). Of the 36 plots in each unit, 5 were selected at random per depth (15 and 30 cm) for this study. However, each unit had two plots at both depths. All of the plots have a "witness tree" located close to the plot center. In each selected plot, probes were placed in the four aspects (NWSE) around the witness tree to accommodate the occurrences of tree-rooting systems in the installation of the probes. Stand density (basal area) measurements were taken relative to the location of the witness trees in each of the plots as well. In most months, measurements were taken on the same day. On a few occasions, measurements were taken on two consecutive days due to weather or time constraints. Once measurements had been recorded, the data were downloaded into a computer and imported into Microsoft Excel, where it was sorted for analysis. The statistical analyses were performed in SAS and SPSS. Where applicable, an alpha level of 0.01 was used.

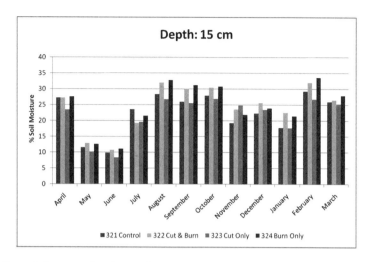

Figure 3. Average soil-moisture values at a depth of 15 cm for four different treatments

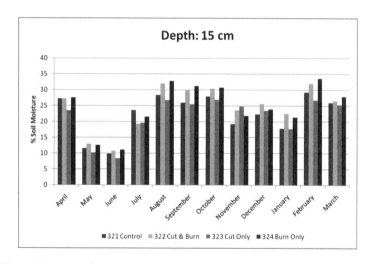

Figure 4: Average soil moisture values at a depth of 30cm for the four different treatments

RESULTS

Due to the high variability between plots in the same treatment unit, the results of the soil moisture measurements show little to no statistical significance. Analyses of depth for concurrently sampled probes were separated by month to interpret *month x depth* interaction. Each depth was also evaluated separately.

Some criticism has been raised about the tendency in the literature to report only statistically significant findings, perhaps biasing our understanding of nature because scientists fail to report the lack of differences. Hence, the results presented here are used to make several empirical observations and interpretations of trends. Figure 3 shows the average soil moisture values at a depth of 15 cm for the four different treatments and Figure 4 shows the average soil moisture values at a depth of 30 cm for the same treatments. Both figures reflect

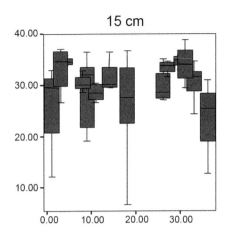 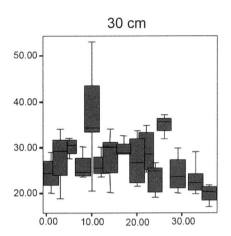

Figure 5. Variability within plots (both 15 and 30 cm) for August (the wettest month).

the metrological cycle in the Southwest. Soil moisture is usually lowest during the dry months of the year (May and June) prior to the onset of the monsoon season in July. In 2006, the North American monsoon started towards the end of July in northern Arizona. The soil moisture measurements in this study were taken after the onset of the monsoon; hence, they reflect an increase in soil moisture compared to the previous months. August is usually the wettest month of the year, which, again, is reflected by the measured soil moisture. The moisture content of the soil then slowly tapers off until the beginning of the snowmelt, which in 2007 began in February.

A comparison of soil-moisture trends in the four different treatments (Figure 3) suggests that the units that had been *burned only* or were *cut and burned* had a tendency to have the highest soil-moisture content at a depth of 15 cm, except during July (when the *control* unit had a tendency to be the wettest) and November (when the *cut only* unit had a tendency to be the wettest). At a depth of 30 cm (Figure 4), however, the *cut and burn* treatment showed a trend of having the highest soil moisture content year round, with the exception of July (when the *cut only* unit had a higher value by less than 1 percent).

As mentioned before, measurements of soil moisture show a high variability between units as well as between plots (Figure 5). The variability at 30 cm depths, however, is considerably less than the variability in soil moisture at 15 cm depth.

Another factor affecting the soil moisture content that this study investigated is the density of overstory. Even though two of the units underwent mechanical thinning, the forest stand density varied from plot to plot in all units, because the silvicultural treatment attempted to mimic the clumpy spatial structure of historical Southwestern ponderosa pine forest stands as described by White (1985). Consequently, the stand density of all 32 plots in this study was measured, expressed in basal area and regressed against the measured soil moisture for six months. As shown in Figures 6 and 7, we found no correlation between soil moisture and overstory stand density. Instead, these figures demonstrate how similar the soil moisture is for each of the displayed months across the different stand densities at depths of 15 and 30 cm.

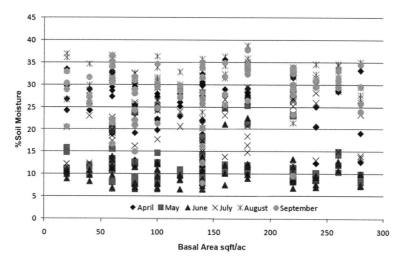

Figure 6. Soil moisture versus stand density expressed in basal area for all 15 cm plots from April until September.

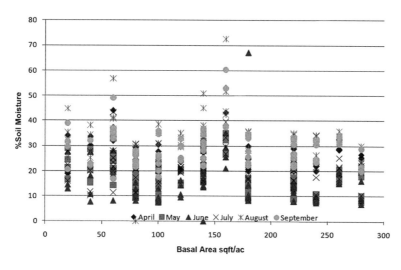

Figure 7. Soil moisture versus stand density expressed in basal area for all 30 cm plots from April until September.

DISCUSSION

Because results were highly variable between plots in the same treatment areas, the study data indicate no statistically significant difference between aspects, treatments or stand density (basal area) as the affect soil-moisture levels. A couple of factors may have contributed to this high variability. The measurements of this pilot study were conducted at only one site (Site 32, Powerline), which has a very rocky soil, and the study was conducted during a historically unusual dry year. Previous studies have shown that hydrological changes in Southwestern ponderosa pine forests after silvicultural treatments only become measurable in wet years (Baker 1986, Neary et al. 2005).

Some soil moisture trends that are noticeable, even in this short pilot study, include higher soil moisture content at a depth of 30 cm in the *cut and burn unit*, especially during the monsoon season. This trend should be investigated further for any correlation with biological observations made during the study period. Block (personal communication 2007), for example, recorded a higher occurrence of native reptiles in the cut and burn unit during the monsoon season as well. Both the *cut and burn* and the *burn only* units showed a tendency to have the highest soil moisture content throughout the year at either depth. Considering that fire used to be a natural part of the Southwestern ponderosa pine ecosystem, this is an important observation that should be also evaluated with other biological observations. A better understanding of the relationship between fire and fire surrogate treatment and soil-moisture levels in ponderosa pine forests might be gained through future studies conducted for longer time periods at other locations in the Southwestern ponderosa pine forests.

REFERENCES

Baker, M. B., Jr. 1986. Effects of ponderosa pine treatments on water yield in Arizona. Water Resources Research 22(1):67–73.

Campbell, R. E., M. B. Baker Jr., P. F. Ffolliott, F. R. Larson, and C. C. Avery. 1977. Wildfire effects on a ponderosa pine ecosystem. USDA Forest Service Research Paper RM-191. Rocky Mountain Forest and Range Experiment Station, Fort Collins, Colorado.

Chen, Z., K. Grady, S. Stephens, J. Villa-Castillo, M. R. Wagner. 2006. Fuel reduction treatment and wildfire influence on carabid and tenebroid community assemblages in the ponderosa pine forest of nothern Arizona, USA. Forest Ecology and Management 225:168–177.

Conner, R. C., J. D. Born, A. W. Green, and R. A. O'Brien. 1990. Forest Resources of Arizona. USDA Forest Service Research Bulletin INT-69. Intermountain Forest and Range Experiment Station, Ogden, Utah.

Covington, W. W., P. Z. Fulé, M. M. Moore, S. C. Hart, T. E. Kolb, J. N. Mast, S. S. Sackett, and M. R. Wagner. 1997. Restoring ecosystem health in ponderosa pine forest of the Southwest. Journal of Forestry 95(4):23–29.

DeBano, L. F., D. G. Neary, P. F. Ffolliott. 1998. Fire's effects on ecosystems. John Wiley & Sons, Inc., New York.

Fox, B. E., W. W. Covington, A. Tecle, J. P. McTague, and M. M. Moore. 1991. An overview of the historical, geographic, social and biophysical characteristics of Southwestern ponderosa pine forests, chapter 2. In Multiresource Management of Southwestern Ponderosa Pine Forests: The status of knowledge, edited by A. Tecle and W. W. Covington, pp. 8–23. USDA Forest Service, Southwestern Region, Albuquerque, New Mexico.

Hart, S. C., T. H. DeLuca, G. S. Newman, M. D. MacKenzie, and S. I. Boyle. 2005. Post-fire vegetative dynamics as driver of microbial community structure and function in forest soils. Forest Ecology and Management 220:166–184.

Haase, S. M. 1986. Effect of prescribed burning on soil moisture and germination of southwestern ponderosa pine seed on basaltic soils. USDA Forest Service, Rocky Mountain Forest Range Experiment Station. Research Note RM-462.

Klock, G. O., and J. D. Hevey. 1976. Soil-water trends following wildfire on the Entiat Experimental Forest. In Annual Proceedings: Tall timber fire ecology, conference 15, pp. 193–200.

Macadam A., and R. Trowbridge. 1988. Effects of prescribed fire on soil moisture and temperature in two sites series of the SBS zone. Canadian Ministry of Forests, Prince Rupert Forest Region, FRDO Memo No. 056.

Milne, M. M. 1979. The effects of burning root trenching, and shading on mineral soil nutrients in southwestern ponderosa pine. Master's Thesis. Northern Arizona University, Flagstaff, Arizona.

Moody J. A., D. A. Martin, T. M. Oakley, and P. D. Blanken. 2007. Spatial and temporal variability of soil properties after a wildfire. United States Geological Survey. Report Series 2007-5015.

Morehouse, K., T. Johns, J. Kave and M. Kave. 2008. Carbon and nitrogen cycling immediately following bark beetle outbreaks in southwestern ponderosa pine forests. Forest Ecology and Management 255:2698–2708.

Neary, D. G. 1995. Effects of fire on water resources: A review. Hydrology and Water Resources in Arizona and the Southwest 22/25:24–53.

Neary, D. G., K. C. Ryan, L. F. DeBano, eds. 2005. Fire effects on soil and water. USDA Forest Service, Rocky Mountain Research Station, General Technical Report RMRS-GTR-42, Volume 4: Fort Collins, Colorado.

Oakley, T. M. 2004. Spatial and temporal

variability of soil moisture and soil temperature in response to fire in a montane ponderosa pine forest. Proceedings from the 26th Conference on Agricultural and Forest Meteorology.

Ower, C. L. 1985. Changes in ponderosa pine seedling growth and soil nitrogen following prescribed burning and manual removal of the forest floor. Master's Thesis. Northern Arizona University, Flagstaff, Arizona.

Patton, D. R. 1991. The ponderosa pine forest as wildlife habitat, chapter 8. In Multiresource Management of Southwestern Ponderosa Pine Forests: The status of knowledge, edited by A. Tecle and W. W. Covington, pp. 361–410. USDA Forest Service, Southwestern Region, Albuquerque, New Mexico.

Paul, E. A., and F. E. Clark. 1996. Soil Microbiology and Biochemistry, 2nd ed. Academic Press Inc., San Diego, California.

Rosgen, D.L. 1996. Applied River Morphology. Wildland Hydrology, Pagosa Springs, Colorado.

Ryan, M. G., and W. W. Covington. 1986. Effect of a prescribed burn in ponderosa pine on inorganic nitrogen concentrations of mineral soil. USDA Forest Service, Rocky Mountain Forest and Range Experiment Station Research Note RM-464.

Sayok, A. K, M. Chang, and K. G. Watterston. 1993. Forest clearcutting and site preparation on a saline soil in East Texas: Impact on sediment loss. In Sediment Problems: Strategies for monitoring, prediction and control. IAHS publ. no. 217:177–184.

USFS. 1989. An Analysis of the Lands Base Situation in the United Sates: 1989–2040. USDA Forest Service General Technical Report RM-181. Rocky Mountain Forest and Range Experiment Station, Fort Collins, Colorado.

Wagner, M. R., W. M. Block, B. W. Geils, and K. F. Wenger. 2000. Restoration ecology: A new forest management paradigm, or another merit badge for foresters? Journal of Forestry 98(10):22–27.

Weatherspoon, C. P. 2000. A proposed long-term national study of the consequences of fire and fire surrogate treatments. In Proceedings Joint Fire Science Conference, pp. 117–126. Boise, Idaho, June 15–17, 1999. University of Idaho Press, Moscow, Idaho.

Weatherspoon, C. P., and J. McIver. 2000. A national study of the consequences of fire and fire surrogate treatments. Joint Fire Sciences Amended Proposal, March 20, 2000.

Wells, C. G., R. E. Campbell, L. F. DeBano, C. E. Lewis, R. L. Fredricksen, E. C. Franklin, R. C. Froelich, and P. H., Dunn. 1979. Effects of fire on soil: A state-of-the knowledge review. USDA Forest Service General Technical Report WO-7, Washington, D.C.

White, A. S. 1985. Presettlement regeneration patterns in a southwestern ponderosa pine stand. Ecology 66:589–594.

Wolfson, B. S. A., T. E. Kolb, C. H. Sieg, and K. M. Clancy. 2005. Effects of post-fire conditions on germination and seedling success of diffuse knapweed in northern Arizona. Forest Ecology and Management 216:342–358.

Wildlife Surveys as a Conservation Framework

MILKSNAKES AT PETRIFIED FOREST NATIONAL PARK, ARIZONA: ADAPTIVE MONITORING OF RARE VERTEBRATES

Erika M. Nowak and Trevor B. Persons

ABSTRACT

Milksnakes (*Lampropeltis triangulum*) are widely distributed across the United States and into South America, and are represented in northern Arizona by a diminutive form. Due to its apparent rarity and unique characteristics, this form is popular with snake collecting enthusiasts. Surveys conducted adaptively at Petrified Forest National Park (NP) between 1997 and 2007 suggest that the park may serve as an important protected site for the species in Arizona. Survey methods included quantitative night driving and bicycling surveys; nocturnal and visual encounter (walking) surveys; pitfall trap and artificial cover arrays and transects; radio-telemetry; and box trap, drift fence, and artificial cover transects. We found a total of 25 individual milksnakes over a 10-year period, and we report additional detections by NPS staff between 1995 and 2007. Controlling for seasonal differences in sampling effort, night-walking surveys during the late summer and use of box trap and drift fence transects in the spring produced the most animals per person-hour of effort, while night driving with an all terrain vehicle (ATV) during the late summer produced the most animals per kilometer. However, detectability was extremely low in all methods. A short-term, late-summer telemetry study indicated extensive use of underground features (burrows) and suggested localized movements. We compare inter-annual detections of milksnakes during night driving surveys with those of other snake species. We conclude by discussing the implications of low detection rates for the status of milksnakes at Petrified Forest NP and make suggestions for future research and monitoring.

INTRODUCTION

The milksnake (*Lampropeltis triangulum*) is one of the most widely distributed snake species in the Western Hemisphere. The species occurs throughout the eastern and central United States and southern Canada to the Rocky Mountain West, and south through Central America into South America (Stebbins 2003, Conant and Collins 1998, Williams 1988, 1994). Twenty-five subspecies have been described rangewide (Williams 1988, 1994), two of which may occur in Arizona: the Utah milksnake (*L. t. taylori*) and the New Mexico milksnake (*L. t. caelanops*). In Arizona, the species is known from only a few disparate localities between Seligman and St. Johns, in the Chino Valley-Prescott Valley area, and in extreme southeastern Arizona in the San Bernardino Valley (Drost et al. 2001, Brennan and Holycross 2006). One noted expert considered the milksnake to be "Arizona's rarest snake" (Lowe 1989).

While the northern Arizona specimens are usually allocated to *L. t. taylori*, they are morphologically distinct (smaller) from *L.*

Figure 1. Adult male milksnake (*Lampropeltis triangulum*) captured during surveys at Petrified Forest National Park, Arizona, in May 2007. This animal exhibits color and banding patterns generally consistent with those described for *L. t. taylori*. Photograph by E. M. Nowak.

t. taylori populations to the north and *L. t. caelanops* populations to the east (Williams 1988, Figure 1). In fact, some professionals consider northern Arizona milksnakes to be a unique form, the "Arizona Milksnake" (A. Holycross, Mesa Community College, personal communication). Southeastern Arizona populations have been placed with *L. t. caelanops* (Stebbins 2003), but this assignment is also questioned (Badman et al. 2003). Although populations of *L. t. taylori* in Utah typically occupy mountain and foothill woodlands (Williams 1988), in Arizona milksnakes appear to be restricted to grasslands (Brennan and Holycross 2006), with sandy clay loam soils suitable for burrowing, often near sedimentary or igneous outcrops (E. M. Nowak unpublished data). Due to the secretive nature of the species, little is known about habitat requirements or abundance in Arizona; however, dietary information indicates that the species feeds on lizards and small mammals in northern Arizona (B. Hamilton, Great Basin National Park, unpublished data).

Due to their apparent rarity, small size, and distinctive red-orange, black, and white-yellow banded coloration, Arizona milksnakes are popular with snake collecting enthusiasts. Milksnakes are easily bred in captivity, and strikingly colored individuals may sell for $250 to $300 each (H. Koenig, retired Arizona Game and Fish Department, personal communication). The species is afforded reduced bag limits compared to more common snake species in the northern part of the state, and is protected from collection in Cochise County (Arizona

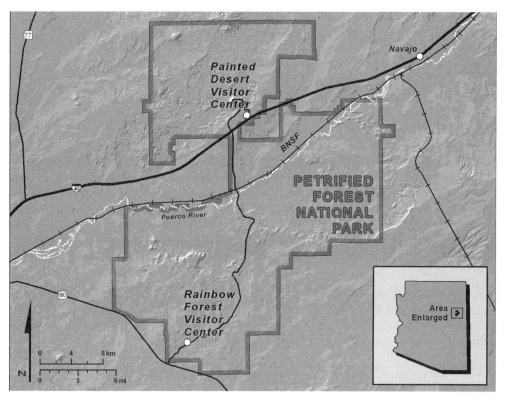

Figure 2. Location of Petrified Forest National Park in Arizona, USA, including location of selected features described in the text.

Game and Fish Department. Arizona reptile and amphibian regulations: 2007 and 2008, Arizona Game and Fish Commission, 2007). Collection pressure on wild populations in northern Arizona is suspected to be low but has not yet been quantified (H. Koenig, retired Arizona Game and Fish Department, personal communication).

Petrified Forest National Park (NP) is well known as a historical and current locality for milksnakes in northern Arizona (Drost et al. 2001, Stebbins 2003). During a two-year reptile and amphibian inventory at Petrified Forest National Park completed in 1999 (Drost et al. 2001), six individuals were found, effectively doubling the number of verified records in the state to that point. The preservation of a large area of healthy, native plains grassland at Petrified Forest NP may be, in part, responsible for the persistence and apparent relative abundance of the species, or the numbers may reflect greater survey effort. Milksnakes may be threatened at Petrified Forest NP by road mortality and illegal collection, especially since the discovery of a predictable population at the park has not escaped the attention of snake collecting enthusiasts (R. Repp, personal communication). Due to these potential threats, the species is of special management concern at Petrified Forest NP.

Building on results of initial surveys conducted in 1997–1998, we conducted a short-term telemetry study in 1998 and monitoring surveys for milksnakes between 1999 and 2007. Our goals were to determine abundance of the species at Petrified Forest NP (Figure 2), develop monitoring protocols based on best detection methods, determine general habitat use, and provide baseline information on a rare and sensitive species. In this paper, we describe detection using different methods over a 10-year period and compare results for other snake species detected during quantitative night road driving and walking surveys. We conclude by discussing implications of detection probabilities for milksnakes, future research, and options for monitoring, including other herpetofauna community members.

METHODS

Sampling Techniques

The research can be divided into three phases between 1997 and 2007, each adopting the most effective techniques of the preceding period, and/or experimenting with new ideas to maximize detections.

1) A general herpetological inventory was conducted from July 1997 to October in 1998, employing quantitative night road driving, visual encounter surveys, and pitfall traps with cover object transects. Short-term focal animal radio telemetry occurred in late summer 1998.

2) Night road driving was conducted during the summer rainy period (July to September) in 1999 and from 2001 to 2005; no surveys were conducted in 2000 due to a lack of personnel. From 2002 to 2005, surveys included two- to four-day intensive "milksnake roundup weekends." During this phase we added night walking surveys, and in 2005, night bicycle surveys.

3) Focused monitoring was conducted from May through September in 2006–2007. During this intensive phase of the project, we conducted night road driving (adding the use of an ATV), night walking surveys, and trapping at box trap and cover object transects.

Quantitative Road Driving Surveys

During quantitative road driving surveys (e.g., Mendelson and Jennings 1992, Shafer and Juterbock 1994), we drove vehicles slowly (32–56 km/hr) along the main paved park road on warm summer nights. Initial surveys indicated that night driving was unproductive from April to June, in part due to cold nighttime temperatures (Drost et al. 2001), so we focused subsequent efforts on the rainy period from mid-July to mid-September. Road survey coverage also varied adaptively. From 1997 to1999, we drove the entire 42 km length of the park road; beginning in 2002, we focused the majority of effort in the most productive northern 19 km of road; however, the entire length of road was driven at least once during every survey.

In August and September 2005 we included surveys by a bicycle equipped with a strong headlamp. This method was not repeated in subsequent years due to a lack of personnel proficient in using the technique, and also because it was difficult to spot animals while riding.

Visual Encounter Surveys

Visual encounter surveys (e.g., Crump and Scott 1994) consisted of walking systematically through a certain micro-habitat for a specified amount of time (time-area constrained searches) or for an unbounded time period (non-time-constrained searches). We searched reasonable microhabitats within that habitat, and recorded reptiles and amphibians encountered.

Diurnal visual encounter walking surveys were used from April to October in 1997 and 1998 as part of the general herpetological survey (Drost et al. 2001). These searches targeted diurnal squamate species, excluding milksnakes.

In 2002, we began non-time-constrained night visual encounter surveys, in which personnel walked with headlamps or spotlights near the road in known milksnake locations. While transects were roughly parallel with the road, walking patterns were random to maximize time spent checking under shrubs and in burrow openings. Night walking surveys were conducted in August and September, typically for 30 to 60 minutes per survey, and typical distances covered were <300m.

Pitfall Traps and Wood Cover Object Transects

During 1997 and 1998, we trapped at seven sampling sites, each consisting of combinations of five-gallon pitfall and drift fence arrays (e.g., Campbell and Christman 1982) with parallel alternating one-gallon pitfall and artificial wood cover object (coverboard) transects (e.g., Fitch 1987, Fellers and Drost 1994). There were a total of 36 five-gallon pitfall buckets in 9 pitfall arrays paired with transects of 40 one-gallon pitfall traps and 50 coverboards (detailed in Drost et al. 2001). These sampling sites were placed in locations throughout the park, including one site within known milksnake habitat. We sampled the sites twice a month between July (1997) or April (1998) and October. Each sampling session lasted about five days, during which pitfall traps were checked at least every other day and coverboards were checked several times (Drost et al. 2001).

Telemetry

In August and September 1998, we opportunistically captured adult individuals for radiotelemetry research. Subjects were large enough so that transmitters would be ≤5% of their body weight (after Reinert 1992). Transmitters (1.8-g; Holohil Systems Ltd., Ontario, Canada) were wrapped in their antennas, coated with vegetable oil, and fed gently to the snakes. The snakes were then released at the point of capture. Transmitters were either regurgitated or passed through the intestinal tract without apparent adverse effect less than two weeks later. We located the snakes at least two times every 24 hours using a TR-4 receiver (Telonics Telemetry Consultants, Mesa, Arizona). When a snake was located, we recorded distance and direction from last location (not geo-referenced), microhabitat association, weather conditions, and behavior if known. After the signals were motionless for four days, we attempted to determine the fate and location of each transmitter.

Box Trap and Cover Object Transects

In 2006 and 2007 between May and September, we conducted trapping using box trap and cover object transects at five monitoring sites in the northern part of park. (Locations are not described per request of Petrified Forest NP.) Box trap and drift fence transect design was adapted from K. Baker and C. Schwalbe (University of Arizona, personal communication). Each fence was a 30-m-long, 0.63 cm mesh piece of hardware cloth 91 cm high, with two box traps of 0.63 cm mesh (59 x 39 x 23 cm with 5–6 cm funnel openings) at each end. Traps were shaded with large pieces of cardboard and closed when not in use. Water was provided in shallow plastic bowls to prevent amphibian mortality; seeds, socks, or cotton balls were provided as food and shelter for small mammals.

Between fences, four artificial cover objects of three alternating types were placed 10 m apart. Cover objects included 60 x 122 x 2 cm plywood pieces (a minimum of 1.9 cm thick), corrugated roofing tin pieces at least 60 x 60 cm, and floor tiles or paving stones at least 30 x 30 cm, intended to mimic rock slabs.

Each monitoring site consisted of three drift fence transects, 12 box traps, and 16 cover objects covering a linear distance of 210 m, for a total of 60 traps and 81 cover objects (27 tin pieces, 27 wood cover objects, and 27 floor or landscape tiles). There were one to two trapping sessions per month in April, May, July, August, and September, with each trapping session lasting four to six

Table 1. Effort and detections of milksnakes during night road driving, walking, and bicycling surveys at Petrified Forest National Park, Arizona, from 1997–2007. Three survey periods are listed, with increasing focus on milksnake detections. Effort and kilometers driven are minimum estimates.

	# Snakes detected	Effort (person-hours)	Kilometers Driven	Snake/hr	Snake/km
1997-1998					
Driving-car	6	359	4450	0.02	0.001
1999-2005					
Driving-car	9	346.5	4554	0.02	0.002
Walking	2	26.2	-	0.08	-
Bicycle	0	2.3	6	0	0
2006-2007					
Driving-car	3	191.6	3755	0.01	0.0008
Driving-ATV	1	28.9	652	0.03	0.001
Walking	1	11.1	-	0.09	-

nights. Traps were checked daily, and the area under each cover object was checked no more than four times per trapping session, generally alternating between mornings and afternoons.

Animal Processing

Each time a milksnake was located, we recorded weather and locality information, body measurements, and snake identification. Through photodocumentation from 1997 to 1999, we determined that individuals can usually be identified by a unique combination of asymmetric dorsal blotch patterns, facial and tail patterns, and wound scars. We drew and photographed the unique features of each animal for individual identification purposes.

Starting in 2002, for permanent and unique identification, we injected a small (8–11 mm) glass-encapsulated passive integrated transponder (PIT tag or microchip) into each animal (Camper and Dixon 1988, Keck 1994, Jemison et al. 1995). Using field-sterile techniques (e.g., Fagerstone and Johns 1987), we injected tags intraperitoneally or subdermally in the posterior third of the snake's body just above the ventral scales.

Statistical Methods

Generally, sample sizes were too small to permit statistical analyses. Chi-square tests (Sokal and Rohlf 1981) were used to examine detection differences (animals captured per hour of person effort) between techniques and seasonality of captures. Two-tailed statistical significance was determined with $\alpha = 0.10$, per convention for small sample sizes (Sokal and Rohlf 1981). Means are reported as ± one standard deviation.

RESULTS

Survey Effort

Inter-annual effort varied over the three phases of monitoring, dependent on survey objectives, available personnel, and funding level (Table 1). During the initial herpetological inventory in 1997 and 1998

(Drost et al. 2001), we drove 4,450 km during night road surveys, over about 359 person hours. About 124 person hours of effort were spent on diurnal visual encounter surveys, and 112 person hours were spent installing and checking pitfall arrays and pitfall trap-coverboard transects. Radiotracking of focal subjects occurred for one to two hours per day over several weeks in August and September 1998.

During 1999 to 2005 (excluding 2000), we conducted a minimum of 4,554 km of road driving over 346 person hours (Table 1). Bicycle surveys in 2005 covered about 10 km, at just over 2 person hours. Night visual encounter (walking) surveys occupied about 26 person hours.

During focused monitoring in 2006 and 2007, we conducted about 4,407 km of road driving, with 3,755 km driven in a car or truck and 652 km driven on an ATV. Of about 232 person hours of survey effort, 192 person hours were spent driving cars or trucks, 29 person hours were spent driving an ATV, and 11 person hours were spent walking (Table 1). Installing, checking, and closing the box traps and cover objects took about 430 person hours.

Detections

During nine years of surveying at Petrified Forest NP over three discrete sampling periods, we found a total of 25 individual milksnakes. Three individuals were recaptured.

Herpetological Inventory (1997–1998)

We did not find milksnakes during diurnal visual surveys, in pitfall traps, nor under cover objects during the original herpetological inventory (Drost et al. 2001). During the initial night road driving surveys, we recorded seven detections, comprising six individuals, including two adult males, one adult female, two unknown sex adults, and one unknown sex juvenile. Average detection rates were 0.001 milksnakes per km of road driving, or 0.02 snakes per hour of searching (Table 1). All milksnakes were found in the northern part of the park, including one snake captured in the middle of the Interstate 40 bridge overpass.

Quantitative Road Driving (1999–2005)

Between 1999 and 2005 (excluding 2000), we recorded 11 detections, comprising 11 adult individuals (six males, four females, and one unknown sex). Nine individuals were found by vehicle, two by walking, and none were seen during bicycle riding.

Average detection rates from 1999 to 2005 were 0.002 milksnakes per km of night surveys, or 0.02 snakes per person hour of searching while driving, and 0.08 snakes per person hour while walking (Table 1). All detections occurred in the northern part of the park.

Focused Monitoring (2006–2007)

We recorded 10 detections comprising eight individual adults (five males and three females) during focused monitoring. One of the males was a recapture from 2005.

Five individuals were found during night road surveys in the northern part of the park: three by vehicle, one by ATV, and one by walking. Average detection rates during night surveys in 2006 and 2007 were 0.0008 milksnakes/km of road driving or 0.01 snakes/person hour of searching while driving, 0.001/km and 0.03/person hour when searching via ATV, and 0.09/person hour when searching on foot (Table 1).

Five detections of three individual adult males occurred in box trap transects at three separate sites: four detections occurred in box traps and one occurred along a drift fence. Detection rates at box trap transects were estimated to be 0.001 snakes/trap-night, or about 0.01 snakes/person hour of effort (Table 1). No milksnakes were found under any wood, tin, or tile cover objects.

Coincidentally, the box trap and coverboard sites were highly productive in sampling small vertebrate diversity. We captured a minimum of 30 species, including

7 lizard species, 7 snake species, 1 amphibian species, 13 or 14 mammal species, and 2 bird species (E.M. Nowak et al. 2009). Rare species were well represented: in addition to milksnakes, 2 uncommon mammal species were detected in box traps: desert shrew (*Notiosorex crawfordi*) and spotted ground squirrel (*Citellus spilosoma*), which constitutes the first record of the latter species for Petrified Forest NP. Regular checking of drift fences at these sites produced 13 species (mostly lizards and snakes) trapped at the fences outside of traps or cover objects.

Detection Rate Comparisons

The number of snakes we detected per person hour varied between comparable methods (night driving, bicycling, and walking surveys) between phases of the research as well as by method (Table 1). Our average detection rate across methods was 0.04 ± 03 snakes/hr, with a high of 0.09 snakes/hr during night walking surveys in 2006–2007 and (excluding bicycle surveys) a low of 0.01 snakes/hr of road driving by car during the same phase of research; however, these results were not statistically different (χ^2 = 0.189, 7 df, $P > 0.10$). There was insufficient data to statistically compare a single method across years or to compare different methods within years. Visual comparison of the data indicates similar detection rates within methods in 1999–2005 and 2006–2007, with walking surveys having the highest detection rates of the methods across years. Detection rates for box trap transects were 0.01 snakes/hr; however, this passive method of trapping is not directly comparable to active driving and walking surveys.

Milksnake Recaptures

We captured at least three milksnakes more than once during nine years of surveys. One recapture was documented through photographic comparison. Twenty-three individuals captured by researchers or National Park Service (NPS) staff were photographed, although three photographs of individuals were either of low quality or did not show unique patterns. Of the remaining 20, only one non-PIT-tagged individual was recaptured; a female captured two weeks apart in 1997. This animal's capture locations were <150 m apart.

Two PIT-tagged individuals were recaptured three times each. We initially captured one adult male during road driving in 2005, recaptured it in a box trap in May 2006, and captured it again in the same drift fence section (different box trap) in April 2007. Photographs of this animal verified the utility of photo-identification over more than two years. We captured a second adult male at a drift fence in May 2007, and recaptured it two days later in a funnel trap at a different drift fence section of the same site.

Temporal Patterns of Captures

Of the 23 detections (22 individuals) during night surveys, 22 occurred during the summer rainy period between the end of July and mid-September, with a peak during August (Figure 3). We found one snake on the road in May. Although detection of snakes/person hour appeared to be double in May (0.08) compared to August (0.04), these results were not statistically different (χ^2 = 0.190, 5 df, $P > 0.10$). Of the five milksnakes we found in box traps, four detections (three individuals) occurred in April or May and one occurred in late July.

Milksnakes detected during night surveys were found after north and south road gates had been closed and the roads cleared of visitors by patrolling law enforcement personnel. Of these, we captured 19 animals (83%) after dark between 2000 and 0130 hours, all in the late summer. Four were captured near dusk (1925–1945 hrs), one in May and the rest in late summer. We found an additional milksnake active at a drift fence before dusk (1811 hrs) during May.

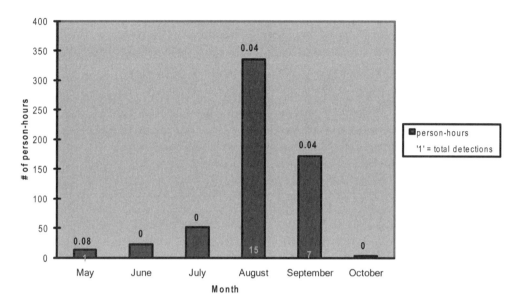

Figure 3. Number of detections of milksnakes by USGS personnel and volunteers by month during quanti-tative night road driving and walking surveys from 1997–2007 at Petrified Forest National Park, Arizona. Numbers at the top of each column indicate detection rates (number of snakes found per person hour of effort) each month. Numbers inside bars indicate snakes captured each month.

TELEMETRY AND HABITAT ASSOCIATIONS

Telemetry

Three adult males were fed radiotransmitters in 1998 and released at the point of capture. Channels 61 and 62 were located two to three times per day between 0815 and 2240 hrs between August 31 and September 12 to determine habitat use and movement distance. Channel 63 was tracked more frequently, from 1135–0003 hrs on September 9 and 10 to determine behavioral patterns. Ambient temperatures varied between 19.3° C and 32° C during the tracking periods.

We released Channel 61 just east of the road at Rattlesnake Butte on the morning of August 31. This snake moved short distances in an area covering less than 50 m around its release site near the road until September 9 (10 days, 21 active locations). The snake was never seen on the surface, although it was found to be moving underground (presumably through burrows) on four diurnal tracking occasions. The snake revisited two of about 10 total sites used. It was located under shrubs and grass clumps on 13 occasions (muhly [*Muhlenbergia* spp.], James' galleta [*Pleuraphis jamesii*], four-wing saltbush [*Atriplex canescens*], rabbitbrush [*Chyrsothamnus* spp.]), and between shrubs presumably in burrows on eight occasions. We found its clean (suggesting regurgitation) transmitter about 1 m underground in a 20 cm diameter burrow.

We released Channel 62 close to Channel 61 on the evening of August 31, and it moved about 150 m to Rattlesnake Butte within the first 12 hours of release. The snake then settled down and made movements of less than five meters on the slope under small sandstone slabs until September 4 (five days, eight active locations). It was never seen on the surface, but moved twice during tracking occasions. This snake used about five sites under the sandstone talus (including one unidentified small bush); we suspect it was using interstitial spaces as well as excavated burrows under the rocks. We could not determine the final location or the fate of this transmitter.

We released Channel 63 on the west side of the park road near Lacey Point on September 9, and tracked eight times throughout that evening and the next day until the evening of September 10. The snake moved a total distance of 288 m, primarily on the night of September 9, during which time it crossed the road. It settled in a burrow complex (visible opening was less than 1 cm in diameter) at the base of a four-wing saltbush before emerging to regurgitate the transmitter. The snake was seen on the surface three times during the tracking period: twice it was seen moving away from the observer, and once it was coiled at the base of 0.3 m tall sagebrush (*Artemesia* spp.). During the observation period between 0003 and 0054 hrs on September 10, the snake was watched from a distance for 51 minutes, during which time it thoroughly covered an area of less than 50 m^2. The snake appeared to investigate the bases of 0.2–0.3 m tall shrubs (four-wing saltbush, Mormon tea [*Ephedra torreyana*], and rabbitbrush), all the while moving continuously and slowly tongue-flicking. The snake did not encounter nor enter any burrows.

Habitat Associations of Non-telemetered Milksnakes

Most animals found during surveys were on the road or the road edge, in sandy loam soils dominated by sand sage (*Artemisia filifolia*), bunchgrasses, and four-wing saltbush. These habitats were the same as those found at the box trap transects (E. M. Nowak, unpublished data). Several locations were near sandstone outcrops or badlands, and one location by NPS staff from the south part of the park was dominated by petrified wood-studded badlands. NPS maintenance staff apparently detected several additional animals in restrooms or other buildings, with surrounding habitats comprised of shrubland associations including one-seed juniper (*Juniperus monosperma*).

Detections of Other Snakes During Quantitative Road Driving Monitoring

Six snake species were found during night quantitative road driving along the park road between 1997 and 2007 (excluding 2000). We excluded striped whipsnake (*Masticophis taeniatus*) from these analyses, as they were only found dead on the road during night surveys. Effort and total detections of each species varied among years, as did detection rates for each species (Table 2). We detected all six snake species in just five of 10 years; only gophersnakes (*Pituophis catenifer*) were detected every year.

Average annual detection per hour of effort across species was highest in 2001 (0.15 ± 0.18 snakes/hr) and lowest in 2004 (0.01 ± 0.01 snakes/hr). During most years, gophersnakes or "Hopi" rattlesnakes (*Crotalus viridis viridis*; Douglas et al. 2002) were the most commonly detected species per hour of effort (Table 2). Rattlesnakes had the highest single-season detection rate, at 0.49 snakes/person hour in 2001. Nightsnakes (*Hypsiglena torquata*) and glossy snakes (*Arizona elegans*) had intermediate detection rates most years (Table 2), while milksnakes and common kingsnakes (*Lampropeltis getula*) had the lowest detection rates (excluding non-detections). The lowest detection rates per hour of effort among the species occurred in 2004, with 0.01 common kingsnakes or glossy snakes detected per person hour.

Table 2. Effort, total number of detections, and detection rate for six species of snakes at Petrified Forest National Park, Arizona, during night road driving surveys (car or truck only) between 1997 and 2007. Effort is given in terms of person hours and total kilometers driven; detection rates are snakes detected per km driven (in italics).

Year	Person hours	km driven	HYTO[1]	LAGE	LATR	AREL	CRVI	PICA
1997	107.4	2681	16 *0.006*	3 *0.001*	4 *0.001*	7 *0.003*	18 *0.007*	28 *0.01*
1998	116.5	3471	17 *0.005*	3 *0.0009*	3 *0.0009*	8 *0.002*	10 *0.003*	19 *0.005*
1999	53.4	391	3 *0.008*	1 *0.002*	0	1 *0.002*	0	4 *0.01*
2001	12.2	549	0	2 *0.004*	2 *0.004*	0	6 *0.01*	1 *0.002*
2002	40.1	584	4 *0.007*	3 *0.005*	1 *0.002*	6 *0.01*	0	3 *0.005*
2003	42.9	533	1 *0.002*	2 *0.004*	2 *0.004*	6 *0.01*	3 *0.006*	2 *0.004*
2004	76.6	1035	0	1 *0.001*	0	1 *0.001*	2 *0.002*	3 *0.003*
2005	90.0	1281	3 *0.002*	0	4 *0.003*	10 *0.008*	9 *0.007*	6 *0.005*
2006	108.5	2242	5 *0.002*	4 *0.002*	2 *0.0009*	3 *0.001*	12 *0.005*	6 *0.003*

Average annual detection per kilometer during night road driving surveys was highest in 2007 (0.006 ± 0.004 snakes/km) and lowest in 2004 (0.001 ± 0.001 snakes/km). Snake detection rates per kilometer ranged from a high of 0.01 snakes/km for rattlesnakes (in 2001 and 2007), glossy snakes (2002, 2003), and gophersnakes (1997), to a low of 0.0009 snakes/km for milksnakes (1998, 2006) and common kingsnakes (1998; Table 2).

Additional NPS Detections of Milksnakes

At least six additional milksnakes have been detected since 1995 by Petrified Forest NP staff in Resource Management, Interpretive, and Law Enforcement Divisions, representing a minimum of five individuals. NPS Resource Management staff conducted an estimated 1,624 km of quantitative road driving between July 1996 and September 1999 over at least 98 person hours (Petrified Forest NP, unpublished data). During these surveys, one adult was found in July 1997. In September 2006 during similar surveys (effort unknown), a second snake was found, PIT-tagged, and photographed. Four additional adults (at least two individuals of unknown sex; one photographed) were found by law enforcement staff between July and September: one in 1995, two during 1996–1998, and one in 2005. Almost all confirmed detections by NPS staff came from the road at the northern end of the park at night. However, one individual was found and photographed on the road in the south part of the park near the Rainbow Forest visitor center in 1995.

Six additional unconfirmed sightings of "red, black, and white snakes" (i.e., milksnakes) exist within Petrified Forest NP. Three were found by maintenance staff in visitor facilities: one from the middle part of the park before 1998, and two from the northern part of the park in 1996 and July 2006, respectively (Petrified Forest NP staff, personal communication). At least three sightings were reported on the road at the northern end of the park by law enforcement rangers between 2002 and 2007.

DISCUSSION

Detections and Detectability

Milksnakes are semi-fossorial and secretive animals across their range in the US (Williams 1988, 1994; Fitch 1999), and the same ecological pattern apparently exists at Petrified Forest NP. Milksnakes were the most difficult to detect snake in the park, even among other small or uncommon nocturnal snake species, an attribute generally noted for the genus *Lampropeltis* by Luiselli (2006). During 10 years of surveys, we detected 28 milksnakes representing 25 individuals (during the same time period, NPS staff detected an additional six animals, with several more unconfirmed sightings).

Contrary to expectation (and unlike other species detected using night road driving), milksnakes did not become substantially easier to find when methods, temporal periods, and sampling locations targeting them were increasingly refined. Detection rates ranged from one to nine milksnakes/100 hours of search effort, and from 0.8 to two snakes/thousand km of driving. Low detection and recapture rates may indicate extreme secretiveness, very clumped distributions, or genuine rarity (MacKenzie et al. 2005, Luiselli 2006, Hamilton 2007). Luiselli (2006) specifically predicts that small, fossorial, dietary specialist species inhabiting narrow ecological niches are more susceptible to rarity. Several if not all of these ecological characters appear to apply to Petrified Forest NP milksnakes. Dietary data indicates that while the species consumes a variety of small vertebrate prey, the majority may be taken in underground nests or burrows (B. Hamilton, unpublished data). We found milksnakes only in certain locations in the northern part of the park, potentially indicating a highly clumped distribution; however, a few detections of animals that were likely milksnakes occurred infrequently at more southerly locations. Our short-term telemetry data suggests that these snakes have localized movements and are primarily active subsurface (e.g., semi-fossorial), at least in the late summer. It is possible that telemetry data was limited by the snakes feeling full; however long-term recapture distances also suggested small ranges and/or localized movements. Further research is needed to assess dietary niche breadth, habitat associations, and seasonal and diel activity patterns.

In addition to suggesting secretiveness, low recapture rates in wildlife populations may also be interpreted to indicate high mortality rates or large population sizes (Nichols 1992). Our recaptures of an adult male over several years may indicate that mortality rates of adults are low; however more life history data is needed from this population. Other studies in which Milksnakes were also particularly difficult to detect under recommended techniques for the species suggest secretiveness is likely (e.g., Luiselli 2006). No milksnakes were detected during a herpetological inventory at Wupatki National Monument, Arizona (a known locality for the species; Persons and Nowak 2006), and they were also not detected during several additional years of quantitative night road driving (over 6,243 km) (Persons 2001). Fitch (1999) noted that milksnakes (*L. t. syspila*) were relatively uncommon over a 50-year study using coverboard transects in Kansas, and estimated an abundance of 0.52-0.70 snakes/ha, compared to an estimated density of 1,040 snakes/ha for the most common species, ringneck snake (*Diadophis punctatus*).

Seasonal Sampling Effort Variation

This long-term monitoring project represents different levels of effort as funding became available. As the project evolved adaptively over the years, our efforts and focal areas became more refined, potentially increasing the probability of detection. However, during years with all-volunteer effort (1999–2005), we focused on maximizing detections using techniques focused in spatial locations and temporal periods previously demonstrated to be successful. The research generated increasing interest by Petrified Forest NP staff, culminating in a request for intensive monitoring for milksnakes in 2006–2007. This unavoidable bias in sampling effort and techniques may have affected our results by ultimately lowering detections.

A majority of research in the first two phases of research (1997–1998, 1999–2005) was conducted during the late summer near the end of the rainy period, based on our initial detections of milksnakes only during this season. We subsequently began sampling in April and May after hearing reports of spring activity for the species in southern Arizona (A. Holycross and T. Brennan, unpublished data, personal communication). As predicted by those findings, sampling in the third and final phase of the project at Petrified NP (2006–2007) indicated that milksnake activity (detectability) is high in May. Future research should occur bi-seasonally to maximize detections, as well as determine potential variation in seasonal activity.

Quantitative night surveys conducted between 1997 and 2007 illustrated that, generally, all snake species detection rates varied from year to year. In contrast to a study in southern Arizona by Mendelson and Jennings (1992), no increasing or decreasing trends in detection rates were obvious for most species, with the possible exception of nightsnakes, which generally exhibited lower detection rates starting in 2001. One possible explanation for a potential decline in this species is ongoing regional drought (Webb et al. 2004), which may negatively effect lizard and insect prey (Brennan and Holycross 2006). Our data illustrate the need for long-term studies to determine population and community trends.

Optimal Monitoring Methods and Timing

No method was found to be particularly effective in detecting milksnakes at Petrified Forest NP, despite extensive, adaptive efforts. Given this, it may be useful to conduct additional telemetric studies of individuals using small implanted transmitters. Such studies could aid in determination of seasonal spatial use and distribution, mortality rates from non-human predators, and proximal causation for patterns of non-detection (e.g., random or co-occurring with habitat features; Gu and Swihart 2004). Results of such a study could be compared to known life and natural history traits for similar-sized colubrid snakes, to estimate what level of removal of adult snakes by collectors would likely impact populations.

Techniques proven for detecting small fossorial snakes in other studies (e.g., Campbell and Christman 1982, Schafer and Juterbock 1994, Fitch 1999) were not equally effective in detecting milksnakes at Petrified Forest NP. However, land managers and resource protection agencies (e.g., other NPS units, Arizona Game and Fish Department, US Forest Service) remain interested in assessing the potential impacts of collecting, road mortality, and Arizona milksnake population status in general. To this end, well-funded, extremely time-intensive studies could be conducted on one or more populations in Arizona. For future monitoring, we suggest using a combination of techniques; availability of trained survey personnel and equipment will dictate their application.

(1) During the late summer rainy period, use night road driving with an ATV or other small, low vehicle such as an electric car to maximize detections of small snakes (E. M. Nowak, unpublished data). These surveys would be paired with repeated walking transects to intensively cover smaller areas.

(2) During late April and May, deploy box trap and drift fence transects in the same areas covered by night driving and walking surveys, to maximize seasonal coverage. Dusk driving and walking may also be effective if warm temperatures permit.

Unfortunately, current methods are unlikely to yield sufficient data for abundance estimates without highly intensive efforts. Uniformly (very) low detection probabilities, in combination with low recapture rates, violate basic assumptions of sample size in information theoretic population modeling (e.g., Burnham and Anderson 2002, Mackenzie and Kendall 2002, MacKenzie et al. 2005) as well as within mark-recapture modeling programs such as MARK (White and Burnham 1999). Instead of estimating actual population abundance, future attempts at understanding the status of milksnakes at Petrified Forest NP might apply habitat-specific site occupancy modeling, or predictions based on snake community characteristics (e.g., richness, local species extinction probability, species turnover), which are more tolerant of low detection probabilities (MacKenzie et al. 2002, MacKenzie et al. 2005). However, detection probabilities may need to increase even for these applications: Bailey et al. (2004) suggest that survey methods be able to detect the target species at least 15% of the time when the species is present. Contrary to the results of MacKenzie and Royle (2005), our data suggests that more intensive, repeated sampling over a smaller area (i.e., walking surveys and use of box traps) may be more effective than less intensive sampling over a larger area (i.e., night road driving): detection rates were highest during walking surveys during the summer rainy period and at box traps in the late spring. Future monitoring efforts could focus exclusively on the intensive seasonal deployment of these methods for the most robust estimates of occupancy.

Implications for Inventory and Monitoring

There is an inherent trade-off in monitoring programs between determining the status of rare or sensitive species, and in detecting measurable trends of more common species which could potentially help to guide future management decisions. Practically speaking, future monitoring for milksnakes within Petrified Forest NP may be most cost effectively conducted as part of an overall monitoring program for small vertebrates, rather than retaining a monotypic focus. A recently completed herpetological inventory of Southern Colorado Plateau National Parks found that only in a few parks were 90% or more of the total expected species documented, even where inventories had been previously completed (Persons and Nowak 2006). This result illustrates the importance of continued monitoring programs in assisting with completion of species lists.

At Petrified Forest NP and similar milksnake localities, a combination of trapping and night surveys (road driving via ATV and walking) may be the most cost-effective strategy for monitoring herpetofauna, with box trap transects added if funding permits. Night driving surveys in particular have been shown to provide a robust sampling regime for long-term inferences about snake and desert amphibian communities, especially nocturnal species that are not easily detected by other methods (e.g., Shafer and Juterbock 1994, Kline and Swann 1998, Drost et al. 2001). Long-term trends may indicate changes in abundance due to habitat or range shifts (Mendelson and Jennings 1992), especially given regional predictions for global climate change (Gibbons et al. 2000; Barnett et al. 2008). Status changes may also occur as a result of changes in road mortality and/or collecting rates (Rosen and Lowe 1994), or from accidental or intentional introductions of non-native species (e.g., H. T. Smith et al. 2007).

CONCLUSIONS AND IMPLICATIONS

Our study at Petrified Forest NP was the most thorough and lengthy attempt to date to assess the status of milksnakes anywhere in Arizona. But the apparently secretive nature of these animals resulted in very low detection probabilities. It is not yet clear whether we are observing a pattern of biological rarity, or simply a hard-to-detect species. However, the park does appear to support an important population of the species in northern Arizona.

Since most milksnakes were found in sandy loam sagebrush grasslands in the northern part of the park, continued protection of those habitats may be necessary to sustain the Petrified Forest NP population. Collectors have recently attempted to (illegally) find milksnakes at Wupatki National Monument (M. Blasing, formerly of Wupatki NM, personal communication), and we have observed illegal collection of other reptiles (eastern collared lizards [*Crotaphytus collaris*]) from Petrified Forest NP (E. M. Nowak, personal observation). Therefore, the threat of illegal collecting of the species remains high, and continuing the current practice of park road closure with gates locked at dusk (especially during the spring months) remains the most important method for protecting milksnakes at Petrified Forest NP. Future studies could compare occurrence and detections on nearby park expansion areas, as well as investigate regional patterns of distribution across Northern Arizona, to improve understanding of these little-known colorful animals.

ACKNOWLEDGEMENTS

We are indebted to the past and present staff of Petrified Forest NP, especially past Superintendents (the late) M. Hellickson and L. Baiza; Resources Management staff M. Depoy, K. Beppler-Dorn, P. Thompson, M. Wilkerson, and many biological technicians; Law Enforcement Rangers G. Caffey, C. Dorn, and D. Kruger; Interpretive Ranger R. Garcia; and many NPS volunteers. This project was funded by the National Park Service's Biological Resource Management Division, and additional funding and accounting assistance were received from the USGS Southwest Biological Science Center and L. Saul. R. Parnell was the Northern Arizona University project proposal lead, and C. Drost was the Principal Investigator during the original inventory. Techniques for animal use were approved through the Institutional Animal Care and Use Committee of Northern Arizona University under permit #05-010. Materials were donated by Ace Homeco, H. Johnson, D. Fisher/Ridgetop Construction, and USGS Colorado Plateau Research Station staff. Field assistants included L. Gilmore, M. Hamilton, AJ Monatesti, B. Parker, M. Santana-Bendix, J. Schofer, and M. Spille. This project would not have been possible without the dedication of many volunteers, especially B. and K. Bezy, K. Boles, C. Drost, E. Enderson,

E. Evans, B. Hamilton, T. Hoffnagle, D. Jex, C. Loughran, J. Meyers, J. Pilarski, J. and T. Pullins, R. Repp, C. and C. Seger, and H. Sweet. B. Hamilton, A. Holycross, and H. Koenig shared unpublished data for comparative purposes. T. Arundel provided the location map. An anonymous reviewer, T. Jones, and B. Wakeling reviewed earlier drafts of this manuscript.

LITERATURE CITED

Badman, J. A., A. T. Holycross, L. Neinaber, and D. F. DeNardo. 2003. Milksnakes (*Lampropeltis triangulum*) from Cochise County: notes on captive breeding and pattern. Sonoran Herpetologist 16

Bailey, L. L., T. R. Simons, and K. H. Pollock. 2004. Estimating site occupancy and species detection probability parameters for terrestrial salamanders. Ecological Applications 14:692–702.

Barnett, T. P., D. W. Pierce, H. G. Hidalgo, C. Bonfils, B. D. Santer, T. Das, G. Bala, A. W. Wood, T. Nozawa, A. A. Mirin, D. R. Cayan, and M. D. Dettinger. 2008. Human-induced changes in the hydrology of the western United States. Science 22 February, 2008, 319:1080–1083.

Brennan, T. C., and A. T. Holycross. 2006. A field guide to amphibians and reptiles in Arizona. Arizona Game and Fish Department, Phoenix, USA.

Burnham, K. P., and D. R. Anderson. 2002. Model selection and multimodel inference. Springer-Verlag Publishers, New York.

Campbell, H. W., and S. P. Christman. 1982. Field techniques for herpetofaunal community analysis. In Herpetological Communities, edited by N. J. Scott, pp.193–200. Wildlife Research Report 13, U. S. Fish and Wildlife Service.

Camper, J. D., and J. R. Dixon. 1988. Evaluation of a microchip system for amphibians and reptiles. Texas Parks and Wildlife Department, Research Publication 7100–159:1–22.

Conant, R., and J. T. Collins. 1998. A field guide to reptiles and amphibians of eastern and central North America. 3rd edition, expanded. Houghton Mifflin Company, New York.

Crump, M. L., and N. J. Scott. 1994. Visual encounter surveys. In Measuring and Monitoring Biodiversity: Standard methods for amphibians, edited by W. R. Heyer, M. A. Donnelly, R. W. McDiarmid, L. C. Hayek, and M. S. Foster, pp. 84–92. Smithsonian Institution Press, Washington D.C.

Douglas, M. E., M. R. Douglas, G. W. Schuett, L. Porras, and A. Holycross. 2002. Phylogeography of the western rattlesnake (*Crotalus viridis*) complex, with emphasis on the Colorado Plateau. In Biology of the Vipers, edited by G. W. Schuett, M. Hoggren, M. E. Douglas, and H. W. Greene, pp.11–50. Eagle Mountain Publishing LC, Eagle Mountain, Utah.

Drost, C. A., T. B. Persons, and E. M. Nowak. 2001. Herpetofauna survey of Petrified Forest National Park, Arizona. In Proceedings of the Fifth Biennial Conference of Research on the Colorado Plateau, edited by C. van Riper III, K. A. Thomas, and M. A. Stuart, pp. 83–102. FRESC Report Series, U. S. Geological Survey.

Fagerstone, K. A., and B. E. Johns. 1987. Transponders as permanent identification markers for domestic ferrets, black-footed ferrets, and other wildlife. Journal of Wildlife Management 51:294–297.

Fellers, G. M., and C. A. Drost. 1994. Sampling with artificial cover. In Measuring and Monitoring Biological Diversity: Standard methods for amphibians, edited by W. R. Heyer, M. A. Donnelly, R. W. McDiarmid, L. C. Hayek, and M. S. Foster, pp. 146–150. Smithsonian Institution Press, Washington D.C.

Fitch, H. S. 1987. Collecting and life-history techniques, chapter 5. In Snakes: Ecology and evolutionary biology, edited by R. A. Seigel, J. T. Collins, and S. S. Novak, pp. 143–164. Macmillan Publishing Co., New York.

Fitch, H. S. 1999. *Lampropeltis triangulum*. In A Kansas Snake Community: Composition and change over 50 years, pp. 79–83. Krieger Publishing Company, Malabar, Florida, USA.

Gibbons, J. W., D. E. Scott, T. J. Ryan, K. A. Buhlmann, T. D. Tuberville, B. S. Metts, J. L. Greene, T. Mills, Y. Leiden, S. Poppy, and C. T. Winne. 2000. The global decline of reptiles, déjà vu amphibians. Bioscience 50:653–666.

Gu, W., and R. K. Swihart. 2004. Absent or undetected? Effects of non-detection of species occurrence on wildlife–habitat models. Biological Conservation 116:195–203.

Hamilton, B., 2007. Rarity as an ecological paradigm, Kansas Journal of Herpetology Number 22, June 2007

Jemison, S. C., L. A. Bishop, P. G. May, and T. M. Farrell. 1995. The impact of PIT-tags on growth and movement of the rattlesnake, *Sistrurus miliarius*. Journal of Herpetology 29:129–132.

Keck, M. B. 1994. Test for detrimental effects of PIT tags in neonatal snakes. Copeia 1994:226–228.

Kline, N. C., and D. E. Swann. 1998.Quantifying wildlife road mortality in Saguaro National Park. In Proceedings of the International Conference on Wildlife Ecology and Transportation, edited by G. Evink, P. Garrett, D. Zeigler and J. Berry, pp. 23–31. Florida Department of Transportation, Tallahassee.

Lowe, C. H. 1989. Arizona's rarest snake *Lampropeltis triangulum*: the milksnake, Tucson Herpetological Society Newsletter 2

Luiselli, L. 2006. Testing hypotheses on the ecological patterns of rarity using a novel model of study: Snake communities worldwide. Web Ecology 6:44–58.

MacKenzie, D. I., and W. L. Kendall. 2002. How should detection probability be incorporated into estimates of relative abundance? Ecology 83:2387–2393.

MacKenzie, D. I., J. D. Nichols, G. B. Lachman, S. Droege, J. A. Royle, and C. A. Langtimm. 2002. Estimating site occupancy rates when detection probabilities are less than one. Ecology 83:2248–2255.

MacKenzie, D. I., J. D. Nichols, N. Sutton, K. Kawanishi, and L. L. Bailey. 2005. Improving inferences in population studies of rare species that are detected imperfectly. Ecology 86:1101–1113.

MacKenzie, D. I., and J. A. Royle. 2005. Designing occupancy studies: General advice and allocating survey effort. Journal of Applied Ecology 42:1105–1114.

Mendelson, J. R., and W. B. Jennings. 1992. Shifts in the relative abundance of snakes in a desert grassland. Journal of Herpetology 26:38–45.

Nichols, J. D. 1992. Capture-recapture models: Using marked animals to study populations. Bioscience 42:94–102.

Nowak, E. M., J. X. Schofer, and T. B. Persons. 2009. Milksnake monitoring at Petrified Forest National Park, Arizona. Final report to Petrified Forest National Park, USGS Southwest Biological Science Center.

Persons, T. B. 2001. Distribution, activity, and road mortality of amphibians and reptiles at Wupatki National Monument, Arizona, Final report to National Park Service, USGS Colorado Plateau Field Station.

Persons, T. B., and E. M. Nowak. 2006. Inventory of amphibians and reptiles in southern Colorado Plateau national parks. U.S. Geological Survey, Southwest Biological Science Center, Colorado Plateau Research Station, Open File Report 2006-1132.

Ramotnik, C. A., and M. A. Bogan. 1998. Baseline surveys for mammals at Petrified Forest National Park, Arizona: Final report of 1996–1997 activities to Petrified Forest National Park.

Reinert, H. K. 1992. Radio-telemetric field studies of pitvipers: Data acquisition and analysis. In Biology of the Pitvipers, edited by J. A. Campbell and E. D Brodie, Jr., pp.185–198. Selva Press, Tyler, Texas.

Rosen, P. C., and C. H. Lowe. 1994. Highway mortality of snakes in the Sonoran desert of southern Arizona. Biological Conservation 68:143–148.

Shafer, H. B., and J. E. Juterbock. 1994. Night driving. In Measuring and Monitoring Biological Diversity: Standard methods for amphibians, edited by W. R. Heyer, M. A. Donnelly, R. W. McDiarmid, L. C. Hayek, and M. S. Foster, pp. 163–166. Smithsonian Institution Press, Washington D.C.

Smith, H. T., W. E. Meshaka Jr., E. Golden, and E. M. Cowan. 2007. The appearance of the exotic green iguana as road-kills in a restored urban Florida state park: The importance of an 11-year dataset, Journal of Kansas Herpetology, 22 June.

Sokal, R. R., and F. J. Rohlf. 1981. Biometry, second edition. W. H. Freeman and Company, New York.

Stebbins, R. C. 2003. A field guide to reptiles and amphibians of western North America, 3rd edition. Houghton Mifflin Company, New York.

Thomas, K. A., M. Hansen, and C. Seger. 2003. Vegetation of Petrified Forest National Park, Arizona, technical report to National Park Service, USGS Southwest Biological Science Center.

Webb, R. H., G. J. McCabe, R. Hereford, and C. Wilkowske. 2004. Climatic fluctuations, drought, and flow in the Colorado River. USGS Fact Sheet 3062-04.

White, G. C., and K. P. Burnham. 1999. Program MARK: Survival estimation from populations of marked animals. Bird Study 46 (Supplement):120–138.

Williams, K. L. 1988. Systematics and natural history of the American Milk Snake. Milwaukee Public Museum, Milwaukee, Wisconsin.

Williams, K. L. 1994. *Lampropeltis triangulum*. Catalogue of American amphibians and reptiles 594.1–594.10.

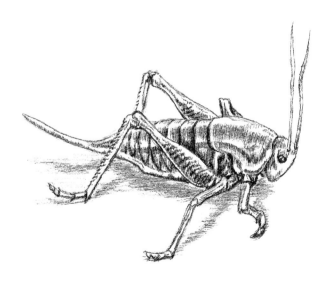

MORMON CRICKET CONTROL IN UTAH'S WEST DESERT: EVALUATION OF IMPACTS OF THE PESTICIDE DIFLUBENZURON ON NONTARGET ARTHROPOD COMMUNITIES

Tim B. Graham, Anne M. D. Brasher, and Rebecca N. Close

ABSTRACT

Grasshopper and Mormon cricket (Orthoptera) populations periodically build to extremely high numbers and can cause significant economic damage in rangelands and agricultural fields of the Great Plains and Intermountain West. A variety of insecticides have been applied to control population outbreaks, with recent efforts directed at minimizing impacts to nontarget fauna in treated ecosystems. A relatively new insecticide for control of Orthoptera is diflubenzuron, which acts to inhibit chitin production, ultimately causing death during the molt following ingestion of the insecticide. All arthropods, including insects, mites, and crustaceans, use chitin to build their exoskeletons and will die if they are unable to produce it during the next molt. Diflubenzuron is not taxon specific—it affects all arthropods that ingest it, except adult insects, which do not molt. Consequently, application of this pesticide has the potential to significantly reduce not only target populations but all terrestrial and aquatic arthropods within treatment zones.

Some research has been done in the Great Plains on the impact of diflubenzuron on nontarget arthropods in the context of grasshopper-control programs, but no work has been done in the Great Basin in Mormon cricket-control areas. This study was instigated in anticipation of the need for extensive control of Orthoptera outbreaks in Utah's west desert during 2005, and it was designed to sample terrestrial and aquatic arthropod communities in both treated and untreated zones. Three areas were sampled: Grouse Creek, Ibapah, and Vernon. High mortality of Mormon cricket eggs in the wet, cool spring of 2005 restricted the agricultural community's need to control Mormon crickets to Grouse Creek. Diflubenzuron was applied (aerial reduced agent-area treatment) in May 2005. Terrestrial and aquatic arthropod communities were sampled before and after application of diflubenzuron in the Grouse Creek area of northwestern Utah in May and June of 2005. In this paper, we discuss only the terrestial results (aquatic results are available online at http://pubs.usgs.gov/of/2008/1305/). In July 2005, U.S. Geological Survey scientists sampled areas in Ibapah and Vernon that had been treated with diflubenzuron in 2004, along with adjacent untreated areas. Pitfall traps at four treated and four untreated sites were used to collect ground-dwelling terrestrial arthropods. One-year post-treatment samples were collected by using the same methods for arthropods at Ibapah and Vernon in July 2005 (treatments applied in June 2004).

More than 124,000 terrestrial arthropods were collected from the three study areas. Direct effects of diflubenzuron on arthropod communities were not apparent in our data from Grouse Creek. Some trends indicate diflubenzuron may affect some terrestrial taxa. Ant communities showed some differences, with possible lag effects at Ibapah and Vernon. *Forelius* was more abundant, while *Tapinoma* and, perhaps, *Formica* declined in treated zones in these two study areas. *Solenopsis* also was more numerous at treated Ibapah sites but varied without pattern at Vernon. Scorpions were abundant at Grouse Creek and Ibapah but rare at Vernon. Numbers did not change during several weeks at Grouse Creek, but at Ibapah, numbers at treated sites were much lower than at untreated sites. The Lygaeidae (in the order Hemiptera) were more abundant in the untreated zones at Ibapah and Vernon, although significantly so only at Ibapah. Lygaeidae were absent from the treated zone at Grouse Creek (before and after treatment) but were present after treatment in the untreated zone. Additional research is recommended to determine more explicitly whether these taxa are sensitive to diflubenzuron applications in the Great Basin.

INTRODUCTION

In rangeland ecosystems of the United States, populations of Orthoptera (including grasshoppers and Mormon crickets) can rapidly build to levels that cause economic damage. Despite efforts to prevent outbreaks, grasshopper (multiple species) and Mormon cricket (*Anabrus simplex*) populations (Figure 1) were at high levels for 5 to 6 years preceding this study in the west desert of northern Utah (U.S. Department of Agriculture 2002). Although much of the area of outbreak was outside of cultivated lands, State and Federal agencies and private landowners were concerned about consumption of crops and range forage during these infestations. The need for rapid and effective suppression of Orthoptera when an outbreak occurs limits the control options available, and the application of an insecticide within all or part of the outbreak area has been the primary response for rapid suppression or reduction of Orthoptera populations to effectively protect rangeland. Control efforts have been implemented in Utah's west desert since 2002 in areas of particularly large Orthoptera populations.

The primary chemicals used for control of grasshoppers and crickets are carbaryl, applied as bran bait, and diflubenzuron, applied as an aerial spray. However, because the use of carbaryl in Utah's west desert has been greatly curtailed and is more localized, we focused on diflubenzuron treatments. Diflubenzuron is a chitin-inhibiting agent, causing arthropods to die during the molting process. Arthropods (including insects, arachnids, and crustaceans) have a hard exoskeleton made of chitin. Since diflubenzuron is a chitin-inhibitor, it affects nontarget arthropods, as well as grasshoppers and crickets. Previous studies have shown that although diflubenzuron is not directly toxic to vertebrates, birds can be indirectly affected when this pesticide reduces availability of key prey items (Sample et al. 1993). Consequently, a major concern in the west desert is that by killing nontarget arthropods, the food base for sensitive, rare, or threatened vertebrates, such as sage grouse and spotted frogs, will be depleted.

Studies on the Great Plains have shown diflubenzuron to have minimal impacts on nontarget arthropods and their vertebrate predators (Wilcox and Coffey 1978, McEwen et al. 1996), reinforcing the decision to use diflubenzuron in a reduced agent-area treatment design (using less pesticide in alternating swaths) instead of carbaryl or malathion. Some nontarget arthropods were affected by diflubenzuron, at least in the short term, in some studies (Catangui et al. 2000, Smith et al. 2006). The generality of previous work has not been established. Information directly applicable to the environment of Utah's west desert is required for assessing potential impacts of diflubenzuron on nontarget arthropods in the Great Basin.

Objectives

This study was designed to help managers improve Orthoptera-control programs by increasing the understanding of how diflubenzuron affects both target and nontarget arthropods. The specific objectives of this study were to (1) compare arthropod community structure (abundance and species composition) in treated and untreated sites in the west desert to determine whether there were changes in either target or nontarget arthropod populations, (2) compare responses at three study areas to determine whether response was similar across the landscape, and (3) compare arthropod community structure over time at each study area following insecticide treatment. The study also yields valuable baseline data on arthropod communities in west desert rangeland ecosystems.

Scope

Three areas of Utah's west desert (Grouse Creek, Ibapah, and Vernon) were chosen for 2005 sampling based on Orthoptera outbreaks in preceding years. However, in 2005, the only area significantly infested, and therefore sprayed with diflubenzuron, was Grouse Creek. We modified our objectives to account for the reduced control effort. We sampled for short-term effects of diflubenzuron at Grouse Creek; at the other two study areas, we tested for lag effects of diflubenzuron application by sampling in untreated zones and zones treated in 2004. Without prespray data or several years of postspray data, our analysis was limited. Topography also proved to be an issue, particularly at Vernon, for both terrestrial and aquatic sampling, because the treatment zone was on the valley floor and sites outside of the treatment zone were approximately 60–65 m higher in elevation.

STUDY AREA

Grouse Creek (Figure 2), Ibapah (Figure 3), and Vernon (Figure 4) in Utah's west desert were chosen for sampling due to large Mormon cricket populations in previous years and large expected populations for 2005. Ibapah and Vernon had large Mormon cricket populations for a number of years but were not treated with diflubenzuron prior to 2004 (G. Abbott, Animal and Plant Health Service, written communication, 2007). Grouse Creek had not been treated before 2005. Patchy application of carbaryl bran bait in previous years was done in all three areas (G. Abbott, Animal and Plant Health Service, written communication, 2007), including some areas we considered untreated relative to diflubenzuron application for this study. Diflubenzuron was applied to the Grouse Creek treatment zone in May 2005. We sampled terrestrial sites in four vegetation associations (Table 1) according to the Southwest Regional Gap analysis (Prior-Magee et al. 2007). Ibapah sites, near the Utah-Nevada border, were all in the same vegetation association, although there was some variability on the ground in abundance of different vegetation components. The Vernon study area, south of Vernon, Utah, was more diverse, encompassing five vegetation communities.

TERRESTRIAL SAMPLING
Study Sites

Terms for the different spatial scales of this study were defined as follows: study area is defined as one of the three major geographic areas studied [that is, Grouse Creek (GC), Ibapah (IB) or Vernon (VE)]; treatment zone refers to the area within a study area that was treated (T) or untreated (U) with diflubenzuron (that is, treated zone, untreated zone); site refers to the individual locations sampled using pitfall traps within a treatment zone in a study area (for example, GC U06, IB T22, and VE U09). In 2005, Mormon cricket-control efforts were concentrated in the mountains east and west of Grouse Creek valley. We established four sites in the eastern Grouse Creek treatment zone; four untreated sites also were established to the west and south of this treated zone (Figure 5). At the other two study areas, Ibapah (Figure 6) and Vernon (Figure 7), we sampled both zones

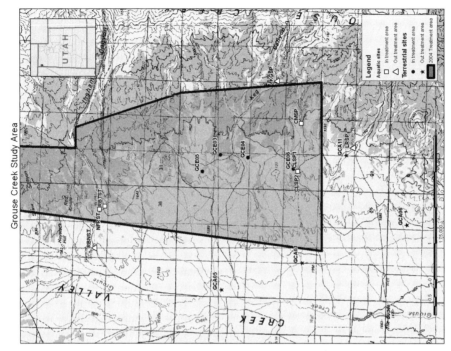

Figure 1. Mormon crickets on the road at Grouse Creek study site in May 2005, Utah (U.S. Geological Survey photograph by Tim Graham).

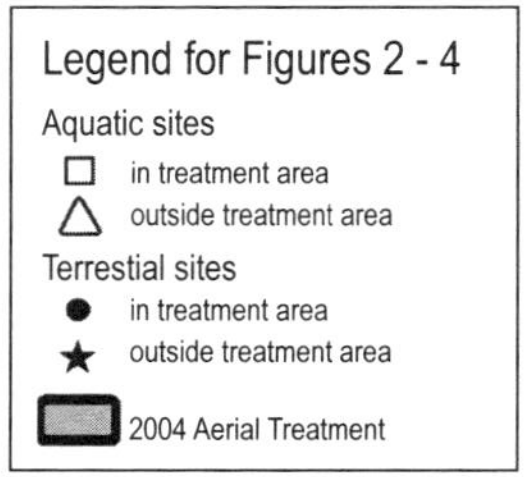

Figure 2. Grouse Creek study area with treated zone shaded; terrestrial and aquatic sampling sites are shown within treated and untreated zones.

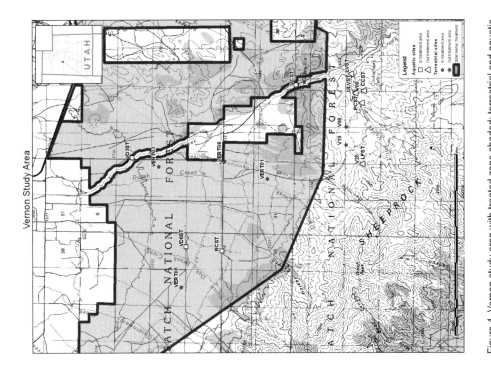

Figure 4. Vernon study area with treated zone shaded; terrestrial and aquatic sampling sites shown within treated zone and untreated zones.

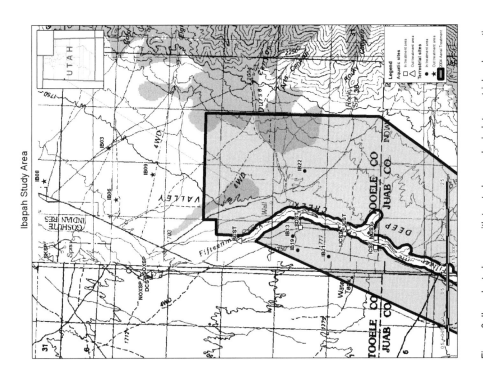

Figure 3. Ibapah study area with treated zone shaded; terrestrial and aquatic sampling sites are shown within treated zone and untreated zones.

Table 1. Terrestrial site and sampling information, west desert study sites, Utah.

Site code	Treatment zone	Elevation in meters	GAP vegetation classification	Sample dates	Sample period	Number of days trap was open
GC U01	Untreated	1,549	Inter-Mountain Basin Mixed Salt Desert Scrub	6/3/2005	pre-treatment	2
				6/20/2005	post-treatment	4
GC U05	Untreated	1,542	Invasive Perennial Grassland	6/3/2005	pre-treatment	2
				6/22/2005	post-treatment	2
GC U06	Untreated	1,573	Inter-Mountain Basin Semi-Desert Grassland	6/6/2005	pre-treatment	3
				6/21/2005	post-treatment	4
GC U11	Untreated	1,669	Inter-Mountain Basin Big Sagebrush Shrubland	5/27/2005	pre-treatment	2
				6/22/2005	post-treatment	5
GC T03	Treated	1,705	Inter-Mountain Basin Big Sagebrush Shrubland	5/26/2005	pre-treatment	2
				6/23/2005	post-treatment	2
GC T04	Treated	1,701	Inter-Mountain Basin Big Sagebrush Shrubland	5/27/2005	pre-treatment	3
				6/21/2005	post-treatment	4
GC T05	Treated	1,665	Inter-Mountain Basin Big Sagebrush Shrubland	5/30/2005	pre-treatment	4
				6/23/2005	post-treatment	2
GC T06	Treated	1,665	Inter-Mountain Basin Big Sagebrush Shrubland	5/26/2005	pre-treatment	2
				6/20/2005	post-treatment	4
IB U3	Untreated	1,734	Inter-Mountain Basin Big Sagebrush Shrubland	7/14/2005	1 year post-treatment	3
IB U6	Untreated	1,707	Inter-Mountain Basin Big Sagebrush Shrubland	7/15/2005	1 year post-treatment	3
IB U8	Untreated	1,670	Inter-Mountain Basin Big Sagebrush Shrubland	7/14/2005	1 year post-treatment	3
IB U9	Untreated	1,747	Inter-Mountain Basin Big Sagebrush Shrubland	7/15/2005	1 year post-treatment	3
IB T3	Treated	1,756	Inter-Mountain Basin Big Sagebrush Shrubland	7/15/2005	1 year post-treatment	3
IB T9	Treated	1,769	Inter-Mountain Basin Big Sagebrush Shrubland	7/15/2005	1 year post-treatment	3
IB T1	Treated	1,784	Inter-Mountain Basin Big Sagebrush Shrubland	7/14/2005	1 year post-treatment	3
IB T2	Treated	1,826	Inter-Mountain Basin Big Sagebrush Shrubland	7/15/2005	1 year post-treatment	3

Table 1 cont.

Site code	Treatment zone	Elevation in meters	GAP vegetation classification	Sample dates	Sample period	Number of days trap was open
VE U08	Untreated	2,082	Southern Rocky Mountain Montane-Subalpine Grassland	7/23/2005	1 year post-treatment	3
VE U09	Untreated	1,998	Great Basin Foothill and Lower Montane Riparian Woodland and Shrubland	7/22/2005	1 year post-treatment	3
VE U19	Untreated	1,879	Great Basin Piñon-Juniper Woodland	7/22/2005	1 year post-treatment	3
VE U20	Untreated	2,117	Inter-Mountain Basin Montane Sagebrush Steppe	7/22/2005	1 year post-treatment	3
VE T01	Treated	1,829	Great Basin Piñon-Juniper Woodland	7/21/2005	1 year post-treatment	3
VE T03	Treated	1,771	Inter-Mountain Basin Big Sagebrush Shrubland	7/23/2005	1 year post-treatment	3
VE T06	Treated	1,801	Inter-Mountain Basin Big Sagebrush Shrubland	7/22/2005	1 year post-treatment	3
VE T08	Treated	1,829	Great Basin Piñon-Juniper Woodland	7/21/2005	1 year post-treatment	3

Figure 5. Typical terrestrial site at Grouse Creek study site, Utah, showing pitfall traps (U.S. Geological Survey photograph by Tim Graham).

Figure 6. Typical terrestrial site at Ibapah study site, Utah (U.S. Geological Survey photograph by Tim Graham).

Figure 7. Typical terrestrial site at Vernon study site, Utah, showing pitfall traps (U.S. Geological Survey photograph by Rebecca Close).

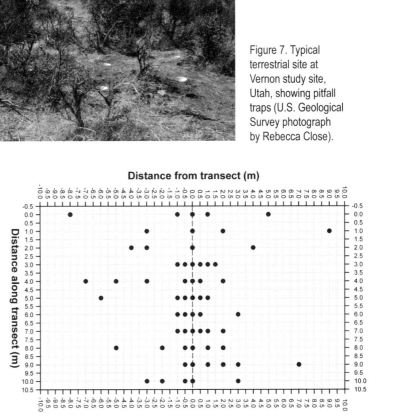

Figure 8. Arrangement of pitfall traps at each terrestrial arthropod sampling site of the study.

that had been treated with diflubenzuron in 2004 and adjacent untreated zones. Here we also established four pitfall sites in each of the treated and untreated zones. All pitfall sites were established at locations randomly selected with a geographic information system. General site characteristics, site-code designations, and sampling dates are shown in Table 1.

Sampling Design

Terrestrial arthropods were sampled by using pitfall traps arranged in a pattern that will allow capture data to be used with DISTANCE software (Buckland et al. 2001) to estimate density of total arthropods and of individual taxa (Lukacs et al. 2004). Pitfall traps at each site were arranged to meet the assumptions of DISTANCE sampling, which are that all invertebrates on the center line are detected (that is, caught) and that distances from the center line are accurately measured. We used 60 traps at each site in the arrangement shown in Figure 8. This pattern was generated by using WebSim (Lukacs 2001, 2002) to simulate a hazard-rate model of invertebrate captures that resulted in estimates with small confidence intervals, and matched trapping results in a pilot study of invertebrate pitfall trapping in Colorado (Lukacs, oral communication, 2005).

Sample Collection and Processing

Pitfall traps were placed by carefully measuring and marking correct locations with flags, then digging in the traps. Pitfall traps were constructed as described by New (1998). For each trap, a 1.5 L plastic jar was buried below ground level and a 500 mL cup containing 125 mL of soapy water was placed in the jar. A 15 cm-diameter funnel was placed over the jar, centered over the cup, with the top of the funnel at ground level.

At Grouse Creek, we sampled in late May/early June, just prior to application of diflubenzuron in both the zone to be treated and in the zone to remain untreated (pre-treatment samples). Traps at all sites in both treated and untreated zones were opened again in late June, about 3 weeks after diflubenzuron application (post-treatment samples). By comparing treated and untreated zones before and after treatment, differences between pre- and post-treatment communities associated with the phenology of the arthropods can be separated from those changes that may be due to exposure to diflubenzuron. Ibapah and Vernon sampling occurred in July 2005, roughly a year following treatment of the treated zones; there was no temporal component to the study in these two areas.

Traps were kept open from 2 to 5 days (Table 1); the time period eventually was standardized at 3 days, but different timeframes were used in early sampling periods due to logistical constraints.

Each trap's contents were washed in the field through a 0.5 mm mesh net three times; everything remaining in the net was placed in a 35 mL vial containing about 25 mL isopropyl alcohol. A paper label with site, date, and trap number was placed inside each vial, and a stick-on label with the same information was affixed to the outside of each vial. Vials were kept in the shade in the field, as cool as possible, and stored at room temperature once they were returned to the lab.

Sample Sorting and Identification

All terrestrial invertebrates were sorted to order. Specimens in the orders Hemiptera and Orthoptera were identified to family; ants (Formicidae) were identified to genus. Taxa were identified following Triplehorn and Johnson (2005), and we followed the taxonomic nomenclature of this source (that is, the order Hemiptera includes Heteroptera and the Homoptera; Thysanura has been split into Microcoryphia and Thysanura). Differences in abundance, or presence/absence of particular taxa that correlated with treatment patterns, were used to identify potential indicator species.

Data Analysis

To allow comparisons among individual sites, treatment zones, and study areas, data are reported as numbers per day (abundance) by taxon, and as relative abundance. Because traps were not kept open for the same number of days during all sampling events, all arthropod numbers were adjusted to average number per day by dividing the number of individuals (both total arthropods and individual taxa) caught in each trap by the number of days the traps at that site were open. Abundance data of terrestrial insects and other arthropods in treated versus untreated zones within a study area were tested for normality and equal variance, then compared using t-tests or Mann-Whitney rank sum tests, using SigmaStat (Systat 2004). Statistical significance was assigned at $\alpha<0.05$; however, several of the observed differences in abundance were large, indicating the potential for biological significance even when statistical tests did not show them to be significant at $\alpha=0.05$. More sampling will tell whether these effects are real (they are masked by high variability, given our sample sizes); for the record, we note these cases with $\alpha<0.20$. Data used for nonmetric multidimensional scaling (NMS) consisted of the average number of individuals per day per taxon for each sampling event. NMS was performed in PC-ORD (McCune and Mefford 2005) using Sorensen's distance measure. Fifty iterations were run with the data, then 250 iterations of a Monte Carlo test were used to estimate the best-fit (least-stress) solution.

TERRESTRIAL RESULTS

In May, June, and July 2005, 1,920 pitfall traps were set at the three study areas. More than 124,000 specimens have been identified to order. The total number of arthropods caught at a single site ranged from 853 at GC T06 (pre-treatment collection) to 36,043 at GC U05 (post-treatment collection). Relative abundance of the 13 orders varied considerably in space and time. The most common taxa were typically Diptera (flies), Hemiptera (true bugs), and Formicidae (ants), with Araneae (spiders), non-ant Hymenoptera (bees and wasps), and Orthoptera (grasshoppers, crickets, and such) fairly common at some sites. The three study areas had very different communities. Additionally, the variability among sites at each study area, even among the "replicate" sites of treated or untreated zones, was considerable. Few indications of the short-term effect of diflubenzuron at Grouse Creek on the relative abundance of any taxon except the Orthoptera (the object of control efforts) were discernible.

Comparison of Proportional Abundance

Grouse Creek.

Flies were most abundant at the Grouse Creek untreated sites in late May and early June, with ants codominant at GC U11 and GC U06 (Figure 9). Spiders, beetles, bees, wasps, Hemiptera, and Lepidoptera also were common at one or more sites. By late June, communities at all sites had changed dramatically. At three sites, Hemiptera were by far the most abundant taxon. Ants and flies were still common at most sites but proportionately less abundant given the increase in numbers of Hemiptera. Spiders and Orthoptera increased at GC U11.

Hemiptera, Diptera, and ants were the most abundant groups at treated sites of Grouse Creek prior to treatment. The three groups accounted for more than 70 percent of all the arthropods caught at the four sites (Figure 10). GC T04 had the lowest proportion of Hemiptera, the highest proportion of ants, and larger numbers of beetles and Orthoptera than at the other sprayed sites prior to treatment. GC T04 had the most Mormon crickets during sampling in late May and early June.

After treatment, Hemiptera were much more abundant at the treated sites, showing a pattern similar to that of untreated sites.

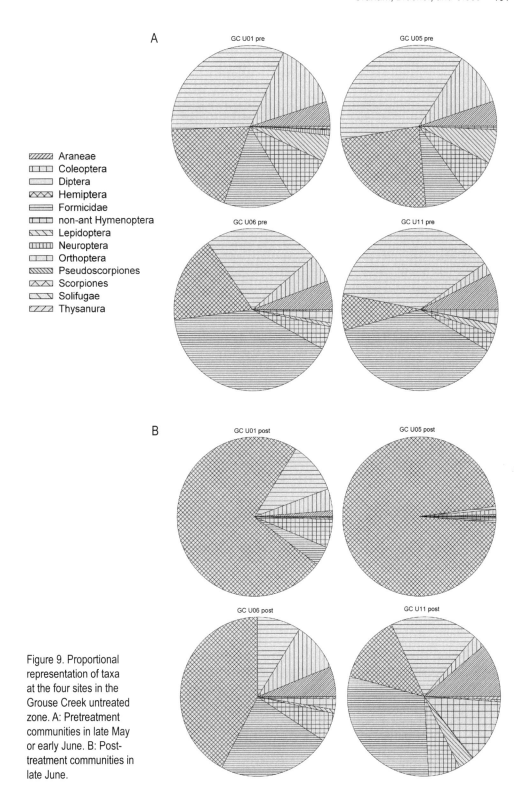

Figure 9. Proportional representation of taxa at the four sites in the Grouse Creek untreated zone. A: Pretreatment communities in late May or early June. B: Post-treatment communities in late June.

Ants and flies accounted for most of the other captures. Numbers of Orthoptera declined following treatment, as expected, showing the largest decline at GC T04. Most differences between pre- and post-treatment communities at the treated sites were similar to the changes observed among the untreated sites, indicating that the differences were likely due to seasonal changes, not treatment effects.

Ibapah.

Ibapah sites were sampled only once, in mid-July 2005 (Table 1). All communities had large Hemipteran components (Figure 11). Two sites, one treated (IB T13, 68 percent) and one untreated (IB U08, 92 percent), were heavily dominated by Hemiptera. Ants and flies also were abundant at all Ibapah sites; spiders were more prevalent than flies in traps at IB T19 and IB T22 (both treated in 2004). Flies were less common at Ibapah than at Grouse Creek, both in treated and untreated sites. Ants dominated all sites if Hemiptera were excluded from the dataset, constituting more than half the individuals at each site. Spider abundance also became more apparent if Hemiptera were excluded.

Vernon.

Vernon was sampled only once in July 2005. At three of four untreated sites and three of four treated sites, about 25–40 percent of all the arthropods caught were Hemiptera; fewer than 15 percent of the arthropods caught at the other two sites (VE U20 and VE T01) were Hemiptera (Figure 12). Compared to Ibapah and Grouse Creek, Hemiptera and flies at Vernon were proportionately less abundant, and the Hymenoptera were relatively more abundant. Specifically, ants and the combined bee and wasp fractions of the Hymenoptera were better represented at Vernon. The non-ant Hymenoptera were more abundant at the untreated sites than at the treated sites, although there was no statistical difference in proportional abundance.

Comparisons of Abundance by Orders in Treated and Untreated Zones

Average abundance (numbers per day) for each taxon was calculated for the four sites within a treatment zone in each study area (Figure 13 A-L). A t test was used if the data passed normality and equal variance tests; the test statistic is represented as a t. If data failed normality or equal variance tests, comparisons were made with the Mann-Whitney rank sum test; the test statistic is represented as T.

Within each study area, abundance for each taxon in the treated zone was compared to abundance in the untreated zone. Data from different study areas were not compared to each other. At Grouse Creek, we also tested whether changes in arthropod abundance following application of diflubenzuron were related to the insecticide, or merely to phenology (seasonal changes in species composition) of the arthropod community. This test was conducted in two ways; the first approach was to compare average abundance before and after treatment within each treatment zone (for example, GC U pre-treatment compared to GC U post-treatment, and GC T pre-treatment compared to GC T post-treatment). Significant differences for a given taxon in pre- and post-treatment numbers in the treated zone were interpreted as indicating a possible effect of diflubenzuron application. The second approach was to compare taxon abundance in the sprayed and unsprayed zones prior to treatment and again with the data collected three weeks after treatment. If there were no differences prior to treatment but treated and untreated average taxon abundance differed following treatment, we assumed diflubenzuron affected that taxon.

Grouse Creek

Pre- versus Post-Treatment Changes in Untreated Zone

Most taxa exhibited an increase in abundance from pre- to post-treatment

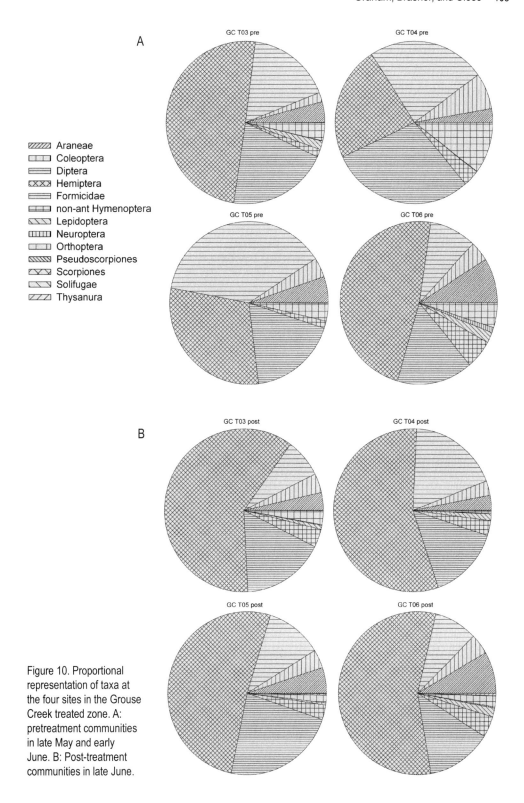

Figure 10. Proportional representation of taxa at the four sites in the Grouse Creek treated zone. A: pretreatment communities in late May and early June. B: Post-treatment communities in late June.

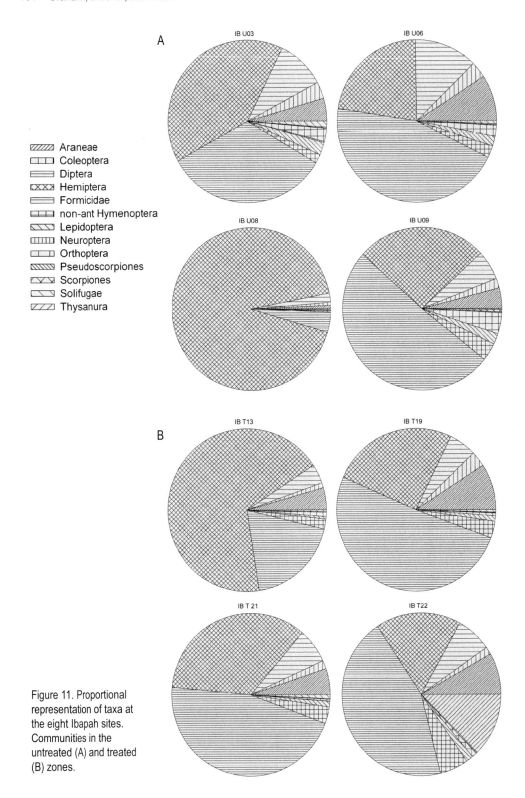

Figure 11. Proportional representation of taxa at the eight Ibapah sites. Communities in the untreated (A) and treated (B) zones.

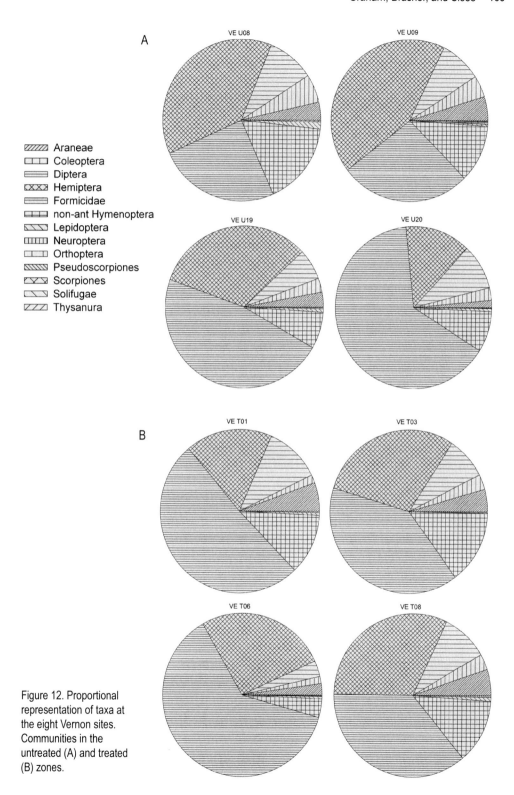

Figure 12. Proportional representation of taxa at the eight Vernon sites. Communities in the untreated (A) and treated (B) zones.

sampling in the unsprayed zone at Grouse Creek (Figure 13 A–L). Spiders ($T_{d.f. 6}=10$; $P=0.029$), non-ant Hymenoptera ($T_{d.f. 6}=10$; $P=0.029$), and total arthropods ($T_{d.f. 6}=10$; $P=0.029$) were significantly more abundant in the post-treatment collections at unsprayed sites. Coleoptera ($t_{d.f. 6}=-2.053$; $P=0.086$), Hemiptera ($T_{d.f. 6}=12$; $P=0.114$), and Scorpiones ($t_{d.f. 6}=-1.926$; $P=0.102$) also showed large increases in average abundance from pre- to post-treatment collections but the differences were not statistically significant.

Pre- versus Post-Treatment Changes in Treated Zone

No significant differences in pre- and post-treatment numbers occurred within the sprayed zone, although Hemiptera ($t_{d.f. 6}=-1.992$; $P=0.093$), non-ant Hymenoptera ($t_{d.f. 6}=-1.482$; $P=0.189$), Orthoptera ($t_{d.f. 6}=2.419$; $P=0.052$) and Scorpiones ($T_{d.f. 6}=12.5$; $P=0.114$) all had average abundance differences that were almost significant statistically. Total arthropods did not differ in the sprayed zone. Only Orthoptera showed a decrease from pre- to post-treatment numbers in the sprayed zone, indicating that diflubenzuron did accomplish the management goal of decreasing Orthoptera numbers in the sprayed zone.

Pre-Treatment Changes in Untreated versus Treated Zones

Most taxa did not differ between unsprayed and sprayed zones prior to application of diflubenzuron. There were statistically significant differences in average abundance for the Hemiptera ($t_{d.f. 6}=-2.726$; $P=0.034$), non-ant Hymenoptera ($T_{d.f. 6}=24.5$; $P=0.035$), and Orthoptera ($t_{d.f. 6}=-2.455$; $P=0.049$). Lepidoptera numbers ($T_{d.f. 6}=23.5$; $P=0.114$) also differed between zones but not to the point of being statistically significant. Hemiptera and Orthoptera were more abundant in the sprayed zone prior to treatment; non-ant Hymenoptera and Lepidoptera were more numerous in the unsprayed zone at the same time.

Post-Treatment Changes in Untreated versus Treated Zones

Post-treatment comparisons of unsprayed and sprayed zones showed that spiders ($t_{d.f. 6}=4.042$; $P=0.007$) and non-ant Hymenoptera ($T_{d.f. 6}=26$; $P=0.029$) were significantly more abundant in the unsprayed zone following application of diflubenzuron. Average numbers of Lepidoptera ($t_{d.f. 6}=2.425$; $P=0.052$), Scorpiones ($t_{d.f. 6}=2.077$; $P=0.083$), and total arthropods ($T_{d.f. 6}=25$; $P=0.057$) also differed markedly in the sprayed and unsprayed zones but not to the point of statistical significance. In all cases, post-treatment numbers were greater in the unsprayed zone. The Lepidoptera decreased somewhat from pre-treatment levels in the unsprayed zone and increased slightly during the same timeframe in the sprayed zone, but Lepidoptera still were more abundant in the unsprayed zone. This post-treatment difference is likely the result of inherent differences in the Lepidoptera communities of the two zones.

Ibapah

Numbers of Orthoptera ($t_{d.f. 6}=2.569$; $P=0.042$) and Scorpiones ($T_{d.f. 6}=25$; $P=0.029$) were significantly lower in the sprayed zone at Ibapah compared to the unsprayed zone. Differences in average abundance that were almost significant were recorded for other taxa, including Coleoptera ($t_{d.f. 6}=1.880$; $P=0.109$), Diptera ($t_{d.f. 6}=1.701$; $P=0.140$), non-ant Hymenoptera ($t=-2.432$; $P=0.051$), and Lepidoptera ($t_{d.f. 6}=0.801$; $P=0.122$). For all taxa showing large differences, more individuals were caught at the unsprayed sites than at sprayed sites.

Vernon

Vernon data from this study provided the strongest indication of diflubenzuron effects on nontarget arthropods. Beetles ($T_{d.f. 6}=26$; $P=0.029$), flies ($t_{d.f. 6}=3.274$; $P=0.017$), Hemiptera ($t_{d.f. 6}=2.458$; $P=0.049$), non-ant Hymenoptera ($t_{d.f. 6}=2.790$; $P=0.032$),

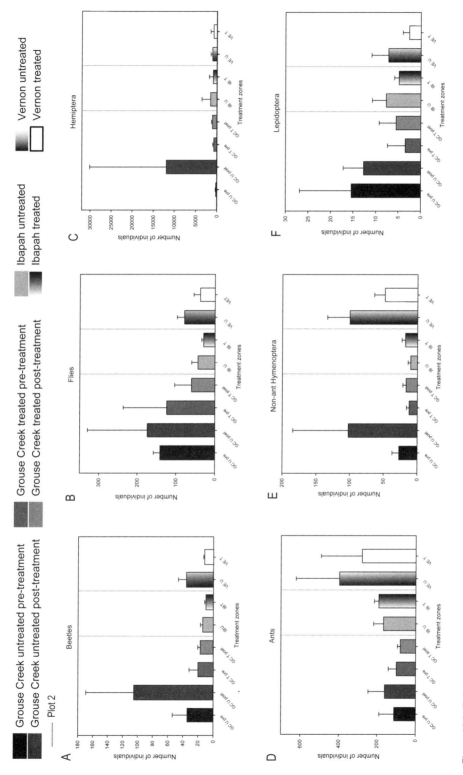

Figure 13 A–F. Average number of individuals (by taxon) in untreated (GC U pre- and post-treated, IB U, VE U) and treated (GC T pre- and post-treated, IB T, VE T) zones at the three study areas.

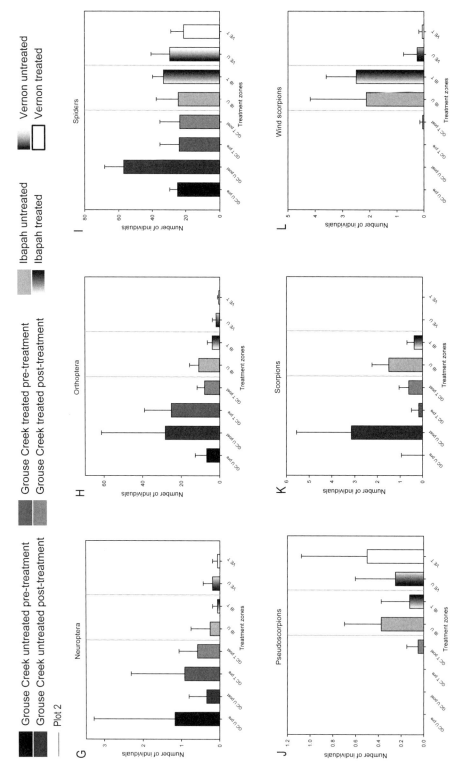

Figure 13 G–L. Average number of individuals (by taxon) in untreated (GC U pre- and post-treated, IB U, VE U) and treated (GC T pre- and post-treated, IB T, VE T) zones at the three study areas.

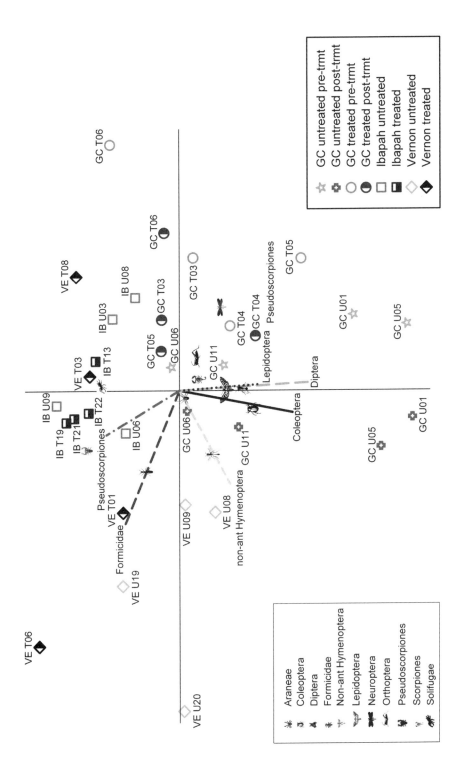

Figure 14. Nonmetric multidimensional scaling ordination for all sites, excluding Hemiptera and Microcoryphia. Vectors represent taxa significantly influencing the spread of sites along the two axes.

and total arthropods ($t_{d.f.\ 6}$=2.650; P=0.038) were all significantly more abundant in the unsprayed zone than in the sprayed zone. Lepidoptera ($t_{d.f.\ 6}$=2.380; P=0.055) and Orthoptera ($t_{d.f.\ 6}$=1.485; P=0.188) also had greater average numbers in the unsprayed zone than in the sprayed zone, although the differences were not quite significant.

Terrestrial Community Structure

Multivariate analysis (NMS) showed no useful ordination solution using either raw data (average numbers caught per day), or log-transformed data. There was no apparent structure in the data; most of the sites were in a single, large cluster with a few (primarily those with very high or very low numbers of Hemiptera) separated individually from the main grouping. Final stress values for the dataset did not differ from analysis with randomized data, indicating that community structure was weak (McCune and Grace 2002). This result may have been due to the influence of particular taxa (for example, Hemiptera) at individual sites. Examination of the data indicated that GC U05 post-treatment and IB U08 were outliers in the raw-numbers dataset, and VE U08 and VE T01 were outliers after the data were transformed, based on the PC-ORD outlier test (McCune and Grace 2002). GC U05 post-treatment and IB U08 were heavily dominated by Hemiptera, with more than 90 percent of all specimens in this order (Figures 9B and 12A). It is not clear why VE U08 was identified as an outlier because no group seemed particularly over- or underrepresented. VE T01 had fewer Hemiptera than any other site, with the exception of the GC U11 pre-treatment collection, and was dominated by ants but not by a higher proportion than at other sites. Removing these outliers did not improve ordination results using either raw numbers or transformed data.

Ordination with the Hemiptera removed from all sites provided a very strong two-axis solution that explained 95.5 percent of the variation in the dataset. This solution left IB T22 isolated from the other sprayed sites at Ibapah, despite what appeared to be very similar communities among the sites. The one significant difference between IB T22 and the other sprayed sites was that no Microcoryphia (silverfish) were found at any Ibapah site except IB T22, where 219 silverfish were collected (no other site in the study had more than 19 silverfish). Ordination (NMS) was then performed, excluding both Hemiptera and Microcoryphia data for all sites (Figure 14).

Ordination with the reduced dataset resulted in a good two-dimensional solution (final stress=10.0747) that explained 92.6 percent of the variation in the dataset. Very little change occurred in the position of the sites, except that IB T22 was brought into close proximity with the other three sprayed sites at Ibapah, and additional separation was achieved between treated and untreated sites at Vernon. The first axis was defined primarily by ant abundance, with some influence from non-ant Hymenoptera numbers. Flies and beetles provided most of the structure on the second axis.

Ibapah and Vernon sites were largely separated from Grouse Creek sites based on abundance of beetles and flies (greater numbers at Grouse Creek) and ants (fewer numbers at Grouse Creek); the former two taxa influenced the position along the second axis, while ants structured sites along the first axis. Abundance of solifugids at Ibapah, beetles and pseudoscorpions at Vernon, and beetle numbers at Grouse Creek also influenced the location of sites in the ordination.

Grouse Creek

Grouse Creek community data indicated a consistent temporal shift for both sprayed and unsprayed sites from pre-treatment (late May and early June) communities to post-treatment (late June) communities, except for GC U05, GC T03, and GC T05 (Figure 14). The basic pattern of data change consisted in a shift to the left and a weak-to-moderate shift down in ordination space. Community structure changed with increases in total arthropod abundance, Coleoptera (beetles), Formicidae (ants), and non-ant Hymenoptera (bees and wasps) from pre- to post-treatment collections. Flies (Diptera) increased at all unsprayed sites and at GC T04 and GC T06 but declined at GC T03 and GC T05. Beetle numbers increased from pre- to post-treatment collections at all sites except GC T05. Sites GC U01 and GC U05 were separated at the bottom of the ordination space and changed primarily along the first axis from pre- to post-treatment collections; these two sites had more beetles than any other site, especially in pre-treatment collections.

The magnitude of increase in abundance differed between sprayed and unsprayed sites, with many more arthropods being caught at unsprayed sites after treatment than at sprayed sites (Table 2a and 2b). The variation among sites, even within a treatment zone or collection period, kept many of these differences from being statistically significant, but the trend is evident when all 16 sampling events are examined (Table 2a and 2b). Additional work is needed to determine whether application of diflubenzuron reduces arthropod abundance, or whether observed differences are related to inherent site differences.

Ibapah

Community structure of Ibapah sites was based primarily upon ant abundance. Without Hemiptera (true bugs and leaf-hoppers) and Microcoryphia (silverfish), the four sprayed sites at Ibapah formed a tight cluster along the second axis, along with IB U09; these five Ibapah sites had similar numbers of ants. The IB U09 ant community resembled communities at the sprayed sites as well, though genus-level data were not included in the dataset used in this ordination. Similarities between IB U09 and the sprayed sites at Ibapah underscore the influence of inherent site properties in structuring arthropod communities. Two of the unsprayed sites (IB U03 and IB U08) were shifted to the right relative to the sprayed sites; these two sites had fewer ants than any other Ibapah site. IB U06, the fourth unsprayed site, had ant numbers similar to the sprayed sites but had more flies, separating it from the sprayed cluster along the second axis. There was little variation along the second axis among the eight Ibapah sites, but what differences did occur were due essentially to differences in abundance of Diptera at each site.

Vernon

As was the case at Ibapah, ant abundance structured the Vernon sites, with more ants at the unsprayed sites than at most of the sprayed sites. Proximity in the multivariate plot to other sites from any study area appeared to be dictated by similar ant numbers. The sprayed sites were scattered across ordination space on the first axis but within a narrow belt on the second axis; large variation in ant abundance at sprayed sites (418 to 2,306 total ants) caused this spread on the first axis. The range of ants collected at unsprayed sites overlapped sprayed sites (870 to 2,826 ants), but the trend was for more ants at unsprayed sites, which accounts for their shift to the left along the first axis (Figure 14). Unsprayed sites also had more beetles and flies than did sprayed sites, and a greater variation in these taxa. The increased numbers of beetles and flies and their greater variation in taxa is reflected in the spread in unsprayed site locations along the second axis and in the small variation in second-axis ordination scores for the sprayed sites, which showed less variation in numbers of flies or beetles (Table 2d).

Range in ordination space was greater for Vernon sites than for Ibapah or Grouse Creek sites, due at least in part to the greater spatial and elevational spread among the sites of the Vernon study area. For example, VE T06 was isolated from all other sites on the ground, being farther west, and was identified as an outlier by using the PC-ORD

Table 2a. Number of arthropods, by order, captured at each site in the untreated area at Grouse Creek, before and after treatment.

	GC U01 pre trmt	GC U01 post-trmt	GC U05 pre trmt	GC U05 post-trmt	GC U06 pre trmt	GC U06 post-trmt	GC U11 pre trmt	GC U11 post-trmt
Coleoptera	162	516	181	355	84	333	29	76
Diptera	400	1220	626	240	352	303	457	549
Hemiptera	233	8581	404	34856	262	1420	88	410
Formicidae	173	453	151	150	602	786	456	869
non-ant Hymenoptera	115	669	123	240	70	197	44	200
Lepidoptera	64	33	114	37	12	25	23	95
Mantodea								
Neuroptera	13			1			1	5
Orthoptera	1	18	12	20	34	73	36	381
Microcoryphia	6	3				1		19
Trichoptera	1	1						1
Araneae	61	139	85	144	86	205	88	325
Scorpiones	2	16			6	14		13
Solifugae								
Pseudoscorpiones								
Chilopoda		1					2	2
Diplopoda								
Total arthropods	1231	11650	1696	36043	1508	3357	1224	2945

outlier analysis routine. VE T06 had more beetles and ants, but fewer flies, than the other treated sites. The other outlier identified (of all 32 communities included in the analysis) was VE U20, which was higher in elevation and slightly farther south than the other unsprayed Vernon sites (Figure 4, Table 1). More ants were collected at VE U20 than at any other site in the study (2,826 ants); flies also were more abundant, and spiders were less common than at other unsprayed sites. Although data on Hemiptera were not included in the ordination analysis, these two sites also differed from the other Vernon sites in numbers of Hemiptera; VE T06 had more Hemiptera than any other sprayed site, and VE U20 had only about half as many Hemiptera as the other unsprayed sites.

Table 2b. Number of arthropods, by order, captured at each site in the treated area at Grouse Creek, before and after treatment.

	GC T03 pre trmt	GC T03 post-trmt	GC T04 pre trmt	GC T04 post-trmt	GC T05 pre trmt	GC T05 post-trmt	GC T06 pre trmt	GC T06 post-trmt
Coleoptera	25	80	93	14	88	65	34	100
Diptera	230	161	314	747	858	204	83	223
Hemiptera	671	1268	295	2264	676	914	407	1401
Formicidae	276	351	381	597	407	408	134	334
non-ant Hymenoptera	21	76	42	121	33	52	48	103
Lepidoptera	23	16		53		7	18	43
Mantodea	1	1						1
Neuroptera	1	3	1	3			9	6
Orthoptera	48	57	133	22	79	27	42	53
Microcoryphia	1							12
Trichoptera	1			1			2	
Araneae	56	73	36	120	118	91	76	206
Scorpiones	2	2		4	2	6		1
Solifugae						1		
Pseudo scorpiones								1
Chilopoda	1			2		1		1
Diplopoda								
Total arthropods	1357	2088	1295	4048	2261	1776	853	2485

Formicidae

Eighteen genera of ants were collected across the three study areas (Table 3). Untreated sites at Grouse Creek had 15 genera; 16 genera were found at sites in the treated zone. Ibapah untreated sites had 11 genera; 14 genera were found at sites in the treated zone. This trend was reversed at Vernon, where 13 genera were found at sites in the untreated zone, and 10 genera were found at sites in the treated zone. Vernon ant communities were slightly less genera rich, but ants were more abundant at Vernon (10,711 ants total) than at the other two study areas. At the eight Ibapah sites, 4,968 ants were collected. Grouse Creek unsprayed sites accounted for 3,622 ants from both collection periods, while 2,887 ants were collected from sprayed sites during the two sampling periods.

Table 2c. Number of arthropods, by order, captured at each site in Ibapah.

	IB U03 un-treated	IB U06 un-treated	IB U08 un-treated	IB U09 un-treated	IB T13 treated	IB T19 treated	IB T21 treated	IB T22 treated
Coleoptera	52	62	64	31	33	50	31	26
Diptera	155	246	206	96	130	90	128	125
Hemiptera	632	428	10289	429	2287	414	638	298
Formicidae	525	868	452	825	631	823	841	773
non-ant Hymenoptera	28	46	27	58	67	54	61	100
Lepidoptera	39	37	13	35	16	17	20	25
Mantodea	1	1			1		2	
Neuroptera	4						1	
Orthoptera	40	42	22	68	29	9	7	17
Microcoryphia								219
Trichoptera								
Araneae	72	177	76	66	148	148	95	143
Scorpiones	3	4	9	8	3	1	2	
Solifugae	19	11	1	3	12	8	15	5
Pseudo-scorpiones	1	2		3		2		
Chilopoda								
Diplopoda								
Total arthropods	1571	1924	11159	1622	3357	1616	1841	1731

Grouse Creek

Ant communities differed in composition among sampling events and sites in the treated and untreated zones at Grouse Creek (Figure 15). Differences appeared to be due to ant phenology and intrinsic site differences, not the application of diflubenzuron. *Formica* were particularly abundant at Grouse Creek, except at GC T03 and GC T05. These two sites were the rockiest sites; GC T03 was on a hillside, GC T05 was at the nose of a small ridge. There were more *Pogonomyrmex* at untreated sites and more *Pheidole* at treated sites, but these differences, which existed prior to treatment, were not related to the insecticide. Seed-harvester numbers increased at nearly every Grouse Creek site from pre- to post-treatment collections. *Forelius* was found only at GC U11 and GC T03; *Forelius* increased at GC U11 from late May to late June but declined over the same period at GC T03.

Ibapah

Ants at Ibapah differed from site to site, but there were some patterns correlated with whether or not the sites had been treated with diflubenzuron in 2004. Untreated sites

Table 2d. Number of arthropods, by order, captured at each site in Vernon.

	VE U08 untreated	VE U09 untreated	VE U19 untreated	VE U20 untreated	VE T01 treated	VE T03 treated	VE T06 treated	VE T08 treated
Coleoptera	198	154	99	110	36	46	53	36
Diptera	336	310	220	397	253	129	117	110
Hemiptera	1326	1516	1162	598	397	453	978	366
Formicidae	870	963	1679	2826	1148	610	2306	418
non-ant Hymenoptera	580	406	261	355	270	222	137	146
Lepidoptera	49	14	25	28	12	3	17	9
Mantodea	4	3	1	1			1	
Neuroptera		1		2			1	
Orthoptera		18	8	4		1	5	
Microcoryphia	12				6			
Trichoptera				1				
Araneae	127	173	107	65	128	68	84	60
Scorpiones			1					
Solifugae		4				1		
Pseudoscorpiones		1	3				4	4
Chilopoda								
Diplopoda		1107		2				
Total arthropods	3504	4670	3566	4389	2256	1533	3703	1162

showed a large proportion of the community composed of ants in the genera *Formica*, *Leptothorax*, and *Tapinoma*, while *Forelius*, *Pheidole*, and *Pogonomyrmex* dominated collections from treated sites (Figure 16). IB U09 was unusual for an untreated site because of the high numbers of *Forelius* found there. It is not clear what features of IB U09 were more similar to the treated sites than to the other untreated sites, except that IB U09 was closer to the treatment zone than the others. Numbers of *Tapinoma* and *Leptothorax* at IB U09 were comparable to the other untreated sites but *Formica* was rare.

Vernon

Community composition of ants at Vernon also was different at each site, but structure again appeared to be correlated with treatment history (Figure 17). *Tapinoma* was found at three of the four untreated sites but not at any treated sites. *Forelius* was present at all four treated sites; one ant found at VE U20 was the only *Forelius* found at any untreated site. *Pogonomyrmex* was rare at Vernon, occurring at only two sites (one treated and one untreated). *Pheidole*, an important seed harvester, also was relatively uncommon at Vernon sites. *Monomorium*,

Table 3. Ant genera found in each west desert treatment zone and during each sampling period, Utah. [Functional group designations assigned from Nash and others (2001, 2004). G, generalist; P, predator; HT, Homoptera tender; SH, seed harvester; LF, liquid feeder; SM, slave maker.] GC = Grouse Creek zone, IB = Ipabah zone, VE = Vernon zone.

Ant genera	Functional group	GC untreated pretreatment	GC untreated post-treatment	GC treated pretreatment	GC treated post-treatment	IB untreated	IB treated	VE untreated	VE treated
Aphaenogaster	G	X	X	X	X		X	X	X
Camponotus	G	X	X	X	X	X	X	X	X
Cardiocondyla	P			X			X		
Crematogaster	G			X	X		X		
Forelius	HT	X	X	X	X	X	X	X	X
Formica	HT	X	X	X	X	X	X	X	X
Lasius	HT	X	X	X	X	X		X	
Leptothorax	G	X	X	X	X	X	X	X	X
Messor	SH	X			X				
Monomorium	SH							X	X
Myrmecocystus	LF	X	X	X	X	X		X	
Myrmica	P	X	X	X	X	X	X	X	X
Pheidole	SH	X	X	X	X	X	X	X	X
Pogonomyrmex	SH	X	X	X	X	X	X	X	X
Polyergus	SM		X	X	X		X		
Prionopelta	P	X							
Solenopsis	G	X	X	X	X	X	X	X	X
Tapinoma	G	X	X	X	X	X	X	X	
Tetramorium	G			X					

Table 4. Total number of ants collected in treated and untreated zones at Grouse Creek after treatment with diflubenzuron, Utah. [Ibapah and Vernon had the nine most common genera; *IB U9 accounted for 541 of the *Forelius* in untreated traps].

Genus	Grouse Creek after treatment		Ibapah after treatment		Vernon after treatment	
	Untreated	Treated	Untreated	Treated	Untreated	Treated
Forelius	127	442	544	1,396	1	2,239
Formica	963	101	536	158	2,700	396
Leptothorax	197	251	273	101	581	181
Monomorium	0	0	0	0	1,055	1,043
Myrmica	7	24	92	103	497	29
Pheidole	122	238	57	257	314	297
Pogonomyrmex	535	300	19	256	29	14
Solenopsis	109	95	44	42	198	102
Tapinoma	3	5	647	24	597	0
Totals	2,063	1,456	2,212	2,337	5,972	4,301

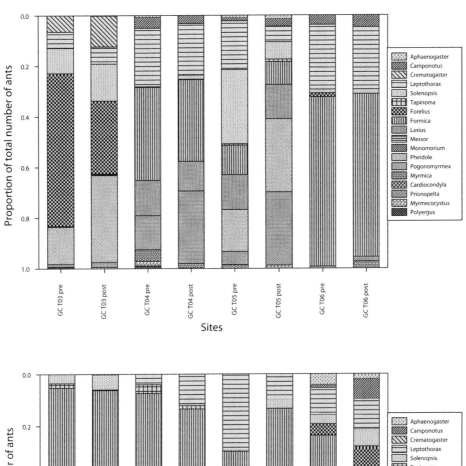
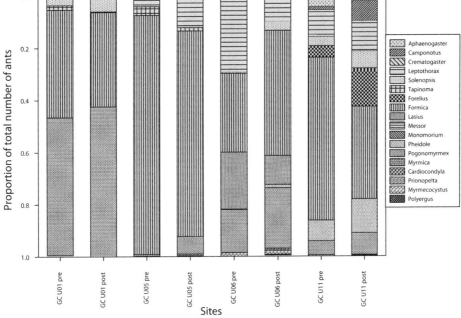

Figure 15. Relative abundance of ant genera at Grouse Creek. Untreated (A) and treated (B) zones before and after treatment.

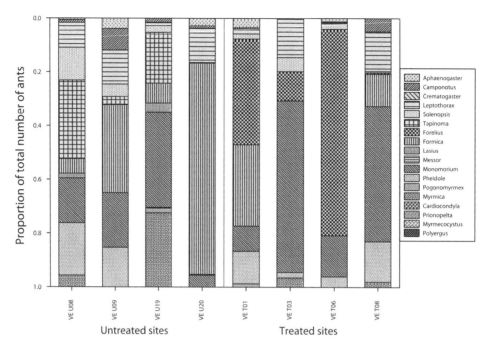

Figure 16. Relative abundance of ant genera at Ibapah untreated and treated zones.

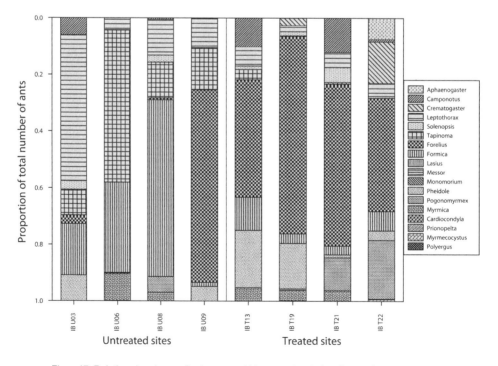

Figure 17. Relative abundance of ant genera at Vernon untreated and treated zones.

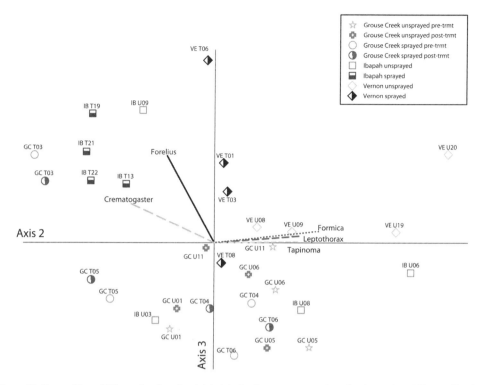

Figure 18. Nonmetric multidimensional scaling joint plot of ant genera as number of ants per day at Grouse Creek, Ibapah, and Vernon. Includes data from before and after treatment at Grouse Creek and the treated and untreated sites of all three areas. Lines indicate genera strongly influencing separation of sites along one or both axes.

another seed-harvesting ant, was found only at Vernon, occurring at all eight sites. On average, *Formica* was more abundant at untreated sites, although numbers varied considerably; it occurred at only two treated sites and ranged from common to uncommon at those sites.

Comparisons Among Study Sites

Most Grouse Creek ant communities were dominated by *Formica* or *Pogonomyrmex* (Figure 15). Ibapah ant communities were characterized by *Tapinoma* and *Formica* at untreated sites and *Forelius* at treated sites (Figure 17). The ant communities at Vernon were distinguished by the presence of *Monomorium* at all eight sites (Figure 17). There were some similarities between Vernon and Ibapah communities that were consistent with treated and untreated zones at both study areas. *Formica* and *Tapinoma* were common to abundant at untreated sites at Vernon and Ibapah, and ants of the genus *Forelius* were very common at most of the treated sites at both study areas.

Ant Community Structure

Ordination (NMS) resulted in a three-dimensional solution providing the best fit and lowest stress (final stress=10.762) and explaining approximately 87 percent of the variation within the dataset. For the sake of graphic simplicity, we present the two-dimensional depiction of axes two and three, which explain about 67 percent of the variation (Figure 18). Axis one was defined primarily by the inverse relationship between *Pogonomyrmex* and *Monomorium*, with some influence from *Solenopsis*. *Monomorium* was found only at

Table 5. Total numbers of ants by genus collected in pitfall traps at each site and sampling event in the three study areas.

Ants	Aphaenogaster	Camponotus	Cardiocondyla	Crematogaster	Forelius	Formica	Lasius	Leptothorax	Messor	Monomorium	Myrmecocystus	Myrmica	Pheidole	Pogonomyrmex	Polyergus	Prionopelta	Solenopsis	Tapinoma	Totals
GC U01pre						72								91			6	3	172
GC U05pre						138								1				5	150
GC U06pre	19	6				200	123	6			8	2		93			2		592
GC U11pre				18	20	285		165	1				36	25			16		456
GC T03pre			1		164	1		47			1	17	41	3			28		275
GC T04pre3	3	16				139	53	18			7			51	4		1		379
GC T05pre	4	5				47	56	88			5	1	67	21			119	2	404
GC T06pre		5			2	88	1	78									2		133
GC U01post						155		34						270			25	1	452
GC U05post						118		1						10		1	1	2	149
GC U06post	18	64			127	380	88	17			12	6	10	182	2	2	27		784
GC U11post	1			42	101	310		77			1	1	112	73			56		866
GC T03post						1		102	1		2	11	120	7			51	1	350
GC T04post		19				189	69	24			1			169	1		10		590
GC T05post	6	12				37	55	120		5		13	118	118			28	4	408
GC T06post		16			1	215		25		3				6			6		341
IB U03		27				34		82					15				33	44	261
IB U06						279	2	107			1	81		1			2	448	858
IB U08		2			2	212		44				11		18			8	43	340
IB U09		2			541	11		44					42				1	112	787
IB T13		57			250	71		78				50	117				7	20	609

Site																Total
IB T19		1	42					25	112	7			10	1		764
IB T21		46		527	24		14	28	3	95			20	3		547
IB T22	56	7		322	12 2	15	16		25	154		1	5			743
VE U08	5	9	108	297	51		34	37	169				105	253		866
VE U09	37	78			48		81		141				45	29		962
VE U19		25			315	57	121		4		460		42	315		1672
VE U20	80	26		1	125	8	22						6			2813
VE T01	39	5		448	2212		357		137	29			22			1140
VE T03		2		66	345		22	21	12	14			32			605
VE T06	26	5		1723			86		87				46			2244
VE T08		20		2	51		13	8	61				2			409
Totals	294	455	210	4594	6165	529	2020	2098	47	772	1429	1438	3	764	1286	22121

the eight sites of the Vernon study area, while *Pogonomyrmex* was common at Grouse Creek and Ibapah but was present at Vernon in small numbers at only two sites. The second axis was structured by the abundance of *Formica*, with *Tapinoma* and *Leptothorax* influencing structure as well. The presence of *Crematogaster*, which occurred at only three sites (GC T03 before and after treatment, IB T19, and IB T22), also influenced scores for NMS axis two and helped separate the latter sites to the left side of the ordination space. Abundance of *Forelius* defined the third axis, with the Vernon and Ibapah treated sites stretched along the third axis primarily according to abundance of *Forelius*. This genus was largely absent from Grouse Creek, being found at only three sites (and rare at two of the three sites) (Table 5). *Forelius* was abundant only at GC T03, and this Grouse Creek site was the only site with positive axis-three scores. Most Grouse Creek unsprayed-zone ant communities were defined by relatively high numbers of *Formica*, *Pogonomyrmex*, and *Leptothorax* and the absence of *Tapinoma*, *Monomorium*, and *Forelius*. Temporal shifts for the Grouse Creek sites did not show any pattern relative to diflubenzuron application, or any other factor that was examined.

Ibapah ant communities separated into sprayed and unsprayed communities primarily on the abundance of *Forelius* at the sprayed sites. IB U09 was near the sprayed-zone sites because it also had a high abundance of *Forelius* (Table 5). The unsprayed sites at Ibapah were scattered throughout the ordination space; each of the four sites had a different ant genus as the most abundant taxon.

The sprayed Vernon sites showed almost no variation along the second axis; their locations were almost entirely based on abundance of *Forelius*. The unsprayed sites had essentially no *Forelius* (one individual at VE U20) and were separated from the sprayed sites on this basis. Abundance

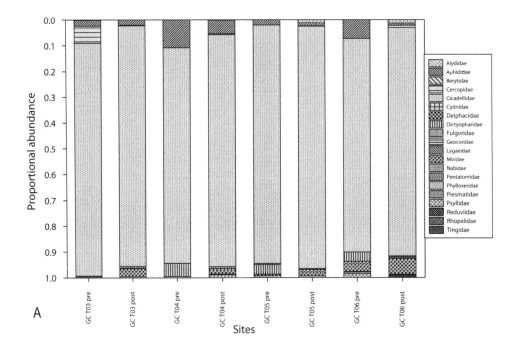

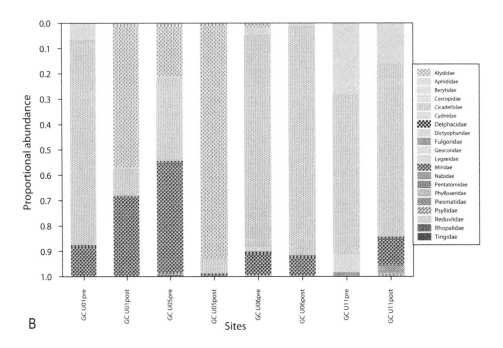

Figure 19. Relative abundances of Hemiptera families at Grouse Creek untreated (A) and treated (B) zones before and after treatment with diflubenzuron.

of *Formica, Tapinoma,* and *Leptothorax* influenced the location of the unsprayed sites at Vernon. All the unsprayed Vernon sites were in the upper right quadrat due to the presence of *Monomorium*, which was found only at Vernon.

The ordination was strongly influenced by the presence of *Forelius*; the location of IB U9 and GC T03 in the same region of ordination space as Ibapah and Vernon sprayed sites strengthens this conclusion. Other than dominance by *Forelius*, there was considerable variation among sites; however, some other genera appear to exhibit trends.

Total numbers of ants differed somewhat between sprayed and unsprayed sites at all three study areas, but the average number of ants in each treatment zone at a study area were not significantly different. However, the abundance of the various genera that made up the "ant" category in the statistical analyses was sometimes quite different among sites and between sprayed and unsprayed zones (Table 4).

Hemiptera

Immature Hemiptera are difficult to identify, even to family, especially by nonexperts. Because most individuals were nymphs, we had large numbers of unidentified Hemiptera in some samples. To address this problem, we estimated total numbers of Hemiptera in each family. The total number of Hemiptera identified in each family and the total of all Hemiptera excluding unknowns were calculated, and the total for each family was divided by the grand total to generate the proportion of identified Hemiptera in each family. The proportion of Hemiptera in each family was multiplied by the total number of unknown Hemiptera caught to estimate the number of nymphs in each family. This value was added to the number of adults to estimate the total number in each family. We recognize that this approach has inherent problems, especially based on the phenology of the different taxa, and that some families might have been represented only by adults, or only by nymphs, at the time of collections. Because the samples at each study area were collected at the same time, we believe any inaccuracies will be consistent across that sampling period.

Grouse Creek

Community composition of Hemiptera varied from site to site and over time at Grouse Creek (Figures 19A and B). Cicadellidae were the most abundant Hemiptera at most sites, but GC U01 and GC U05 sites were dominated by the Alydidae in late June (post-treatment period), representing a dramatic increase in this family from the early June sampling event. Most other families were relatively rare at Grouse Creek sites; Miridae were uncommon to abundant at some untreated sites and increased from pre- to post-treatment collections. The only other family that occurred with any frequency was the Aphididae. The GC U11 Hemipteran community appeared to differ from communities at the other untreated sites.

The treated sites at Grouse Creek also were dominated by the Cicadellidae, even more strongly than in the untreated zone, during both sampling periods (Figure 19*B*). The Alydidae were absent from all treated sites prior to treatment; they were captured in post-treatment samples but in much lower numbers than at post-treatment unsprayed sites. Aphids were again present in all collections and increased across the two collection periods. The Miridae were less frequent in the treated zone and showed only a slight increase from pre- to post-treatment collections, while the increase in Miridae abundance in the untreated zone was large.

The Lygaeidae were absent from both untreated and treated zones prior to treatment at Grouse Creek. No Lygaeids were detected in the treated zone 3 weeks after treatment, but low numbers (averaging 1.9 individuals per day) were detected at each untreated site in the post-treatment collection period. Although there was no difference in numbers of Lygaeidae in the treated zone before and

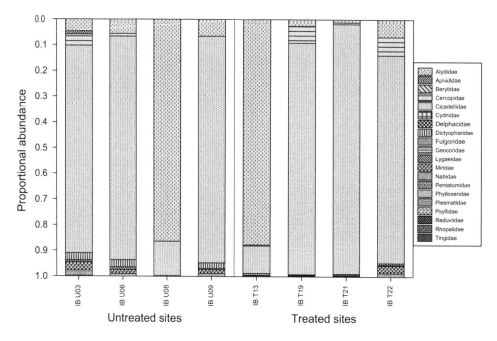

Figure 20. Relative abundances of Hemiptera families at Ibapah treated and untreated sites.

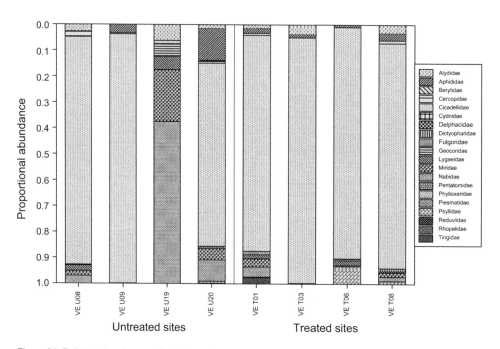

Figure 21. Relative abundances of Hemiptera families at Vernon untreated and treated sites.

after treatment (no individuals were detected), the comparison between post-treatment untreated and treated zones was significantly different ($T_{d.f.\ 6}=26$; $P=0.029$). The phenology of the Lygaeidae may account for the lack of any individuals being detected in pre-treatment samples from either treatment zone; they may emerge later in the season and were not active during the early sampling period.

Ibapah

Cicadellidae dominated six of eight Ibapah sites, with the Alydidae most numerous at IB U08 and IB T13 (Figure 20). The Alydidae were present at all eight Ibapah sites, but each treatment zone had one site where the Alydidae dominated the collection of Hemiptera. IB U08 and IB T13 appeared to have finer soils; greasewood, saltbush, and other shrubby Chenopodeaceae made up a significant portion of the shrub component of the vegetation at these two sites. Sagebrush was relatively scarce at IB U08 but common at IB T13. The Lygaeidae had significantly more individuals in untreated sites than treated sites ($T_{d.f.\ 6}=26$; $P=0.029$).

Vernon

Vernon hemipteran communities differed among sites and were more diverse than Ibapah or Grouse Creek communities (Figure 21), with Cicadellidae dominating some sites; other families, such as the Lygaeidae, Miridae, Nabidae, and Psyllidae, were common. As at Ibapah and Grouse Creek, no pattern in the community structure appeared to be associated with the treatment history at Vernon. The Lygaeidae, which showed indications of sensitivity to diflubenzuron at Grouse Creek and Ibapah, were highly variable at Vernon, but total numbers hinted at sensitivity to diflubenzuron (63 caught in the untreated zone, 32 caught in the treated zone).

Community Structure of Hemiptera

Ordination of numbers per day for hemipteran families did not provide a good solution because the Cicadellidae (at most sites) and the Alydidae (at a few sites) overwhelmed the variation among other families. Data were adjusted by adding 0.01 to each value to eliminate zeros and then log transformed in PC-ORD (McCune and Mefford 2005) to compress high values (as suggested by McCune and Grace 2002) of the Cicadellidae and Alydidae; the Euclidean distance measure was used. Ordination of transformed data resulted in a three-axis solution that explained 89.2 per-cent of the dataset variation; we present the first two axes, which explain 69.9 percent of the variation, for graphic simplicity (Figure 22). The Alydidade, Cercopidae, Miridae, and Nabidae influenced the first axis structure; the second axis was structured by the Aphididae and Psyllidae (negative values) and the Alydidae (positive values). Grouse Creek sites were spread across ordination space with no differences correlated with insecticide application. Seven of the eight Grouse Creek sites showed similar shifts in pre- to post-treatment community structure that were correlated with large increases in Miridae and Cicadellidae. GC U11 had only modest increases in these families (Table 6), and it moved in ordination space differently than the other seven sites (Figure 22). The Ibapah sites were relatively isolated from most Vernon and Grouse Creek sites by the abundance of the Cicadellidae. Within the Ibapah cluster, three treated sites and two untreated sites formed a tight group. IB T13 and IB U08 were somewhat isolated on the basis of abundant Alydidae. The Vernon sites were scattered throughout ordination space with no apparent pattern (Figure 22).

The abundant Hemiptera in this study are phloem-feeding herbivores, such as Alydidae, Cicadellidae, and Aphididae. Many species of Cicadellidae and Aphididae are tended by ants (Buckley 1987, Fischer and Shingleton 2001, Offenberg 2001) in commensal to symbiotic relationships. The Geocoridae and Nabidae consist mostly of generalist predators, but some species may

Table 6. Total numbers of Hemiptera by family collected in pitfall traps at each site and sampling event in the three study areas.

	Alydidae	Aphidae	Berytidae	Cercopidae	Cicadellidae	Cydnidae	Delphacidae	Dictyopharidae	Fulgoridae	Geocoridae	Lygaeidae	Miridae	Nabidae	Pentatomidae	Phylloxeridae	Piesmatidae	Psyllidae	Reduviidae	Rhopalidae	Tingidae	Totals
GC U01pre	83	16			192							29									237
GC U05pre		2		2	126							174	1				1		2		391
GC U06pre	4	5			152			4				17					1				183
GC U11pre		35			77			7	13				2								123
GC U01post	4485	53		28	767			3		2	19	2470	1	1			11				7852
GC U05post	35852	40		3	2349						1	450	15			1				1	38711
GC U06post	3	9		1	749			1			2	66	1	1			1				834
GC U11post		36			147			1		1	1	24	7	1			2				221
GC T03pre	2	24		73	979							4					4				1086
GC T04pre		59			462			28									3				552
GC T05pre		13			610	3		25				4					5				660
GC T06pre		45			515			22		1		24	2		4		9				621
GC T03post	3	25		3	1149			9				40	5								1232
GC T04post		65	1	4	1143			9				27	4				15				1270
GC T05post	11	12			909			2	1			25	5				1				967
GC T06post	11	7		7	734	2		4	1	1		51	3				3	5			829
IB U03	29	8		27	510			17			4	21	13					1			630
IB U06	24	4		4	371			12			4	7	4								426
IB U08	3958				612			1			5		5					1			4582
IB U09	29				391			10			2	6	5								443
IB T13	2006			8	245			16				5	6	1							2287
IB T19	11			27	372			1				1	1								413

specialize on particular taxa or a particular habitat, spider webs, for example (Readio and Sweet 1982, Schuh and Slater 1995). The abundance of Nabidae at some Vernon sites is surprising since predators are usually less abundant than potential prey. Total numbers of arthropods captured at Vernon sites were not particularly high (Table 2d).

Solifugae and Scorpiones

Solifugae were rare at Grouse Creek (1 individual in 960 traps) and Vernon (7 individuals in 480 traps) but were common at Ibapah (74 individuals in 480 traps) and occurred at all 8 sites. Four species of Solifugae (wind scorpions) were recorded in this study. The single specimen at Grouse Creek was *Eremobates ascopulatus*; Vernon's seven specimens included two species, *Hemerotrecha handfordana* and *Eremobates actenidia*. Fifty-five of the wind scorpions from Ibapah were sent for identification, and three species were identified: *H. handfordana* (41), *E. ascopulatus* (10), and *E. corpink* (1), with three immature specimens that could not be identified. The remaining 19 specimens were sent to Jack Brookhart (Denver Museum of Natural Sciences) for identification.

Scorpions, another large, cursorial predator, were common at both Grouse Creek and Ibapah (66 individuals found in 960 traps at Grouse Creek and 30 individuals found in 480 traps at Ibapah) but rare at Vernon (1 individual found in 480 traps). Captures at Ibapah indicated that the association between scorpions and solifugids differed in treated and untreated zones. Regression of the average number of solifugids caught per day compared to the mean number of scorpions caught per day showed a highly significant negative relationship at the untreated sites and a strong positive relationship, though not significant, among the sites treated in 2004 (Figure 23).

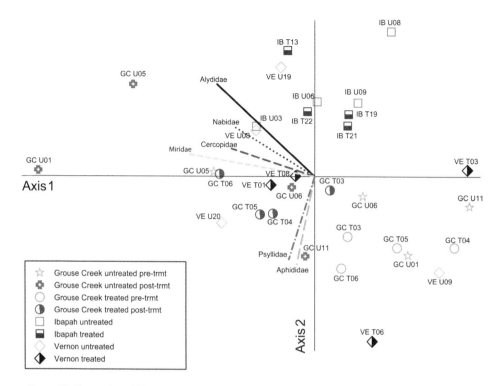

Figure 22. Nonmetric multidimensional scaling ordination of Hemiptera families at the three study areas.

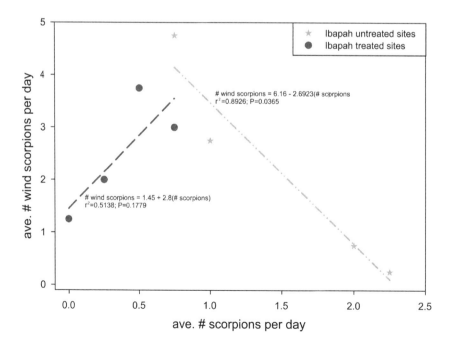

Figure 23. Number of wind scorpions per day versus number of scorpions per day at Ibapah study site.

DISCUSSION

Our results indicate that nontarget arthropods vary considerably in susceptibility to diflubenzuron when used to control Mormon cricket populations in the Great Basin. Sensitivity to diflubenzuron is not necessarily predictable on a taxonomic basis or on an assessment of life history traits. Our data indicate that some taxa assumed not to be susceptible did have population differences that correlated with treatment zones. The importance of identifying specimens to the lowest possible taxonomic level in analyzing results should be emphasized, as conclusions may be very different depending on how the arthropods are classified (for example, counts at family versus genus level). A total of 42 comparisons of untreated abundance versus treated abundance were possible for the three study areas. Nine of these comparisons showed significant differences between treatment zones; another nine comparisons indicated large, but not statistically significant, differences. In all 18 instances, more individuals were collected in the untreated zone.

At the order level, no consistent patterns of difference in proportional representation between treated and untreated sites at any of the three study areas indicate that treatment with diflubenzuron affects nontarget arthropods. Our results are not conclusive, however, because there is tremendous variation between individual sites within each treatment type, and because order-level resolution may mask changes in communities at lower taxonomic levels. For example, shifts between genera or from one family to another may not be detected if one taxon replaces another with a similar number of individuals. In addition, there may have been confounding effects from topography and other issues with the spatial distribution of treated and untreated sites due to the nature of Mormon cricket treatment applications in general.

Studies conducted in South Dakota and Wyoming found diflubenzuron had minimal impacts on nontarget arthropods and their vertebrate predators (Wilcox and Coffey 1978, McEwen et al. 1996), indicating that use of diflubenzuron in a reduced agent-area treatment design (by using less pesticide in alternating swaths) would have less environmental impact than carbaryl or malathion. However, some taxon-specific effects have been recorded. Catangui et al. (2000) reported temporary declines in ants, "flying predators," and "parasites." Weiland et al. (2002) reported that Diptera increased significantly and spider numbers were reduced in treated plots in a Wyoming study. Smith et al. (2006) found that for some applications, numbers decreased at low pesticide-application rates, but abundance was greater at higher pesticide-application rates. Studies indicating diflubenzuron had limited impacts on nontarget arthropods and was effective at low-application rates were conducted in the Great Plains, but additional information that is directly applicable to the environment of Utah's west desert is needed.

Catangui et al. (2000) found that ants in pitfall traps declined by 43 percent 49 to 55 days after treatment, but subsequent sampling periods showed a rebound to pre-treatment numbers. Smith et al. (2006) reported that Formicidae as a family showed mixed responses in a Wyoming study of nontarget arthropod responses to treatments of carbaryl and diflubenzuron at different dose rates and carrier oils. Ants were reduced in diflubenzuron treatments relative to carbaryl treatments using the same carrier oil in 2001, but no significant differences between the two treatments were found in 2002. Weilund et al. (2002) reported no response at the order level by Hymenoptera to diflubenzuron applications in Wyoming.

Grouped at the family level, ants did not show declines in treated zones at any of our study areas. In fact, more ants were collected at Ibapah sprayed sites than at unsprayed sites, and at Grouse Creek more ants were collected in the sprayed zone 3 weeks after diflubenzuron application than were collected prior to treatment. However, when examined at the genus level, differences in abundance were correlated with diflubenzuron treatment. Some genera (for example, *Forelius*) had higher numbers in sprayed zones, while the abundance of other genera (for example, *Tapinoma*) was lower in sprayed zones. Potential indicators in the ant communities we sampled include genera that appear to decrease in response to diflubenzuron treatment and those that increase. *Formica* and *Tapinoma* tended to have lower numbers in treated zones, while *Forelius* and perhaps *Pheidole* tended to increase in treated zones.

Ants perform a number of important ecological functions, especially in arid and semiarid ecosystems (Greenslade 1976, Risch and Carroll 1982, Lobry de Bruyn and Conacher 1990, Lobry de Bruyn and Conacher 1994, Nash et al. 1998, Whitford et al. 1999). Predictions of the sensitivity of ants to environmental perturbations and restorations and, thus, their value as bioindicators have not been consistently supported (Bestelmeyer and Wiens 1996, Whitford et al. 1999, Read and Andersen 2000, Nash et al. 2001, Andersen et al. 2002, Andersen and Majer 2004, Nash et al. 2004). Because ants perform significant functions in ecosystems, it is important to understand how individual species and collective units at functional group and community levels respond to disturbance, and how they recover from perturbation in the short and long term.

Diflubenzuron may be a factor in developing ant community structure at Ibapah and Vernon. However, pesticide treatments could be correlated with other environmental factors we did not measure that made these sites favorable for *Forelius* and, perhaps, less favorable for *Formica* and *Tapinoma*. There are potential confounding effects of topography based on where and how diflubenzuron was applied. Ibapah treated sites were all south of untreated sites, and most were a little higher in elevation. These conditions were reversed at Vernon, however, where untreated sites were higher and south of the treated area; elevation differences were greater at Vernon than at Ibapah. Similarities among ant communities from treated zones at Ibapah and Vernon, despite environmental differences between them, suggest a role for diflubenzuron in structuring ant communities in the treated zones. Untreated-zone communities at Ibapah and Vernon study areas differed; each untreated site within the Ibapah and Vernon study areas had different ant communities (Figure 18). Convergence of communities in the treated zones indicates a possible link to diflubenzuron treatment. Additional research is needed, especially a study designed to address the possibility of lag effects of diflubenzuron on the ant communities, and ants should be identified to species level to elucidate ecological functions that might be affected by changes in ant community structure in response to diflubenzuron application.

Non-ant Hymenoptera (including bees and predatory and parasitic wasps) were significantly lower in treated zones at Grouse Creek and Vernon. Coleoptera demonstrated the same pattern as non-ant Hymenoptera, with significantly fewer beetles at treated sites at Vernon and a trend toward more beetles at untreated sites in the Ibapah study area. Flies also were significantly reduced at Vernon and showed a trend toward declining numbers at treated sites at Ibapah. Catangui et al. (2000) reported that diflubenzuron reduced the number of flying predators by 59 percent 15 to 20 days after treatment, however, numbers rebounded in subsequent samples. Parasite numbers also were reduced (by about 18 percent) 35 to 41 days after treatment, but abundance returned to near-control levels after 41 days (Catangui et al.

2000). Flying insects classified as parasites included both Hymenoptera and Diptera; flying insects classified as predators included Hymenoptera, Diptera, and Coleoptera (Catangui et al. 2000).

Hemiptera showed large differences in abundance among sites, and large changes over time were observed at Grouse Creek. Some differences were correlated with treatment zone; more Hemiptera and more of particular families were found in the untreated zones in all three study areas. Smith et al. (2006) found the Hemiptera had mixed responses to various treatments, but some diflubenzuron treatments did reduce Hemiptera numbers.

Within the Hemiptera, there was some indication that the Lygaeidae (seed bugs), in particular, were sensitive to diflubenzuron treatments. Lygaeidae were absent at Grouse Creek from all late May and early June samples, probably because of phenology. No seed bugs were found in any traps from treated sites during the late June collection period, but they were caught consistently, albeit at low numbers, in the untreated zone during this sampling period. The seed bugs at Grouse Creek and Ibapah showed a trend toward reduced numbers in the treated zones; differences in numbers of seed bugs between zones were significant at both study areas. There were large differences in total numbers of seed bugs at Vernon (63 in the untreated zone, 32 in the treated zone), but the differences were not significant because of high variability among sites. Seed bugs represented a relatively small component of the arthropod community at all three study areas. The ecological impact of changes in Lygaeidae numbers, if they are affected by diflubenzuron, is probably relatively minor given the small population of bugs that will be affected.

Scorpions and solifugids are large, nocturnal predators that forage on the ground. Scorpions and solifugids compete for food (Polis and McCormick 1986a, Polis et al. 1989). Polis and McCormick (1986a) found that scorpions and solifugids also prey on each other. Thus, it is not surprising to find an inverse correlation between scorpion and solifugid abundance. This correlation was seen among the untreated sites at Ibapah. The association is reversed among plots treated with diflubenzuron in 2004. The data indicate that, rather than some interaction between scorpions and solifugids in the presence of diflubenzuron, scorpions may simply be more sensitive to diflubenzuron than solifugids. There was a dramatic decline in scorpion numbers in treated sites compared to untreated sites, while solifugid abundance remained unchanged. In the four untreated areas, 34 solifugids were found in traps, and 40 solifugids were caught in the traps in the treated zone. Only 6 scorpions were caught within the treated zone, while 24 scorpions were trapped in the untreated zone.

Scorpion life histories may make them more susceptible to exposure to diflubenzuron. Young scorpions molt more frequently than older individuals (Polis and McCormick 1986b); if more young individuals are present during treatment periods, the numbers could be affected. Another possibility is that scorpions' prey may have been affected more than the solifugids' prey. Thus, scorpion numbers declined, but solifugid numbers remained relatively unchanged in the treatment zone. At least one group of large, potential prey showed concomitant declines at the treated sites at Ibapah; Orthoptera were roughly one-third as abundant at treated sites as at untreated sites (62 total at the four treated sites compared to 172 total at the four untreated sites, Figure 16H). The relative importance of Orthoptera in scorpion and solifugid diets should be explored to test this hypothesis.

Scorpions eat solifugids and solifugids eat scorpions. In the Coachella Valley of California, Polis and McCormick (1986a) found the diet of the scorpion *Paruroctonus mesaensis* consisted of up to 14.4 percent solifugids and up to 65.4 percent intraguild prey (spiders, scorpions, and solifugids). However, in an experiment in which scorpions were removed from experimental plots and the numbers of solifugids and spiders were sampled spring and fall for two years, no significant increases in numbers or size of solifugids were observed in the removal plots compared to the control plots where scorpions were not removed (Polis and McCormick 1986b). In our study, solifugid numbers did not differ in treated and untreated zones despite a fourfold difference in scorpion numbers, which is consistent with the findings of Polis and McCormick (1986b) that indicate solifugids do not show a numerical response to reduction or removal of a potential competitor/predator.

The importance of scorpions in Great Basin sagebrush communities has not been well studied, but they can be extremely abundant in some systems (Polis and McCormick 1986a) and may play a significant role in population regulation of other arthropods in arid and semiarid ecosystems. Research is needed to clarify the impact of scorpions on the arthropod community, especially experiments in which scorpions are removed from large plots and other arthropods are monitored.

Numbers of beetles in the families Carabidae and Tenebrionidae were reduced shortly after diflubenzuron treatment (Catangui et al. 2000). Smith et al. (2006) reported that tenebrionid beetles (collected in pitfall traps) and Coleoptera (collected with sweep nets) showed some differences in response to a variety of carbaryl and diflubenzuron treatments. Coleoptera collected in sweep nets had the most consistent response to diflubenzuron and carbaryl, with fewer beetles caught in all insecticide applications compared to the control plots. We found significant reductions in beetles at Vernon and markedly fewer beetles at Ibapah, indicating there may be long-term effects on beetles from diflubenzuron application in the Great Basin.

Ants and beetles are major components of sage grouse chick diets, as are many Orthoptera (Klebenow and Gray 1968, Peterson 1970, Pyle and Crawford 1996, Drut et al. 1994). Few food-habit studies of sage grouse chicks have been done, and many lump invertebrates found in bird stomachs into very basic categories (for example, "beetles" or "worms"); thus, it is difficult to assess from these studies the potential effect of changes in arthropod communities from diflubenzuron application on sage grouse population dynamics. Because chicks concentrate on insects and other arthropods during the first month after hatching, even a temporary decline in numbers, such as those reported by Catangui et al. (2000), could have a significant impact on sage grouse chick survival. Sample et al. (1993) found that five species of songbirds in diflubenzuron-treated sites had significantly different and potentially less nutritious diets compared to songbirds in untreated sites, and two species in treated sites had significantly less invertebrate biomass in their stomachs. Whitmore et al. (1993) found seven of nine bird species tested on diflubenzuron-treated plots had lower fat levels than those on untreated plots due to reductions in invertebrate prey populations, increased foraging costs, reduction in food quality, or some combination of the three. Bell and Whitmore (1997) reported lower numbers of birds of most species in plots that had been treated with diflubenzuron, and they attributed this to a reduction in habitat quality related to prey availability in treated forest plots.

Our temporal comparisons, while covering the critical insectivore phase of sage grouse chick life history (Drut et al. 1994), did not yield definitive results because of the extreme variability in sites and taxa response to diflubenzuron and the uncertainty of sage grouse chick food preferences. However, diflubenzuron application to control Orthoptera has the potential to affect sage grouse chick foraging by altering prey-species composition and/or abundance. Survival of chicks has been shown to be critical for sustainable sage grouse populations (Johnson and Boyce 1990). Additional work is needed to clarify whether diflubenzuron affects sage grouse chick survival and, thus, sage grouse population dynamics in areas where Mormon cricket or grasshopper control occurs.

CONCLUSIONS

Our study design included several treated and untreated zones to facilitate statistical comparisons. Because a limited outbreak of Orthoptera occurred in 2005, only one area was treated with diflubenzuron, thus, severely limiting our ability to detect the effects of the pesticide spraying.

The effects of diflubenzuron on arthropod communities are not apparent in our data from Grouse Creek, the only area treated in 2005. The treatment was designed to avoid spraying pesticide on water bodies, and no measurable impacts on aquatic community composition, richness, or abundance on either springs or streams were observed, with the exception of reduced taxa richness at Vernon (a result confounded by elevational differences in the treatment and nontreatment zones). Our study did indicate that treatment with diflubenzuron was correlated with changes in abundance for some terrestrial taxa, notably some ant genera, the Lygaeidae (Hemiptera), non-ant Hymenoptera, beetles (Coleoptera), and scorpions (Scorpiones).

Important ecosystem functions (for example, seed predation and pollination) are performed by the arthropods showing reduced abundance during this study, and, thus, ecological function could be adversely affected by declines caused by diflubenzuron application. Many of these taxa are used by sage grouse chicks at a critical stage in their development; sustainability of sage grouse populations could be indirectly affected by use of diflubenzuron in sage grouse brood habitat. Differences between sprayed and unsprayed zones were greater at Ibapah and Vernon when sampled a year after diflubenzuron application, suggesting that the effect may lag behind application. Although direct effects may still occur, the potential for indirect effects increases greatly. Differences in abundance may not be observed at higher taxonomic levels (for example, order or family) for some taxa; thus, work to evaluate the effects of diflubenzuron on nontarget arthropods should include identification of arthropods to at least family, and for ecologically or taxonomically diverse groups (for example, ants), identification should be to genus and species when possible.

Although few apparent short-term effects of diflubenzuron on terrestrial arthropods at Grouse Creek were observed that were statistically significant, mean abundances of some taxa at Ibapah and Vernon were significantly different at untreated sites than at sites treated with diflubenzuron the previous year, and nearly significant differences were observed at all three study areas. The same taxa differed over several study areas. Sometimes the differences were statistically significant and sometimes they were not. These taxa, which included Coleoptera, Diptera, Hemiptera, non-ant Hymenoptera, Lepidoptera, Orthoptera, and Scorpiones, may be more susceptible to diflubenzuron. Additional research targeting these taxa would be informative. Funding should be sought to identify specimens of the taxa collected in this study to the lowest taxonomic level possible (at least to genus, preferably to species). This finer resolution may show which taxa are actually affected by pesticide spraying. For example, the mean number of ants (as a group) did not differ for any comparison of the treated and untreated zones, but there were significant differences for some genera. The same could be true for Coleoptera, Diptera, or other taxa.

To determine whether diflubenzuron application caused the observed differences in arthropod communities, a study should be designed to control for environmental differences. Ideally, an area where Mormon cricket-control is judged to be needed should be divided into randomly assigned treatment and nontreatment blocks and sampled extensively the year prior to treatment to provide quantitative baseline data. Sampling should also occur just before treatment and at several intervals after treatment from about 3 weeks to at least 18 months.

Control efforts will continue to affect nontarget arthropods as long as diflubenzuron is the insecticide of choice; the current application methods are likely the most effective for Orthoptera control and do not lend themselves to avoiding particular patches in the treated areas. Although additional research is needed to clarify the suspected relationships identified in this study, we recognize that efforts to control Orthoptera on rangelands in the Intermountain West will continue. We suggest that the potential impacts of diflubenzuron treatment discussed here be considered in future decisions regarding control efforts. The value of Orthopteran population control must be weighed against the potential direct and indirect effects on ecosystem structure and functioning that may result from changes in arthropod community structure through shifts in sensitive taxa.

ACKNOWLEDGMENTS

We would like to thank Mike Freeman, Annie Caires, Patrick Milliman, Tyler Cuff, James Hereford, Lisa Bryant, and Lovina Abbott for their help in digging pitfall traps, collecting trap contents, and collecting aquatic samples. Evelyn Cheng, Julius Schoen, Jessie Doherty, Kristen Shelburg, Rebecca Allmond, April Johnson, John Barry, and Steve Deakins sorted and identified terrestrial arthropod samples. Sandy Brantley identified some spiders to start a reference collection; Jack Brookhart identified the Solifugae. Aquatic samples were identified and enumerated by Rhithron in Missoula, Montana. Gery Wakefield and Karen Hansen provided GIS support, including generating random points and making many maps. Paul Lukacs generated the pitfall trap layout in WebSim. Project design benefited from discussions with Greg Abbott, Lisa Bryant, Clint Burfitt, and comments from two reviewers of the proposal. Funding for fieldwork was provided by the Bureau of Land Management. Funding to process samples, identify specimens, and analyze data was provided by the Utah Department of Agriculture and Food, and the U.S. Geological Survey Southwest Biological and Utah Water Science Centers.

REFERENCES

Andersen, A. N., B. D. Hoffmann, W. J. Mueller, and A. D. Griffiths. 2002. Using ants as bioindicators in land management: Simplifying assessment of ant community responses. Journal of Applied Ecology 39(1):8–17.

Andersen, A. N., and J. D. Majer. 2004. Ants show the way down under: Invertebrates as bioindicators in land management. Frontiers in Ecology and Environment 2(6):291–298.

Bell, J. L., and R. C. Whitmore. 1997. Bird populations and habitat in *Bacillus thuringiensis* and dimilin-treated and un-treated areas of hardwood forest. American Midland Naturalist (137):239–250.

Bestelmeyer, B. T., and J. A. Wiens. 1996. The effects of land use on the structure of ground-foraging ant communities in the Argentine Chaco. Ecological Applications 6(4):1225–1240.

Buckland, S. T., D. R. Anderson, K. P. Burnham, J. L. Laake, D. L. Borchers, and L. Thomas, eds. 2001. Introduction to distance sampling: Estimating abundance of biological populations. Oxford University Press, Oxford, United Kingdom.

Buckley, R. 1987. Ant-plant-homopteran interactions. Advances in Ecological Research 16:53–85.

Catangui, M. A., B. W. Fuller, and A. W. Walz. 2000. VII.3 Impact of Dimilin® on nontarget arthropods and its efficacy against rangeland grasshoppers. In Grasshopper Integrated Pest Management User Handbook, edited by G. L.

Cunningham and M. W. Sampson, pp. 1–5. U.S. Department of Agriculture, Animal and Plant Health Inspection Service, Washington, D.C.

Drut, M. S., W. H. Pyle, and J. A. Crawford. 1994. Diets and food selection of sage grouse chicks in Oregon. Journal of Range Management 47:90–93.

Fischer, M. K., and A. W. Shingleton. 2001. Host plant and ants influence the honeydew sugar composition of aphids. Functional Ecology 15(4):544–550.

Greenslade, P. J. M. 1976. The meat ant *Iridomyrmex purpureus* (Hymenoptera: Formicidae) as a dominant member of ant communities. Australian Journal of Entomology 15(2):237–240.

Johnson, G. D., and M. S. Boyce. 1990. Feeding trials with insects in the diet of sage grouse chicks. Journal of Wildlife Management 54:89–91.

Klebenow, D. A., and D. M. Gray. 1968. Food habits of juvenile sage grouse. Journal of Range Management 21:80–83.

Lobry de Bruyn, L. A., and A. J. Conacher. 1990. The role of termites and ants in soil modification: A review. Australian Journal of Soil Research 28:55–93.

Lobry de Bruyn, L. A., and A. J. Conacher. 1994. The bioturbation activity of ants in agricultural and naturally vegetated habitats in semi-arid environments. Australian Journal of Soil Research 32:555–570.

Lukacs, P. M. 2001. Estimating density of animal populations using trapping webs: Evaluation of web design and data analysis. Colorado State University, Fort Collins, Colorado.

Lukacs, P. M. 2002. WebSim: Simulation software to assist in trapping web design. Wildlife Society Bulletin 30: 1259–1261.

Lukacs, P. M., A. B. Franklin, and D. R. Anderson. 2004. Passive approaches to detection in distance sampling. In Advanced Distance Sampling: Estimating abundance of biological populations, edited by S. T. Buckland et al., pp. 260–280. Oxford University Press, Oxford, United Kingdom.

McCune, B., and J. B. Grace. 2002. Analysis of ecological communities. MjM Software, Gleneden Beach, Oregon.

McCune, B., and M. J. Mefford. 2005. PC-ORD, multivariate analysis of ecological data, version 5.0, 5th ed. MjM Software Design, Gleneden Beach, Oregon.

McEwen, L. C., C. M. Althouse, and B. E. Peterson. 1996. Direct and indirect effects of grasshopper integrated pest management (GHIPM) chemicals and biologicals on nontarget animal life. U.S. Department of Agriculture, Animal and Plant Health Inspection Service Technical Bulletin 1809, Section III.2.

Nash, M. S., D. F. Bradford, S. E. Franson, A. C. Neale, W. G. Whitford, and D. T. Heggem. 2004. Livestock grazing effects on ant communities in the eastern Mojave Desert, USA. Ecological Indicators 4(3):199–213.

Nash, M. S., W. G. Whitford, D. F. Bradford, S. E. Franson, A. C. Neale, and D. T. Heggem. 2001. Ant communities and livestock grazing in the Great Basin, U.S.A. Journal of Arid Environments 49(4):695–710.

Nash, M. S., W. G. Whitford, Z. J. Van, and Kris Havstad. 1998. Monitoring changes in stressed ecosystems using spatial patterns of ant communities. Environmental Monitoring and Assessment 51(1-2):201–210.

New, T. R. 1998. Invertebrate Surveys for Conservation. Oxford University Press, Oxford, United Kingdom.

Offenberg, Joachim. 2001. Balancing between mutualism and exploitation: The symbiotic interaction between *Lasius* ants and aphids. Behavioral Ecology and Sociobiology 49(4):304–310.

Peterson, J. G. 1970. The food habits and summer distribution of juvenile sage grouse in central Montana. Journal of Wildlife Management 34:147–155.

Polis, G. A., and S. J. McCormick. 1986a. Scorpions, spiders, and solpugids: Predation and competition among distantly related taxa. Oecologia 71(1): 111–116.

Polis, G. A., and S. J. McCormick. 1986b. Patterns of resource use and age structure among species of desert scorpion. Journal of Animal Ecology 55(1):59–73.

Polis, G. A., C. A. Myers, and R. Holt. 1989. The ecology and evolution of intraguild predation: Potential competitors that eat each other. Annual Review of Ecology and Systematics 20:297–330.

Prior-Magee, J. S., K. G. Boykin, D. F. Bradford, W. G. Kepner, J. H. Lowry, D. L. Schrupp, K. A. Thomas, and B. C. Thompson, eds. 2007. Southwest Regional Gap Analysis Project final report. U.S. Geological Survey, Moscow, Idaho.

Pyle, W. H. and J. A. Crawford. 1996. Availability of foods of sage grouse chicks following prescribed fire in sagebrush-bitterbrush. Journal of Range Management 49:320–324.

Read, J. L., and A. N. Andersen. 2000. The value of ants as early warning bioindicators: Responses to pulsed cattle grazing at an Australian arid zone locality. Journal of Arid Environments 45(3):231–251.

Readio, J., and M. H. Sweet. 1982. A review of the Geocorinae of the United States east of the 100th meridian (Hemiptera: Lygaeidae). Miscellaneous publications of the Entomological Society of America 12:1–91.

Risch, S. J., and C. R. Carroll. 1982. The ecological role of ants in two Mexican agro-ecosystems. Oecologia 55:114–119.

S-PLUS, 2000. Seattle, MathSoft, Inc.

Sample, B. E., R. J. Cooper, and R. C. Whitmore.

1993. Dietary shifts among songbirds from a diflubenzuron-treated forest. The Condor 95:616–624.

Schuh, R. T., and J. A. Slater. 1995. True Bugs of the World (Hemiptera: Heteroptera): Classification and natural history. Cornell University Press, New York.

Smith, D. I., J. A. Lockwood, A. V. Latchininsky, and D. E. Legg. 2006. Changes in non-target arthropod populations following application of liquid bait formulations of insecticides for control of rangeland grasshoppers. International Journal of Pest Management 52:125–139.

Systat. 2004. SigmaStat. Richmond, California, Systat Software, Inc.

Triplehorn, C. A., and N. F. Johnson. 2005. Borror and DeLong's introduction to the study of insects, 7th ed. Thomson Brooks/Cole, Belmont, California.

U.S. Department of Agriculture. 2002. Rangeland grasshopper and Mormon cricket suppression program. Animal and Plant Health Inspection Service, U.S. Department of Agriculture.

Weiland, R. T., F. D. Judge, Teun Pels, and A. C. Grosscurt. 2002. A literature review and new observations on the use of diflubenzuron for control of locusts and grasshoppers throughout the world. Journal of Orthoptera Research 11(1):43–54.

Whitford, W. G., J. Van Zee, M. S. Nash, W. E. Smith, and J. E. Herrick. 1999. Ants as indicators of exposure to environmental stressors in North American desert grasslands. Environmental Monitoring and Assessment. 54(2):143–171.

Whitmore, R. C., R. J. Cooper, and B. E. Sample. 1993. Bird fat reductions in forests treated with dimilin. Environmental Toxicology and Chemistry 12: 2059–2064.

Wilcox, H. III, and T. Coffey Jr. 1978. Environmental impacts of diflubenzuron (Dimilin) insecticide. Forest Service, U.S. Department of Agriculture.

YELLOW-BILLED CUCKOO DISTRIBUTION AND HABITAT ASSOCIATIONS IN ARIZONA, 1998–1999: FUTURE MONITORING AND RESEARCH IMPLICATIONS

Matthew J. Johnson, Robert T. Magill, and Charles van Riper III

ABSTRACT

In 1998 and 1999, we surveyed throughout Arizona areas of suitable habitat in all known riparian drainages historically occupied by Western Yellow-billed Cuckoos (*Coccyzus americanus occidentalis*). Of the 30-odd drainages identified with historical Yellow-billed Cuckoo detections, 26 drainages had at least one Yellow-billed Cuckoo detection during the 1998–1999 surveys. In 1998 we completed 97 individual surveys resulting in 166 Yellow-billed Cuckoo detections within 22 drainages. Yellow-billed cuckoos were detected in that field season were mainly located along 5 drainages: Cienega Creek, Sonoita Creek, San Pedro River, Bill Williams River, and the Verde River. In 1999, our survey effort increased, and we completed 169 individual surveys resulting in 404 Yellow-billed Cuckoo detections within 47 drainages. The majority of Yellow-billed Cuckoo detections in 1999 were along the San Pedro River, Verde River, and Cienega Creek. Although the Agua Fria River was not surveyed in 1998, we detected a high number of Yellow-billed Cuckoos there in 1999. Sonoita Creek had a high number of Yellow-billed Cuckoo detections in 1998, whereas in 1999 it only accounted for 6 percent of all detections. In 1999, we had 14 drainages with no Yellow-billed Cuckoo detections. To evaluate our Yellow-billed Cuckoo survey method, we visited sites either once, twice, or three times. During the 1998 and 1999 Yellow-billed Cuckoo breeding season, we found that if a Yellow-billed Cuckoo was detected during the first survey, the probability of detecting Yellow-billed Cuckoos during the second and third surveys was very high.

During both years of this study we found 85 percent of all Yellow-billed Cuckoo detections in native habitat (>75% native species), dominated by cottonwood (*Populus* spp.), willow (*Salix* spp.), and mesquite (*Prosopis* spp.). Yellow-billed cuckoo detections in the mixed native habitat (51–75% native species) were dominated by cottonwood, mixed with willow and tamarisk (*Tamarix* spp.). A smaller percent of Yellow-billed Cuckoo detections (5%) occurred in the mixed-exotic category (51–75% exotic species), which was dominated by tamarisk; however, cottonwood was present at all but two Yellow-billed Cuckoo detection sites within this category. In addition, riparian habitat at sites with Yellow-billed Cuckoo detections of >5 had a greater surface area (100 m wide) at its widest point of the drainage than did sites with <5 cuckoo detections.

INTRODUCTION

The Yellow-billed Cuckoo (*Coccyzus americanus*), a neotropical migrant, spends early June through early September in northern Mexico, the United States, and southern Canada and winters primarily in South America (Hughes 1999). Yellow-billed cuckoos begin arriving in Arizona in late May and early June with the majority arriving in mid to late June (Bent 1940, Hughes 1999, Cormon and Wise-Gervais 2005). Nesting activities usually occur between late June and early July, but can begin as early as late May and continue through late September (Hughes 1999, Laymon et al. 1997). In Arizona, nesting peaks between mid July and early August, later than for most co-occurring bird species (Cormon and Wise-Gervais 2005). The timing of nesting may be triggered by an abundance of cicadas (Cicadidae), katydids (Tettigoniidae), caterpillars (Lepidoptera), and other large prey items (e.g., tree frogs [*Hyla* spp.]), which are the bulk of the species' diet (Hamilton and Hamilton 1965, Rosenberg et al. 1982, Hughes 1999).

Western Yellow-billed Cuckoos (*C. a. occidentalis;* i.e., populations west of the continental divide) have historically bred in riparian zones from western Washington to northern Mexico, including Oregon, southwestern Idaho, California, Nevada, Utah, western Colorado, Arizona, New Mexico, and western Texas (American Ornithologists' Union 1983, 1998). Comparisons of historical and current information suggest the Yellow-billed Cuckoo's range and population numbers have declined substantially across much of the western United States over the past 80 years (USFWS 2002). Analysis of population trends is difficult because quantitative data, including historic population estimates, are generally lacking. However, rough extrapolations based on both observed densities of Yellow-billed Cuckoos and historical habitat distribution indicate that western populations were once substantially larger than they are today (USFWS 1985, USFWS 2002).

In a 25 July 2001 finding by the U.S. Fish and Wildlife Service (USFWS), the western Yellow-billed Cuckoo was declared to represent a distinct population segment, and as such warranted protection under the Endangered Species Act as "threatened." However, the Yellow-billed Cuckoo was classified as a Candidate Species under the Endangered Species Act, because there was "sufficient information on their biological status and threats to propose them as endangered or threatened under the ESA, but for which development of the proposed listing regulation is precluded by other higher priority listing activities" (USFWS 2002). The Arizona Game and Fish Department (AGFD) has designated the Yellow-billed Cuckoo as Wildlife of Special Concern in the state (AGFD 2002), and the U.S. Forest Service Regional Forester designated it a Sensitive Species on National Forests (USDA 2000) within Arizona. In addition, the species is considered likely to become an endangered species throughout its range on the Navajo Nation (Navajo Nation 2005).

Factors that probably contributed to population declines in the West are the loss, fragmentation, and alteration of native riparian breeding habitat; the possible loss of wintering habitat; and pesticide use on breeding and wintering grounds (Gaines and Laymon 1984, Franzreb 1987, Laymon and Halterman 1987, Hughes 1999). Local extirpations and low colonization rates may have contributed to the declines (Laymon and Halterman 1989). Populations may be further restricted by limited food availability, and birds may not nest if food supply at the breeding grounds is inadequate (Veit and Petersen 1993). Food availability for the Yellow-billed Cuckoo is likely affected by drought conditions (Newton 1980, Durst 2004, Scott et al. 2004).

Western Yellow-billed Cuckoos require structurally complex riparian habitats with tall trees and a multi-storied vegetative understory (Hughes 1999, Johnson et al. 2008). This bird is known to breed in large blocks of riparian habitat (5–20 ha), particularly in woodlands dominated by cottonwoods (*Populus* spp.) and willows (*Salix* spp.) (Ehrlich et al. 1988, USFWS 2002a, Johnson et al. 2008).

Arizona Historical Abundance

In Arizona, the Yellow-billed Cuckoo was once considered a fairly common species throughout the state, breeding within riparian forests that were dominated by cottonwood, willow, and/or mesquite (*Prosopis* spp.) (Stephens 1903, Swarth 1905, Swarth 1914, Visher 1910, Phillips et al. 1964). A 1977 statewide Arizona survey of suitable habitat found an estimated 205–214 pairs, with more than half of these along the lower Colorado River (Gaines and Laymon 1984). Past estimates suggested that <200 pairs remained in 1986 (Layman and Halterman 1987) and that <50 pairs were present five years later (Ehrlich et al. 1992).

Prompted by continued concern regarding severe population declines, habitat loss, and the lack of statewide data, the USFWS asked the U.S. Geological Survey, in conjunction with the AGFD, to initiate Yellow-billed Cuckoo surveys to (1) determine statewide distribution of Yellow-billed Cuckoos in Arizona on public lands and (2) identify habitats associated with the presence of Yellow-billed Cuckoos.

METHODS

Study Site Selection

We selected 107 specific Yellow-billed Cuckoo survey sites prior to the initial survey season using an *a priori* knowledge of the bird, including habitat preferences, expert opinion, and basic biology of this species. This survey method identifies suitable habitats before conducting surveys (Bibby et al. 1992). We selected sites primarily where historical detections occurred prior to 1998, which is a preferred method for surveying rare birds (Dawson 1981) when resources (e.g., time) are limited. To help identify riparian areas where potentially suitable Yellow-billed Cuckoo habitat might occur we used information from the AGFD's Habitat Data Management System.

Because breeding Yellow-billed Cuckoos require large continuous areas of intact habitat (Laymon et al. 1997), we focused our 107 survey efforts within 29 large drainages (>0.2 ha) with dense broadleaf deciduous vegetation or large intact mesquite bosques. However, because of the fragmented nature of many Arizona riparian zones and because there are records of Yellow-billed Cuckoos occupying smaller patches (Johnson et al. 2008), smaller areas were also considered. We did not survey areas smaller than two ha, as this was considered an absolute minimum size for Yellow-billed Cuckoo occupancy; no Yellow-billed Cuckoos have been detected attempting to nest in patches that size or smaller in Arizona or California (Laymon et al. 1997). Dominant broadleaf deciduous species of interest included Fremont cottonwood (*Populus fremontii*) and Goodding willow (*Salix gooddingii*). Co-dominant species included Arizona sycamore (*Plantanus wrightii*), velvet ash (*Fraxinus pennsylvanica* ssp. *velutina*), seep willow (*Baccharis salicifolia*), netleaf hackberry (*Celtis reticulata*), Arizona alder (*Alnus oblongifolia*), and Arizona walnut (*Juglans major*). Exotic species included tamarisk (*Tamarix* spp.) and Russian olive (*Eleagnus angustifolia*). We identified survey sites using topographical maps and aerial photography.

A museum and literature search was conducted to determine where Yellow-billed Cuckoos had been historically documented. Curators of museum and university collections throughout the United States were contacted for information on Yellow-billed Cuckoo specimens collected in Arizona.

Yellow-billed Cuckoo Survey Methods

We modified a survey protocol used in California (Laymon, S. A., Yellow-billed cuckoo survey and monitoring protocol for California, unpublished report, 1998) to determine the statewide distribution of Yellow-billed Cuckoos in Arizona. We reduced the number of within-site surveys (three surveys per site) at some of the sites to increase the number of overall locations that could be surveyed within such a large geographic area and because our study was basically designed to collect baseline data. Surveys were conducted along 29 drainages historically associated with the presence of Yellow-billed Cuckoos in addition to those with no records of previous historical surveys. All counts were conducted between 31 May and 29 August 1998, and between 5 June and 2 September 1999. The timing of the surveys coincided with the peak of the Yellow-billed Cuckoo's breeding season in Arizona (Hamilton and Hamilton 1965, Nolan and Thompson 1975, Hughes 1999). Surveyors were trained prior to the breeding season to standardize field techniques.

We used a taped recording of the Yellow-billed Cuckoo's *kowlp* call (Hughes 1999) during surveys. Playback equipment was capable of projecting this call at least 100 m with a minimum of distortion. Surveys were conducted from half an hour before sunrise until 11:00 a.m., and were terminated if shade temperatures exceeded 41° C or during steady rainfall. One transect (i.e., a series of points, 100 m apart, from which the tape was broadcast) was made through the habitat for every 200 m of habitat width. We bypassed areas of unsuitable habitat (e.g., a monoculture of young, short tamarisk, or an extensive cobble bar) between patches. To be excluded from the survey, unsuitable habitat had to be at least 300 m in length.

The surveyor initially stopped at a survey point and remained quiet for one minute to acclimate to the ambient noise. If no Yellow-billed Cuckoos were heard in this one-minute period, the surveyor then played the *kowlp* call once, followed by one minute of silence to listen for a response. If no detections occurred, this playback-listen sequence was repeated an additional four times. The surveyor then moved 100 m along the transect (by foot or by boat) and repeated the playback-listen protocol.

At each of the 107 survey sites we recorded Universal Transverse Mercator coordinates of the survey site boundaries (including start and stop points) and provided a description of the habitat and surrounding area. When a Yellow-billed Cuckoo was detected, the surveyor recorded its location and its use of the habitat patch, types of vocalization, and any apparent presence of nesting, and, if so, the stage of nesting. The surveyor also attempted to observe and record any other Yellow-billed Cuckoos present, possible interactions between individuals, and any apparent breeding behavior (e.g., nest building, active nest, food delivery to young). The interpretation of these behaviors was later used to ascertain breeding status.

Habitat Measurements

Visual estimates of the dominant vegetation and habitat class along the survey site were made at the time of the Yellow-billed Cuckoo survey to identify associations between the habitat type and presence of Yellow-billed Cuckoos. Habitats were classified according to the percentage of native and exotic dominant tree species: native habitats contained >75 percent native species; mixed native had 51–75 percent native species; mixed exotic habitats had 51–75 percent exotic species; and exotic habitats had >75 percent exotic species. In addition, vegetation classification according to the dominant tree species com-prising the canopy and sub-canopy were recorded.

RESULTS
Museum and Literature Search

Museum and literature searches identified a number of major drainages in Arizona where Yellow-billed Cuckoos had been detected prior to the initiation of this study (Table 1). The majority of the records documented Yellow-billed Cuckoos in the central, western, and southeastern portions of the state along perennial drainages below the elevation of 1,500 m. Of the drainages with historically known detections of Yellow-billed Cuckoos, only the Teec Nos Pos

Table 1. Yellow-billed cuckoo historical (prior to 1998) detection locations along riparian corridors in Arizona. This information was compiled from the published literature and museum records.

Historical location	Historical location
San Juan River	Gila River - upper cont.
Colorado River – above Lake Mead	Pima
Colorado River – below Lake Mead	Ft. Thomas
Bill Williams River	Geronimo
Big Sandy River	San Francisco River
Santa Maria River	San Pedro River – lower
Hassayampa River	Bass Canyon
Agua Fria River	Cascabel
Verde River	lower Aravaipa Canyon
Oak Creek	San Manuel Crossing
Wet Beaver Creek	Aravaipa Creek confluence
Tonto Creek	Cook's Lake
Little Colorado River	PZ Ranch
Salt River	Dudleyville Crossing
Gila River – lower	San Pedro River – upper
Salt confluence	Garden Canyon
Agua Fria confluence	Ramsey Canyon
W. of Airport Road	Palominas
Buckeye	Hwy. 92 – Hereford
W. of Hwy. 85	Hwy. 90
Hassayampa confluence	Lewis Springs
Gila Bend	Charleston
Painted Rock Dam	Boquillas Ranch
Dome Valley	Curtis Windmill
W. of Fortuna Wash	Saint David
Gila River – middle	Cave Creek Canyon
Winkelman	San Bernardino Valley
near Grayback Mtn.	Sulphur Springs Valley
Whitlow Dam – Queen Cr.	Babocomari River
Picacho Reservoir	Sonoita Creek
Sacaton	Sycamore Canyon
Santa Cruz confluence	Altar Valley
Gila River – upper	Arivaca Cienega
Bonito Creek	Arivaca Creek near Arivaca
Sanchez Road	Arivaca Creek
Safford	Santa Cruz River
Thatcher	Cienega Creek

Wash and Sulphur Springs Valley were not surveyed, since we determined that they lacked adequate Yellow-billed Cuckoo habitat. Of the 30-odd major drainages identified with historical Yellow-billed Cuckoo detections, 26 had at least one Yellow-billed Cuckoo detection during our 1998–1999 surveys.

1998–1999 Yellow-billed Cuckoo Detections

During the 1998 field season, 72 sites in 22 drainages were surveyed between 31 May and 29 August (Table 2; Figure 1). We completed 97 individual surveys resulting in 166 Yellow-billed Cuckoo detections at 46 of the 72 sites (64%). The mean detection

Table 2. Total number of yellow-billed cuckoos detected during 1998–1999 surveys along riparian corridors in Arizona (NS = No survey conducted).

Drainage	1998	1999	Total
San Pedro River	28	86	114
Cienega Creek	35	39	74
Verde River	20	40	60
Agua Fria River	NS	45	45
Sonoita Creek	20	15	35
Gila River	5	28	33
Bill Williams River	21	11	32
Clear Creek	3	20	23
Hassayampa River	NS	23	23
Big Sandy River	5	11	16
Wet Beaver Creek	2	11	13
Arivaca Creek	1	11	12
Santa Cruz River	NS	10	10
Santa Maria River	5	2	7
Arivaca Cienega	NS	6	6
Roosevelt Lake	NS	6	6
San Bernardino Valley	NS	6	6
Penitas Wash	NS	5	5
Picacho Reservoir	NS	4	4
San Luis Wash	NS	4	4
Lower Colorado River	3	1	4
San Francisco River	4	0	4
Babocomari River	4	NS	4
Oak Creek	1	2–23	3
O'Donnell Creek	3	NS	3
Sycamore Canyon	NS	3	3
Granite Creek	NS	2	2
Pajarito Canyon	NS	2	2
Champurrado Wash	0	2	2
Lindberg Tank Wash	NS	2	2
Blue River	2	NS	2
Little Colorado River	NS	2	2
Salt River	NS	2	2
Burro Creek	1	NS	1
Bitter Well Wash	NS	1	1
Tonto Creek	1	0	1
Pinal Creek	1	NS	1
Virgin River	1	NS	1
Huachuca Creek	NS	0	0
Red Tank Draw	NS	0	0
Fossil Creek	NS	0	0
Waddell Dam	NS	0	0
Queen Creek	NS	2	2
Cerro Prieto Wash	NS	0	0
Sabino Creek	NS	0	0
Sawmill Canyon	NS	0	0
Date Creek	NS	0	0
Leslie Canyon	NS	0	0
Chevelon Crossing	NS	0	0
Cave Creek, Chiricahuas	NS	0	0
Seven Springs	NS	0	0
Romero Wash	NS	0	0
Total	166	404	570

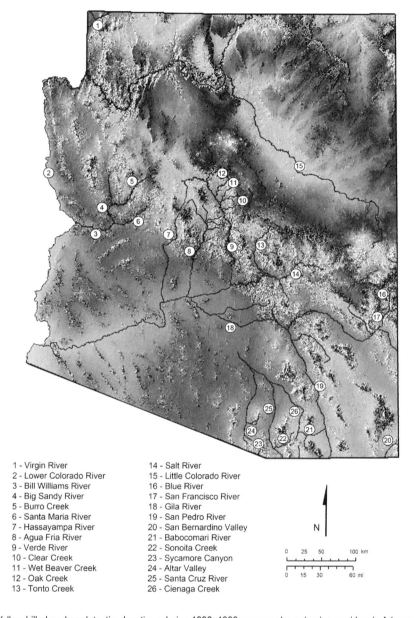

Figure 1. Yellow-billed cuckoo detection locations during 1998–1999 surveys along riparian corridors in Arizona.

date in 1998 was 26 July. The majority of Yellow-billed Cuckoos detected in 1998 were located along five drainages; Cienega Creek (22%) San Pedro River (17%), Bill Williams River (13%), Sonoita Creek (12%), and Verde River (12%). All drainages surveyed in 1998, except for Champurrado Wash, had at least one Yellow-billed Cuckoo detection.

During the 1999 field season our survey effort increased and we covered 134 sites in 46 drainages between 5 June and 2 September 1999 (Table 2; Figure 1). We completed 169 individual surveys resulting in 404 Yellow-billed Cuckoo detections at 88 of the 134 sites (65%). The mean detection date in 1999 was 9 July. Yellow-billed cuckoo detections in 1999 were also centered along the San Pedro River (21%), Verde River (10%) and Cienega Creek (10%). Agua Fria was a site not surveyed in 1998, but in 1999 we detected a high number of Yellow-billed Cuckoos, accounting for 11 percent of all detections that year. Sonoita Creek had a high number of Yellow-billed Cuckoo detections in 1998, whereas in 1999 it only accounted for 6 percent of all Yellow-billed Cuckoo detections. In 1999, we found 14 (31%) drainages with no Yellow-billed Cuckoo detections.

Combining both 1998 and 1999, we surveyed 53 drainages of which 39 drainages had at least one Yellow-billed Cuckoo detection. During this study, 16 drainages were surveyed in both years. Of those 16 drainages, 13 had Yellow-billed Cuckoo detections in both years. Those 13 drainages accounted for 75 percent of all Yellow-billed Cuckoo detections during this study.

Yellow-billed Cuckoo Detection Rates

In our evaluation of the survey protocol, we found the majority (72%) of Yellow-billed Cuckoo detections were solicited through broadcast. The number of solicited detections peaked during the middle of July and then declined as the breeding season progressed. We found that 74 percent of Yellow-billed Cuckoo detections, solicited or unsolicited, were aural, and 26 percent were visual detections. We found that areas with the largest Yellow-billed Cuckoos populations had the highest rate of vocalizations.

During the 1998 and 1999 breeding season, we found that if a Yellow-billed Cuckoo was detected during the first survey, the probability of detecting Yellow-billed Cuckoos during the second and third surveys was high. Eighty-three percent of the sites with Yellow-billed Cuckoo detections during the first survey had Yellow-billed Cuckoos detections during the second survey, and 78 percent of those sites with detections during the first survey had detections during the third survey. In contrast, if a Yellow-billed Cuckoo was not detected during the first survey, but detected during the second survey, only 43 percent of those sites had Yellow-billed Cuckoo detections during the third survey. Overall, if detections occurred during the first visit, 71 percent of the time detections occurred during all three visits ($n = 26$). In general, Yellow-billed Cuckoos tend to occupy a site and remain in that location for the breeding season.

Yellow-billed Cuckoo Habitat Type

Yellow-billed cuckoos were detected primarily within three habitat classes: native, mixed native, and mixed exotic habitats. These classes were subsequently divided into 17 habitat types according to the dominant vegetation (Table 3). We combined both the 1998 and 1999 data to summarize Yellow-billed Cuckoo detections per habitat type. We found 85 percent of all Yellow-billed Cuckoo detections within the native habitat category (Table 3). Within the native habitat category, we found that habitat dominated by cottonwood with a strong association of willow and mesquite (45%) was the most common habitat composition in which Yellow-billed Cuckoo were detected, followed by cottonwood with willow, but without mesquite (17%), followed by a cottonwood-willow-ash mix (14%). We found 10 percent of all Yellow-billed Cuckoo detections were within mixed native habitats (Table 3). Within the mixed native habitat category, 89 percent of the Yellow-billed Cuckoo detections occurred in areas dominated by cottonwood with a willow and tamarisk mix. Only 5 percent of all Yellow-billed Cuckoo detections in this study occurred within mixed exotic habitats, which were dominated by tamarisk; yet cottonwood was present at all but two Yellow-billed Cuckoo detection sites within this category. No Yellow-billed Cuckoo

Table 3. Habitat classes and vegetation dominance, comprising the canopy and sub-canopy associated with Yellow-billed Cuckoo detections and cuckoo surveys with zero detections for all sites surveyed in 1998 and 1999.

Habitat Class	Yellow-billed Cuckoo Detections	Yellow-billed Cuckoo Surveys with No Detections
Native	1998–1999	1998–1999
Cottonwood-Willow-Mesquite	216 (45%)	27
Cottonwood-Willow	82 (17%)	28
Cottonwood-Willow-Ash	66 (14%)	8
Cottonwood-Willow-Sycamore	48 (10%)	12
Cottonwood Dominated Mix	23 (5%)	12
Mesquite Dominated Mix	20 (4%)	12
Willow Dominated Mix	12 (2%)	8
Ash Dominated Mix	9 (2%)	7
Sycamore Dominated Mix	7 (1%)	9
Black Walnut Dominated Mix	3 (<1%)	0
Subtotal	485 (85%)	123
Mixed Native		
Cottonwood-Willow-Tamarisk	51 (89%)	15
Cottonwood-Russian Olive	2 (3.5%)	0
Cottonwood-Willow-Mesquite-Tamarisk	2 (3.5%)	3
Willow-Tamarisk	2 (3.5%)	3
Subtotal	57 (10%)	21
Mixed Exotic		
Tamarisk-Cottonwood-Willow-Mesquite	24 (86%)	8
Tamarisk-Cottonwood	2 (7%)	0
Tamarisk-Willow	2 (7%)	0
Subtotal	28 (5%)	8
1998–1999 Total	570	152

detections occurred in exotic habitats in our 1998 or 1999 surveys. We are unable to report specific selections because we did not enumerate the availability of each habitat type.

DISCUSSION

Our 1998–1999 surveys included almost all areas historically occupied by Yellow-billed Cuckoo and other potentially suitable sites in Arizona. During these surveys we had 532 Yellow-billed Cuckoo detections along 38 drainages in Arizona. The majority of Yellow-billed Cuckoo detections were on the San Pedro River, Cienega Creek, Verde River, Sonoita Creek, and the Agua Fria River. Previous studies documented Yellow-billed Cuckoos along many of the same drainages that we surveyed (Hamilton and Hamilton 1965, Groschupf 1987, Halterman 1998, Hughes 1999); however, the detections were not all along the same reaches. Rea (1983) reported Yellow-billed Cuckoos present along the Gila River from the confluence with the Santa Cruz River to its confluence with the Hassayampa River. Groschupf (1987) reported Yellow-billed Cuckoos were present along the length of the Santa Cruz River from 1970 to 1986. Yellow-billed cuckoos were once considered abundant throughout the riparian floodplain along the lower Colorado River. Grinnell and Miller (1944) cited only Stephen's (1903) observations of several cuckoos near Needles in 1902.

Surveys in mid-June 1964 along the lower Colorado River near Laguna Dam indicated the abundance of Yellow-billed Cuckoos was similar to, and possibly higher than, that on the San Pedro River in southeastern Arizona (Hamilton and Hamilton 1965).

Given the lack of consistent long-term Yellow-billed Cuckoo survey information and the challenges of interpreting historical Yellow-billed Cuckoo detection data, we cannot make precise comparisons between the historical and current numbers of breeding Yellow-billed Cuckoos with this current study. However, there are probably substantially fewer Yellow-billed Cuckoos and fewer breeding sites (less habitat) now than what occurred in Arizona during the early 1900s, and possibly fewer than during the 1970s and 1980s.

Historical Water Diversion and Landscape Changes along the Lower Colorado River

Somewhere between 85 and 98 percent of Arizona's native riparian habitat has been reduced or degraded since Euro-American settlement (Noss et al. 1995). This began with the construction of a series of dams: Laguna Dam in 1907, Hoover Dam in 1936, Parker Dam and Imperial Dam in 1938, and Davis Dam in 1954. Dam operations changed the natural flows of the lower Colorado River by ending the cycle of annual flooding, except when heavy runoff from local rains produced floods from the larger tributaries (e.g., Bill Williams River). Without these floods, new, rich alluvial seedbeds were no longer formed and the replacement cycle of cottonwoods, willows, and mesquites were changed, ultimately eliminating the vast majority of Yellow-billed Cuckoo habitat. With floods controlled and irrigation water readily available, large stands of natural floodplain vegetation were converted to agricultural uses. In the 1940s and 1950s wide portions of the floodplain near Yuma, Blythe, Parker, and Needles were cleared for agriculture. Extensive farm tracts, "clean" farming practices, and shifts to crops such as cotton and lettuce resulted in the removal of large tracts of cottonwood-willow forests and mesquite bosques and greatly reduced the extent of wildlife habitat, including habitat required by Yellow-billed Cuckoos to breed (Rosenberg et al. 1991). The only large tracts of natural riparian vegetation that remained through the 1970s were in five Native American nations and four national wildlife refuges.

Historical Habitat Changes in Arizona

Knowledge of habitat-selection patterns and identification of potential breeding habitat is essential to guide conservation efforts (Laymon 1998, Hughes 1999). Our study has demonstrated that Yellow-billed Cuckoos are found in sites containing relatively large areas of native deciduous riparian habitat, at least 100 m wide, with dominant tree species comprising mainly cottonwood, willow, and mesquite. In addition, Yellow-billed Cuckoos are more likely to occupy riparian habitat with adjacent patches of mesquite (Holmes et al 2008).

Habitat within areas where Yellow-billed Cuckoos were absent, few, or solitary—such as some sites along Tonto Creek, Salt River, Gila River, and many sites along the lower Colorado River—lacked a multi-structure understory with a dominant cottonwood canopy, and were dominated by exotic vegetation, had no standing water, or was noticeably fragmented or influenced by adjacent land practices (e.g., agriculture, urbanization). Agriculture practices and urban development have greatly contributed to the fragmentation of the remaining riparian areas in Arizona (Anderson and Ohmart 1984, Younker and Anderson 1986).

Arizona's native riparian habitat has been reduced over the last century (Noss et al. 1995), resulting in the fragmentation of riparian forest tracts, which are progressively reduced to smaller and more isolated patches embedded within a relatively permanent matrix of largely unsuitable habitat (Saab 1999). Western Yellow-billed Cuckoos may be especially sensitive to fragmentation, as it appears that they require tracts of large contiguous and unfragmented patches. In California, sites larger than 80 ha in extent and wider than 600 m were found to be the optimal patch size for Yellow-billed Cuckoos (Laymon and Halterman 1989).

Agricultural lands currently dominate much of the riparian landscape within many regions in Arizona, particularly along the lower Colorado River, and agricultural development on adjacent lands affects riparian bird communities. While studying habitat use by breeding birds in cottonwood riparian forests along the South Fork of the Snake River in southeastern Idaho, Saab (1999) found riparian patches surrounded by an agriculture matrix supported different bird assemblages than did patches surrounded by a natural habitat matrix. In addition, avian nest predators, brood parasites, and exotic species all prospered in the human-altered landscapes resulting from agricultural development, fragmentation, residential areas, or all three factors (Saab 1999). Of 55 sites we surveyed for Yellow-billed Cuckoos along the lower Colorado River in Arizona in 2006 and 2007, 65 percent were bordered on at least one side by agriculture fields (Johnson et al. 2008).

Changes to avian and vegetative communities have the potential to influence Yellow-billed Cuckoo habitat use. In arid regions, Yellow-billed Cuckoos are restricted to river bottoms, ponds, swampy areas, and damp thickets with relatively high humidity (Gaines and Laymon 1984, Hughes 1999). Most breeding pairs of western Yellow-billed Cuckoos have been found nesting in riparian patches within 100 m of water (Laymon and Halterman 1987; Johnson et al. 2006a, 2006b, 2007). Surface water in these cottonwood-willow groves may help lower the air temperature via evaporative cooling (Laymon and Halterman 1987, Hughes 1999), which provides the optimal microclimatic breeding conditions for Yellow-billed Cuckoos.

One factor that drastically changed riparian vegetation structure and composition throughout many parts of Arizona was the introduction of exotic tamarisk, changing riparian ecosystem processes that initially promoted its establishment and persistence. In 1894, Mearns (1907) estimated that native riparian vegetation covered between 160,000 and 180,000 ha of alluvial bottomland between Fort Mohave and Fort Yuma (lower Colorado River). As of 1986, only about 40,000 ha of riparian vegetation remained (Anderson and Ohmart 1984, Younker and Anderson 1986). About 40 percent of the area remaining in 1986 was dominated by tamarisk, 16 percent by honey mesquite and/or native shrubs, and only 0.7 percent by mature cottonwood or willow habitat (Ohmart et al. 1988). Tamarisk and Russian olive are currently the third and fourth most frequently occurring woody riparian plants in the Southwest (Friedman et al. 1998). Although tamarisk was the most common tree in all of our study sites, we found that Yellow-billed Cuckoo occupancy rates in Arizona were highest in sites dominated by native tree species and lower in habitats consisting of mixed native or >75 percent tamarisk cover.

In summary, multiple factors such as water diversion and redistribution as well as agricultural and urban development have impacted riparian habitats and landscapes throughout Arizona in the previous 50 to100 years. While riparian habitats have been lost, fragmented, and degraded, we found that Yellow-billed Cuckoos persist in the region mainly in riparian habitats that are still dominated by native vegetation. Other factors, such as the presence of surface water, microclimate conditions, and landscape-level habitat features, may also play a role in Yellow-billed Cuckoo habitat selection.

Probability of Detecting Yellow-billed Cuckoos using Tape Playback Recordings

Even though the probability of detecting Yellow-billed Cuckoos on any given survey of a site is less than 1.0, our data confirm that conducting multiple surveys (3 surveys) using tape playback recordings within an area increases the probability of detecting resident Yellow-billed Cuckoos (i.e., individuals that are consistently in an area throughout the breeding season) as the breeding season progresses. Even though broadcasting vocalizations generally increases Yellow-billed Cuckoo detections, no survey methodology can guarantee an absolute determination of presence or absence, especially on any given survey. Evidence that birds may be present but not responding includes our observation that on some occasions when no detections were made at a survey point during the broadcast and listening period, a Yellow-billed Cuckoo vocalized several minutes after a surveyor left that point (but was still close enough to detect the bird calling). Also, Halterman (2005) conducted a test of the playback survey method, in which one researcher carried out a normal survey while another researcher with telemetry equipment located and monitored the response of birds to playback. Only 50 percent of birds responded to the broadcast calls, suggesting that any single survey may miss birds present at a site.

The probability of detecting secretive species such as the Yellow-billed Cuckoo may depend on season, stage of nesting, and sex of the bird(s) present. In a study of Bicknell's thrushes (*Catharus bicknelli*), birds called and sang consistently during the day in early to mid June, but later in the season songs were infrequent and calls were concentrated at dawn and dusk (Rimmer et al. 1996). Some researchers have found that birds may be less likely to vocalize in response to playback after nesting has started (Sogge et al. 1997, Legare et al. 1999, Bogner and Baldassarre 2002). We found that aural Yellow-billed Cuckoo detections (both solicited and unsolicited) peaked somewhat during the middle of the breeding season, but declined thereafter.

The type of song a bird uses may vary through the breeding season (Ritchison et al. 1988) and using different calls to elicit response may be one way to increase detection probabilities. The Yellow-billed Cuckoo has a diverse repertoire, which most commonly includes the *kowlp* call but also includes a variety of *kuks*, *coos*, and *rattles*. Playback during our surveys only used recordings of the *kowlp* call, and although this helps to ensure consistency across survey areas, it may fail to elicit responses from some birds.

RESEARCH NEEDS

Continue Yellow-billed Cuckoo Presence-Absence Surveys

If our study were repeated at least once every 10 years the data gathered would update current knowledge on Yellow-billed Cuckoo distribution in Arizona. Including measurements on available habitat in future surveys would help identify the species' habitat selection. These data could be used to identify selected riparian habitat for Yellow-billed Cuckoos in Arizona. One way to improve the reliability of the surveys, would be to conduct three Yellow-billed Cuckoo surveys within the study year using established protocols (Halterman, M. D., et al., Western Yellow-billed Cuckoo natural history summary and survey methodology, unpublished report, 2006), including at least one survey during the August or September breeding season.

Using Occupancy Models to Identify Core Breeding Yellow-billed Cuckoo Habitat

There is a need to identify core breeding habitat areas and their characteristics for use as a basis for future habitat expansion through riparian habitat enhancement and restoration efforts. Occupancy might be a reliable method of habitat-quality assessment (MacKenzie et al. 2002; Halterman, M. D., et al., Western Yellow-billed Cuckoo natural history summary and survey methodology, unpublished report, 2006). Particular sites could be classified based on duration of occupancy (i.e., occupancy rate) and monitored across years. Occupancy could then be correlated with productivity and/or with some other measure of site or habitat quality (Sergio and Newton 2003).

Identifying Yellow-billed Cuckoo Habitat Model Variables

Knowledge of habitat selection patterns and identification of potential breeding habitat is essential to guide conservation efforts (Laymon 1998, Hughes 1999). Including additional habitat characterization measures during future Yellow-billed Cuckoo surveys would be helpful for planning processes. Data on plant species composition, vegetation structure within riparian patches, riparian patch size, and characterization of the surrounding landscape matrix could be analyzed for selection patterns and used to develop predictive models for breeding-season habitat occupancy. This information could help predict the effects of riparian restoration and other management options.

ACKNOWLEDGEMENTS

We would like thank the Arizona Game and Fish Department, the Colorado Plateau Research Station, the U. S. Fish and Wildlife Service, and the University of Arizona for funding and logistical support. We thank all of the museums that provided us with information on the Yellow-billed Cuckoos in their collections or granted us the opportunity to visit their collections, including: the American Museum of Natural History; California Academy of Sciences; Cornell University; Los Angeles County Museum; Museum of Vertebrate Zoology, Berkeley; National Museum of Natural History, Smithsonian Institution; San Bernardino County Museum; San Diego Natural History Museum; University of Arizona; University of Kansas, Lawrence; University of Michigan, Ann Arbor; University of New Mexico, Albuquerque; and Western Foundation of Vertebrate Zoology, Los Angeles. We also thank all of our field assistants for their dedication and hard work during the surveys throughout Arizona, including: Alison Cariveau, Richard Bush, Kristen Pearson, Jan Hart, Marg Lomow, L.B. Meyers, Chris O'Brien, Susanne Oehlschlaeger, Joelle Viau, Angela Wartell, and Brian Wooldridge.

LITERATURE CITED

American Ornithologists' Union. 1983. Check-list of North American Birds: The species of birds of North America from the Arctic through Panama, including the West Indies and Hawaiian Islands. American Ornithologists' Union, Washington, D.C.

American Ornithologists' Union. 1998. Checklist of North American Birds, 6th ed. American Ornithologists' Union, Washington, D.C.

Anderson, B. W., and R. D. Ohmart. 1984. A vegetation management study for the enhancement of wildlife along the lower Colorado River. Final report submitted to United States Bureau of Reclamation, Boulder City, NV.

Arizona Game and Fish Department. 2002. Heritage Data Management System. http://www.gf.state.az.us/frames/fishwild/hdms.

Bent, A. C. 1940. Life Histories of North American Cuckoos, Goatsuckers, Hummingbirds, and their Allies. Dover Publications, New York.

Bibby, C. J., N. D. Burgess, and D. A. Hill. 1992. Bird Census Techniques. Academic Press, New York.

Bogner, H. E., and G. A. Baldassarre. 2002. The effectiveness of call-response surveys for detecting least bitterns. Journal of Wildlife Management 66:976–984.

Corman, T. E., and C. Wise-Gervais. 2005. Arizona Breeding Bird Atlas. University of New Mexico Press, Las Cruces.

Dawson, D. G. 1981. Experimental design when counting birds. Studies in Avian Biology 6:392–398.

Durst, S. L. 2004. Southwestern willow flycatcher potential prey base and diet in native and exotic habitats. Master's thesis, Northern Arizona University, Flagstaff.

Ehrlich, P. R., D. S. Dobkin, and D. Wheye. 1988. The Birders Handbook: A Field Guide to the Natural History of North American Birds. Simon and Schuster, New York.

Ehrlich, P. R., D. S. Dobkin, and D. Wheye. 1992.

Birds in Jeopardy. Stanford University Press, San Francisco, California.

Franzreb, K. 1987. Perspectives on managing riparian ecosystems for endangered bird species. Western Birds 18:10–13.

Friedman, J. M., W. R. Osterkamp, M. L. Scott, and G. T. Auble. 1998. Downstream effects of dams on channel geometry and bottomland vegetation: regional patterns in the Great Plains. Wetlands 18:619–633.

Gaines, D., and S. A. Laymon. 1984. Decline, status, and preservation of the yellow-billed cuckoo in California. Western Birds 15:49–80.

Grinnell, J., and A. Miller. 1944. The distribution of the birds of California Pacific Coast. Avifauna 18.

Groschupf, K. 1987. Status of the yellow-billed cuckoo (*Coccyzus americanus occidentalis*) in Arizona and West Texas. Contract 20181-86–00731, U.S. Fish and Wildlife Service, Phoenix, Arizona.

Halterman, M. D. 1998. Population status, site tenacity and habitat requirements of the yellow-billed cuckoo at the Bill William's River, Arizona: Summer 1998. Report for U.S.D.I. Bureau of Reclamation Lower Colorado Regional Office, Boulder City, NV, and U. S. Fish and Wildlife Service, Bill Williams River National Wildlife Refuge, Parker, Arizona.

Halterman, M. D. 2005. Surveys and life history studies of the yellow-billed cuckoo: Summer 2004. Administrative report to the Bureau of Reclamation, Boulder City, Nevada.

Hamilton, W. J., and M. E. Hamilton. 1965. Breeding characteristics of yellow-billed cuckoos in Arizona. In Proceedings of the California Academy of Sciences. Fourth Series, pp. 405–432. Vol. XXXII, No.14.

Holmes, J. A., M. J. Johnson, and C. Calvo. 2008. Yellow billed cuckoo distribution, habitat use and breeding ecology in the Verde Watershed of Arizona, 2003–2004. Final Report, Arizona Game and Fish Heritage Program, Phoenix, AZ.

Hughes, J. M. 1999. Yellow-billed cuckoo (*Coccyzus americanus*). In The Birds of North America, edited by A. Poole and F. Gill, The Birds of North America, Inc., Philadelphia, Pennsylvania.

Johnson, M. J., J. A. Holmes, and R. Weber. 2006a. Final report: Yellow-billed cuckoo distribution and abundance, habitat requirements, and breeding ecology in select habitats of the Roosevelt Habitat Conservation Plan, 2003–2006. Report submitted to Salt River Project (SRP), Phoenix, AZ; USGS, Southwest Biological Science Center, Colorado Plateau Research Station, Flagstaff, Arizona.

Johnson, M. J., J. A. Holmes, R. Weber, and M. Dionne. 2006b. Yellow-billed cuckoo distribution, abundance and habitat use along the lower Colorado and Gila Rivers in La Paz and Yuma Counties, 2005. Report submitted to Arizona Game and Fish Heritage Program, Bureau of Land Management, Bureau of Reclamation, and Northern Arizona University, Flagstaff, Arizona.

Johnson, M. J., S. L. Durst, C. M. Calvo, L. Stewart, M. K. Sogge, G. Bland, and T. Arundel. 2008. Yellow-billed cuckoo distribution, abundance, and habitat use along the lower Colorado River and its tributaries, 2007 Annual Report: U.S. Geological Survey Open-File Report 2008–1177.

Laymon, S. A. 1998. Yellow-billed cuckoo (*Coccycus americanus*). In The Riparian Bird Conservation Plan: A strategy for reversing the decline of riparian-associated birds in California. California Partners in Flight. http://www.prbo.org/calpif/htmldocs/riparian.

Laymon, S. A., and M. D. Halterman. 1987. Can the western subspecies of the yellow-billed cuckoo be saved from extinction? Western Birds 18:19–25.

Laymon, S. A., and M. D. Halterman. 1989. A proposed habitat management plan for yellow-billed cuckoos in California. In U.S. Department of Agriculture Forest Service General Technical Report PSW–110, pp. 272–277.

Laymon, S. A., P. L. Williams, and M. D. Halterman. 1997. Breeding status of the yellow-billed cuckoo in the South Fork Kern River Valley, Kern County, California: Summary report 1985–1996. Prepared for U.S. Department of Agriculture, Forest Service, Sequoia National Forest, Cannell Meadow Ranger District. Challenge Cost-Share Grant 92–5–13.

Legare, M. L., W. R. Eddleman, P. A. Buckley, and P. Kelly. 1999. The effectiveness of tape playback in estimating black rail density. Journal of Wildlife Management 63:116–125.

MacKenzie, D. I., J. D. Nichols, G. B. Lachman, S. Droege, J. A. Royle, and C. A. Langtimm. 2002. Estimating site occupancy when detection probabilities are less than one. Ecology 83:2248–2255.

Mearns, E. A. 1907. Mammals of the Mexican boundary of the United States. A descriptive catalogue of the species of mammals occurring in that region, with a general summary of the natural history, and a list of trees. Smithsonian Institution, U.S. Natural Museum Bulletin 56, U.S. Government Printing Office, Washington, D.C.

Navajo Nation. 2005. Navajo Endangered Species List. Navajo Fish and Wildlife Department. http://nnhp.navajofishandwild life.org/nnhp_nesl.pdf.

Newton, I. 1980. The role of food in limiting bird numbers. Ardea 68:11–30.

Nolan, V., Jr., and C. F. Thompson. 1975. The occurrence and significance of anomalous reproductive activities in two North American non-parasitic cuckoos *Coccyzus* spp. Ibis 117:496–503.

Noss, R. F., E. T. Laroe III, and J. M. Scott. 1995. Endangered ecosystems of the United States: A preliminary assessment of loss and degradation. National Biological Service, Biological Report 28.

Ohmart, R. D., B. W. Anderson, and W. C. Hunter. 1988. The ecology of the lower Colorado River from Davis Dam to the Mexico-United States International Boundary: A community profile. U.S. Fish and Wildlife Service Biological Report 85 (7.19).

Phillips, A., J. Marshall, and G. Monson. 1964. The Birds of Arizona. University of Arizona Press, Tucson.

Rea, A. 1983. Once a River. University of Arizona Press. Tucson.

Rimmer, C. C., J. L. Atwood, K. P. McFarland, and L. R. Nagy. 1996. Population density, vocal behavior, and recommended survey methods for bicknell's thrush. Wilson Bulletin 108:639–649.

Ritchison, G., P. M. Cavanagh, J. R. Belthoff, and E. J. Sparks. 1988. The singing behavior of eastern screech-owls: Seasonal timing and response to playback of conspecific song. Condor 90:648–652.

Rosenberg, K. V., R. D. Ohmart, and B. W. Anderson. 1982. Community organization of riparian breeding birds: Response to an annual resource peak. Auk 99:260–274.

Rosenberg, K. V., R. D. Ohmart, W. C. Hunter, and B. W. Anderson. 1991. Birds of the lower Colorado Valley. University of Arizona Press, Tucson.

Saab, V. 1999. Importance of spatial scale to habitat use by breeding birds in riparian forests: A hierarchical analysis. Ecological Applications 9:135–151.

Scott, M. L., M. E. Miller, and J. C. Schmidt. 2004. The structure and functioning of riparian ecosystems of the Colorado Plateau: Conceptual models to inform the vital-sign selection process. Prepared for National Park Service, Southern Colorado Plateau Network, Northern Arizona University, Flagstaff.

Sergio, F., and I. Newton. 2003. Occupancy as a measure of territory quality. Ecology 72:857–865.

Sogge, M. K., R. M. Marshall, S. J. Sferra, and T. J. Tibbitts. 1997. A southwestern willow flycatcher natural history summary and survey protocol. National Park Service Technical Report NPS/NAUcprs/NRTR–97/12.

Stephens, F. 1903. Bird notes from eastern California to western Arizona. Condor 5:75–78.

Swarth, H. S. 1905. Summer birds of the Papago Indian Reservation and of the Santa Rita Mountains, Arizona. Condor 7:23–28, 47–50, 77–82.

Swarth, H. S. 1914. A distributional checklist of the birds of Arizona. Pacific Coast Avifauna 10:1–33.

USDA Forest Service Region 3. 2000. Regional Forester's Sensitive Species List.

U. S. Fish and Wildlife Service. 1985. Sensitive species management plan for the western yellow-billed cuckoo. Portland, Oregon.

U. S. Fish and Wildlife Service. 2002. Yellow-billed cuckoo candidate listing on Endangered Species List. Federal Register 67:114.

Veit, R., and W. Petersen. 1993. Birds of Massachusetts. Massachusetts Audubon Society, Lincoln, Massachusetts.

Visher, S. S. 1910. Notes on the birds of Pima County, Arizona. Auk 27:279–288.

Younker, G. L., and C. W. Andersen. 1986. Mapping methods and vegetation changes along the lower Colorado River between Davis Dam and the border with Mexico. Final report to U.S. Bureau of Reclamation, Lower Colorado Region, Boulder City, Nevada.

A HISTORICAL ASSESSMENT OF CHANGES IN AVIAN COMMUNITY COMPOSITION FROM MONTEZUMA CASTLE NATIONAL MONUMENT, WITH OBSERVATIONS FROM THE CAMP VERDE REGION OF ARIZONA

Charles van Riper III, Mark K. Sogge, and Matthew J. Johnson

ABSTRACT

This chapter is a historical assessment of changes within the 211 bird species that have been recorded as occurring in the National Park Service areas of Montezuma Castle National Monument, including its Montezuma Well unit. There are also observations from the Camp Verde region of the Verde Valley in central Arizona. Information in this assessment is based on an intensive six-year bird-inventory project conducted from 1991 through 1997, entailing more than 600 field hours. While our formal surveys were focused at the Montezuma Castle National Monument (MCNM) complex, we made observations throughout the Camp Verde region. Some of the information collected on this project is detailed in Johnson and van Riper (2004), Sogge and Johnson (1998), and Schmidt et al. (2006). We have also used the published literature (e.g., North American Birds), museum specimens, and information generously provided by individuals with many years of experience birding in the Camp Verde region and at the MCNM complex (see acknowledgments). Based on comparisons with historical data, the abundance and status of several species appear to have changed substantially. We discuss species that have increased in numbers, those that appear to have declined, and possible explanations for those changes. We found that most avian species that increased or decreased over the past several decades have done so over the entire southwestern United States, not only within the National Park Service areas and the Camp Verde region.

INTRODUCTION

The historical record for individual birds in the Camp Verde region began in the late nineteenth century, but the first complete listing of birds from Montezuma Castle National Monument did not appear until the mid-twentieth century with observations by Betty Jackson from 1926 to1940 (Jackson 1941). A more extensive compilation of birds in that area was completed by C. B. Frost (1947), building upon the information from Jackson. A wider geographic area was covered by the bird list created by Sutton (1954), including not only Camp Verde but the entire Verde Valley. Additional information on the birds from the Monuments and the Camp Verde region came with Christmas bird counts that were conducted between 1973 and 1980 and published in the journal American Birds by the National Audubon Society, and the later Arizona Breeding Bird Atlas (Corman and Wise-Gervais 2005).

Our avian surveys were conducted from 1991 through 1997, focusing on the

Figure 1. Montezuma Castle at Montezuma Castle National Monument, Aizona. The monument contains diverse habitats, and hosts many species of breeding birds.

birds within Montezuma Castle National Monument, including the Montezuma Well unit, comprising the Montezuma Castle National Monument (MCNM) complex, with additional observations made during that period and through 2007 in the Camp Verde region, northeastern Yavapai County, Arizona. In this chapter we use all the sources noted above, as well as field notes of observations from birders associated with the Northern Arizona Audubon Society, and a summary of the status of Verde Valley birds provided by Roger Radd. Sightings listed in North American Birds and Audubon Field Notes that we use are specifically referenced, as are records from museum specimens.

This chapter provides a list of all known birds recorded from the MCNM complex, and notes from observations made within the Camp Verde region of Central Arizona. The species entries include notes on historical records and pertinent information on each species as it relates to their presence in the area. The listing may also be used as a general guide to birds of similar habitats in northern Arizona, but the status of many may differ outside of the immediate Camp Verde region. The wide variety of birds found at MCNM is largely due to the diversity of habitats within and around the monument, complemented by other habitat-types throughout the Verde Valley.

More than seventy species of birds are known or suspected to breed (or have bred historically) within the MCNM complex boundaries. A number of species have undergone range expansions into the region, while other species seem to have disappeared from the Camp Verde region. The status of some species changed during the time period that we surveyed, and bird species presently undocumented may well be found in the monument in the foreseeable future as habitat and climatic changes continue, while some current species may be lost.

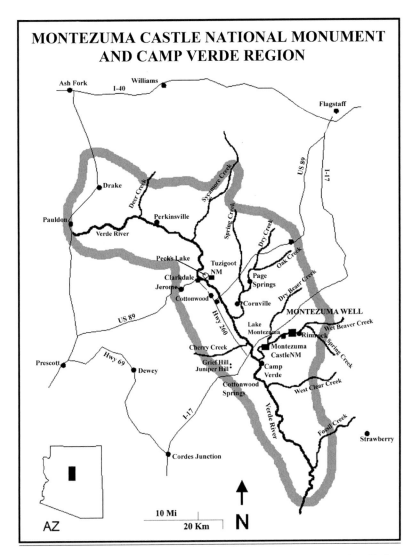

Figure 2. Map of the Verde Valley, Yavapai County, Arizona. The boundary of the Verde Valley, as defined in this chapter, is indicated by the broad shaded line which approximates the 5,000 ft elevation contour.

METHODS

Study Area

The Camp Verde region is located within the Verde River valley, in the northeastern portion of Yavapai County, Arizona (34o36' N, 111o50' W), about in the center of the state (Figure 2). The MCNM complex is about 6 km northeast of the town of Camp Verde. Comprising two subunits (Montezuma Castle and Montezuma Well), the complex totals 340 ha, of which about 116 ha encompasses riparian habitat (including mesquite bosque) along Beaver Creek, and about 190 ha of upland habitat types (Figure 3). For the purposes of this report we define the "Camp Verde region" as that area surrounding the

Figure 3. Montezuma's Well, a deep limestone sink in Montezuma Castle National Monument, Arizona. Photo by Chris O'Brien.

town of Camp Verde, bounded on the north by Peck Lake and the town of Clarkdale, on the northeast along the I-17 corridor to Rimrock and Montezuma Well in the Montezuma Castle National Monument, on the west by Grief and Juniper Hills, and stretching to Wingfield Mesa and Fossil Creek on the south.

Climate within this region is semiarid, with mild winters and hot summers. Elevation ranges from 963 m along the riparian area south of Camp Verde to 1048 m elevation in the Montezuma Castle unit and 1110 m at Montezuma Well, and then up to 2000 m elevation around Clarkdale and Peck's Lake, the upper limit of our area of study. The Montezuma Well is a flooded limestone sinkhole with open water (O'Brien et al. 2003), and mean annual precipitation is 30 cm. Surface water and groundwater resources sustain aquatic and riparian habitats within this region, while winter and summer rains provide water for upland habitats.

Bird Census Techniques

Point Counts (Breeding Season) – We established point count stations in each of six habitat types (Schmidt et al. 2006). The six habitat types were later reduced to four general vegetation types (after Ellison and van Riper 1998). Count stations were placed systematically a minimum of 150 m apart, so that as much of the count radius as possible was included within a single major habitat type. The number of census points varied from five to nine per habitat type. In some cases, count stations were removed or relocated following destruction or habitat loss from floods, or to eliminate practical sampling biases such as background noise from roads or water.

During the breeding season (from April through July), we conducted point counts once or twice per month at each census point. Point counts were made following the recommendations of Reynolds et al. (1980) and Ralph et al. (1993). We used a 50-meter count radius and a minimum six minute count period. Point counts began as soon as the observer reached the point count station. At each point, the observer stood quietly and recorded all birds seen or heard within the 50 m radius. For each detection, the observer recorded time of detection, bird species, number of individuals, type of detection (aural or visual), distance from the observer, and activity that the bird was engaged in when first detected.

Walking Counts (Nonbreeding Season) – Although point counts are an effective bird survey technique during the breeding season, they are not well suited for inventory or monitoring birds during the migration and wintering seasons (Reynolds et al. 1980, Ralph et al. 1993). Nonbreeding birds are often found in flocks, meaning that many individuals may be clumped together in a small area, decreasing the probability they will be found within a randomly situated count station during any given count period. Furthermore, nonbreeding birds and flocks generally do not maintain territories, and so often move about the landscape during short periods of time. Thus, they are easily double counted which can inaccurately inflate abundance estimates (especially if large flocks are encountered more than once). Finally, nonbreeding birds usually do not sing and are much less conspicuous than territorial breeders, and are not as readily detected by a stationary observer, particularly in dense habitats.

Because of these limitations, we used walking surveys during the nonbreeding season (August through March). Walking surveys were conducted by moving slowly and quietly through an area, recording the species and number of each bird detected. This allowed observers to efficiently sample large areas, move closer to noticeable bird activity and to obtain better views of inconspicuous individuals, and even flush quiet birds that would otherwise have been overlooked. Survey route and duration were not standardized so that survey effort could be flexible to meet the challenges of nonbreeding season surveys. For example, if several large mixed species flocks were encountered, surveyors could follow the flock and/or spend as much time as necessary to determine what species were present. Similarly, if birds were seen in particularly dense habitat or behaving secretively, surveyors could spend the time needed to obtain a positive identification. Although less repeatable and structured than point counts (and hence less useful for formal monitoring), walking surveys provide a better tool for documenting the presence and abundance of birds during the nonbreeding season (Verner 1985).

Bird list format and contents – Bird species entries in this chapter present the seasonal status of each species (based largely on numbers observed during surveys and dates on museum collections) and the preferred habitat(s). The list of birds is presented in taxonomic order, with all English bird names and listing sequence conforming to the American Ornithologists' Union's Checklist of North American Birds (1998 seventh edition, and subsequent supplements). Following the English and scientific names, we provide bulleted information for each species that includes: 1) residency status, 2) dates known to be present, 3) breeding status, 4) abundance, and 5) preferred habitat type. Below this information is a "Notes" section with additional comments and clarifications for each bird species.

The following definitions and explanations describe the bulleted summary categories we use throughout:

1) **Residency:** We use four residency status groupings; however, in referencing these assignments, keep in mind that bird species can sometimes be observed beyond identified time frames in which we have found them, and species will sometimes leave prior to the

end of an identified "season." Therefore, the identified residency status should be viewed as a general guide to when one would expect to see the species in the Camp Verde region, and especially at the MCNM complex. Some species occupy multiple residency categories, and in those cases each category will be listed, with the most commonly recorded period listed first. Neotropical migratory species are designated with a "•" symbol following the common name.

The notations used for the four residency status groups include:

Breeding: This term within the residency status category indicates bird species that are known to breed within the MCNM complex. Species in this category are usually observed during the spring and early summer, the principal breeding period. Notes are also provided for those species that are found in the Camp Verde region.

Migrant: Birds assigned to this residency category are usually observed during the spring (March to May) and/or fall (September to November) migration periods. These birds will often use the Camp Verde region, and specifically the MCNM complex, as stopover sites on their annual northward spring migration, or during their southward fall migration

Winter: Birds in this residency category are usually observed over-wintering, sometimes throughout the entire winter period.

Yearlong: Bird species in this residency category are found throughout the year in the MCNM complex and Camp Verde region. These birds breed in and do not migrate out of the area. An observer can, therefore, expect to find the species present during any month of the year.

a. *Dates known to be present:* This is the range of dates (in months) for which we have records of the presence of these species in the MCNM complex. There is undoubtedly a wider range of dates for the entire Camp Verde region. In the case of spring and fall migrants, this may include a split range of dates for each migratory season. A date of "January – December" is given for species that are present yearlong. "Unknown" is given for species with few or no recent records.

b. *Breeding status:* This section reflects documented evidence for nesting activity, or rationale for those bird species that are strongly suspected to breed in the Monuments. Additional nesting information is aprovided for the Camp Verde region. The notation "not known to breed in area" is used for species with no published records or substantiated evidence of breeding. The notes accompanying each species account will contain specific information, such as confirmed breeding dates, locations where we found an active nest, or observations of adults feeding just-fledged young.

2. Abundance: This section of the account indicates how certain an observer could be of seeing a bird species during the appropriate season. It does not provide a true quantitative assessment of bird numbers, but a relative index of their abundance. Many or most of the generally "uncommon" species are conspicuous and, therefore, often observed even though few individuals are present. Definitions for terms that we use in the "abundance" groupings throughout are the following:

Abundant: Usually observed or present in large numbers (possibly more than 100 individuals of this species in one day).

Common: Usually observed or present in moderate numbers (10 to 100 individuals of this species in one day).

Uncommon: Usually observed in low numbers (1 to 10 per day).

Rare: One to a low number observed annually (1 to 10 per year).

Irregular-casual: This notation describes

species fairly common to abundant some years, but totally absent in others. Casual species could be observed in low numbers some years, or not at all.

3. Preferred Habitat Type: This notation indicates recorded habitat associations for each of the described bird species. Determination of preferred habitat was based on the relative frequency of detections that we had for the species in each of nine habitats. We then consolidated all detection locations into four general habitat types:

Open water: Aquatic habitat characterized by flowing or open standing water. This habitat is found in Montezuma Well, the sewage reclamation ponds, along the Verde River, and along Beaver and Fossil Creeks.

Marsh: Wetland habitat that typically includes bulrushes (*Scirpus* spp.), cattails (*Typha* spp.), and seep willows (*Baccharis salicifolia*). In the Montezuma Castle National Monument complex, Peck's Lake and the Wet Beaver Creek shoreline contains marsh habitat, while in Camp Verde it is found within backwater areas along the Verde River.

Riparian: Wetland-associated habitat that consists largely of dense associations of sycamores (*Platanus wrightii*), cottonwoods (*Populus fremontii*), willows (*Salix* spp.), tamarisk (*Tamarix ramosissima*), and seep willow. For purposes of this list, mesquite (*Prosopis* spp.) bosque is considered a riparian habitat. Riparian habitat occurs primarily along the edges of Beaver and Fossil Creeks and the Verde River.

Mixed desert upland shrub: Upland habitat dominated by shrubs such as creosote bush (*Larrea tridentata*), acacia (*Acacia* spp.), and several species of cacti and seasonal grasses are included in this habitat type. It also includes cliffs and rocky outcrops, the predominant habitat surrounding riparian areas.

RESULTS

Species accounts – The following accounts provide details for all bird species that have been known to occur at the MCNM complex and generally within the Camp Verde region.

Greater White-fronted Goose (*Anser albifrons*)
Residency: winter; migrant
    Dates present: December – April
    Breeding status: not known to breed in area
Abundance: rare
Preferred habitat type: open water
The Greater White-fronted Goose is seen more frequently (though still rare) in winter and fall within the Verde Valley (Johnson and Sogge 1995). One record of this species occurred in February 1949, at Montezuma Well (Sutton 1954).

Snow Goose (*Chen caerulescens*)
Residency: winter
    Dates present: December – April
    Breeding status: not known to breed in area
    Abundance: rare
Preferred habitat type: open water
The Snow Goose is a very casual winter visitor, seen more frequently (though still rare) at Lake Montezuma and in other portions of the Camp Verde region (Stejskal et al. 1995, Johnson and Sogge 1995, van Riper personal observation 1999). We did observe one at Montezuma Castle in December 1995.

Canada Goose (*Branta canadensis*)
Residency: yearlong
    Dates present: January – December
    Breeding status: known to breed in area
Abundance: common
Preferred habitat type: open water
Sutton (1954) lists Canada Geese as common in the Verde Valley during winter, and describes them in Montezuma Well. The species is now a yearlong resident of the Verde Valley and has a breeding population of birds on the lower Verde River south of the bridge on Highway 260, and especially south on Salt Mine Road.

Wood Duck *(Aix sponsa)*
Residency: yearlong
   Dates present: January – December
   Breeding status: known to breed in area
Abundance: uncommon
Preferred habitat type: open water; riparian
This is an uncommon breeding resident usually seen on open water associated with cottonwood and sycamore riparian systems, especially the Verde River. Between 21 February and 19 October, we detected 24 individuals during our surveys along Wet Beaver Creek. We found breeding evidence of this cavity-nesting duck, observing adults and ducklings along Beaver Creek and the Verde River. Wood Ducks are generally believed to be increasing in the region, as they were not noted by Jackson (1941) or Sutton (1954), and only recently have been verified as breeding in the Verde Valley (Witzeman and Stejskal 1984). Now these ducks are often seen in pairs and with ducklings elsewhere in the Verde Valley (Radd in litt.).

Green-winged Teal *(Anas crecca)*
Residency: winter; migrant
   Dates present: January – May
   Breeding status: not known to breed in area
Abundance: uncommon
Preferred habitat type: open water
There are two recent records of Green-winged Teal in Montezuma Well (February 1997 and March 1997). Sutton (1954) states, "Uncommon winter visitor... noted several times at Montezuma Well and Castle." The species is an uncommon and brief winter visitor to the Camp Verde region.

Mallard *(Anas platyrhynchos)*
Residency: yearlong; migrant
   Dates present: January – December
   Breeding status: known to breed in area
Abundance: common
Preferred habitat type: open water; marsh
We found the Mallard present 11 February – 6 June; 11 September; and 5 December during 12 of our surveys. The species breeds locally in the Verde Valley (Radd in litt.), but is only a rare winter visitor, casual spring-fall migrant at the MCNM complex, found in open water associated with marsh (e.g. in Montezuma Well) or riparian areas. Mallards are a fairly common migrant and wintering species in other parts of the Verde Valley including Peck's Lake and Tavasci Marsh (Sutton 1954, Johnson and Sogge 1995)

Northern Pintail *(Anas acuta)*
Residency: winter
   Dates present: November – April
   Breeding status: not known to breed in area
Abundance: rare
Preferred habitat type: open water
This diving duck was historically more common in the Camp Verde region (Sutton 1954), but is currently uncommon in winter (Radd in litt.). No occurrences have been recorded at the MCNM complex since 1954, but occasionally vagrants occur at Peck's Lake and larger bodies of open water.

Blue-winged Teal *(Anas discors)*
Residency: winter; migrant
   Dates present: November – April
   Breeding status: not known to breed in area
Abundance: uncommon
Preferred habitat type: open water; marsh
Sutton (1954) notes only Verde Valley records from Clarkdale (Peck's Lake), so it was apparently rare then. The species is still rare to uncommon fall through spring elsewhere in Verde Valley area (Johnson and Sogge 1995). We had two recent detections, 9 May and 26 May 1992, during our surveys at the MCNM complex.

Cinnamon Teal *(Anas cyanoptera)*
Residency: winter; migrant
   Dates present: January – April; August – September
   Breeding status: known to breed in area
Abundance: rare
Preferred habitat type: open water; marsh
Sutton (1954) lists this species as casual, noted from Montezuma Well and Castle. Johnson and Sogge (1995) found it uncommon to

fairly common from fall to spring elsewhere in Verde Valley, with a 1998 breeding record at Tavasci Marsh (Radd in litt.). In the Camp Verde region we found it to be a rare transient, and detected six birds from 22 January to 23 March and from 30 August to 11 September.

Northern Shoveler *(Anas clypeata)*
Residency: winter
   Dates present: January – March
   Breeding status: not known to breed in area
Abundance: rare
Preferred habitat type: open water; marsh
Sutton (1954) lists this as an occasional winter visitor in Verde Valley, but it is now uncommon from fall through spring (Johnson and Sogge 1995) and is a casual winter transient. We detected only four birds during our surveys.

Gadwall *(Anas strepera)*
Residency: winter
   Dates present: November – March
   Breeding status: not known to breed in area
Abundance: uncommon
Preferred habitat type: open water; marsh
Historically, this bird was "rarer" at Montezuma Well (Sutton 1954), with only a single 1985 sighting (Thornburg in litt.), but our detections from 14 February 1995 and 3 and 5 March 1997 suggest that the species is probably now more abundant throughout the Camp Verde region. The Gadwall is abundant in winter at Peck's Lake and along the Verde River.

Eurasian Wigeon *(Anas penelope)*
Residency: winter
   Dates present: October – February
   Breeding status: not known to breed in area
Abundance: rare
Preferred habitat type: open water; marsh
Eurasian Wigeons at Montezuma Well have been seen mixed with flocks of American Wigeons. There are three records from 16 January 1985 (Thornburg in litt.), and 17 November 1988 (Witzeman and Stejskal 1988), and 27 October 1996 (V. Gilmore in litt.) at Montezuma Well. There is also a record outside the MCNM complex at the nearby Village of Oak Creek during fall migration (October 1998, Rosenberg and Stejskal 1990).

American Wigeon *(Anas americana)*
Residency: winter
   Dates present: November – April
   Breeding status: not known to breed in area
Abundance: common
Preferred habitat type: open water; marsh
American Wigeon is the most frequently seen waterfowl in the MCNM complex and along Verde River, a fairly common winter visitor and migrant. They were present during most of our late fall surveys and during our early winter surveys in Montezuma Well, usually in flocks of 20–40 birds. We detected them in 22 surveys with a date range of 2 January to 25 March and 11 September to 29 November.

Canvasback *(Aythya valisineria)*
Residency: winter
   Dates present: November – April
   Breeding status: not known to breed in area
Abundance: rare
Preferred habitat type: open water
The Canvasback was listed as uncommon in winter at "ponds like Montezuma Well" (Sutton 1954), but there are no records after that listing. Currently this species is uncommon to fairly common fall through spring in the Verde Valley (Johnson and Sogge 1995, Radd in litt.).

Redhead *(Aythya americana)*
Residency: winter
   Dates present: August – March
   Breeding status: not known to breed in area
Abundance: uncommon
Preferred habitat type: open water
We had eight detections on 21 February and between 30 August and 24 September. There is also one recent winter record within the MCNM complex from a sighting in February 1997. This species is an uncommon visitor fall through spring in the Verde Valley

(Johnson and Sogge 1995), occurring on large expanses of open water.

Ring-necked Duck *(Aythya collaris)*
Residency: winter
    Dates present: November – April
    Breeding status: not known to breed in area
Abundance: common
Preferred habitat type: open water
Sutton (1954) listed them as occasional at Montezuma Castle and rare at Montezuma Well. We detected six individuals on open water associated with marsh and rivers during surveys from 2 January to 7 March and on 26 November. Currently the Ring-necked Duck is a fairly common fall-through-spring visitor to the Verde Valley (Johnson and Sogge 1995).

Lesser Scaup *(Aythya affinis)*
Residency: winter
    Dates present: November – March
    Breeding status: not known to breed in area
Abundance: rare
Preferred habitat type: open water
The species is listed as historically rare in winter by Sutton (1954) at Montezuma Well and at pools along Beaver Creek at Montezuma Castle. There are no recent records at Montezuma Well. Currently this scaup is an uncommon fall and winter visitor in the Verde Valley (Johnson and Sogge 1995).

Common Goldeneye *(Bucephala clangula)*
Residency: winter
    Dates present: November – April
    Breeding status: not known to breed in area
Abundance: rare
Preferred habitat type: open water
A "sparse to uncommon" winter visitor to central Arizona (Monson and Phillips 1981). Two records for the Verde Valley region date from 13 January 1989 (Thornburg in litt.) and 11 April 1998 (Radd in litt.).

Bufflehead *(Bucephala albeola)*
Residency: winter
    Dates present: November – May
    Breeding status: not known to breed in area
Abundance: uncommon
Preferred habitat type: open water
The Bufflehead is fairly common on open water during fall through spring in the Verde Valley, but is not uniformly distributed over the region (Johnson and Sogge 1995, Radd in litt.). Sutton (1954) lists this as an occasional winter migrant (November through April) in the Verde Valley, and "known from Montezuma Well." Thornburg (in litt.) provided four sightings from 1986 to 1989. We detected five individuals of this species at Montezuma Well during our surveys from 8 January to 4 March and on 8 May. Jackson (1941) lists the species present in February and March within the Camp Verde region.

Hooded Merganser *(Lophodytes cucullatus)*
Residency: winter
    Dates present: November – March
    Breeding status: not known to breed in area
Abundance: rare
Preferred habitat type: open water
The Hooded Merganser was not listed as present by Jackson (1941) or Sutton (1954). There is one record on 26 November 1992, and the species is a rare fall through winter visitor to the Verde Valley (Johnson and Sogge 1995).

Common Merganser *(Mergus merganser)*
Residency: yearlong
    Dates present: January – December
    Breeding status: known to breed in area
Abundance: uncommon
Preferred habitat type: open water
Sutton (1954) listed the Common Merganser as a yearlong resident in the Verde Valley (noting a nest in a hole in the bank of Beaver Creek near the Castle in 1940) and considered the bird rare at Montezuma Well. Jackson (1941) recorded it throughout most of year and considered it a breeding species (April through May) within the MCNM complex. Christmas bird counts from 1938–1939 recorded up to 10 Common Megansers (Jackson 1939). We detected 18 individuals during our surveys between 10 February and 23 August and found adults with young on Beaver Creek.

Red-breasted Merganser *(Mergus serrator)*
This species was listed in Frost's (1947) bird checklist for the MCNM complex, based on an April 1935 sighting reported in the monthly park report (Superintendents of MOCA 1932). The MCNM complex is far outside the Red-breasted Mergansers typical breeding and winter range, and Phillips et al. (1964) note it as "decidedly uncommon." In winter, the appearance of this species is quite similar to the Common Merganser, which is regularly detected in the MCNM complex. Therefore, the Red-breasted Merganser seen in 1935 might possibly have been a Common Merganser in winter plumage, and this sighting should be considered "possible," pending additional Red-breasted Merganser records.

Ruddy Duck *(Oxyura jamaicensis)*
Residency: winter
　　Dates present: November – April
　　Breeding status: not known to breed in area
Abundance: uncommon
Preferred habitat type: open water
The Ruddy Duck has historically been uncommon in winter (Sutton 1954). We found the species on open water and associated with the marsh in Montezuma Well, detecting five individuals from 5 December to 13 March. Although this species breeds from 46 m up to 2835 m elevation throughout much of Arizona (Corman and Wise-Gervais 2005), we did not find evidence of breeding in the Camp Verde region.

Ring-necked Pheasant *(Phasianus colchicus)*
Residency: yearlong
　　Dates present: January – December
　　Breeding status: known to breed in area
Abundance: rare
Preferred habitat type: mixed desert upland shrub
Ring-necked Pheasants were introduced into the Verde Valley by the Arizona Game and Fish Department as a game bird in the 1930s. Although the introduction was initially considered successful with "large numbers in and around Camp Verde" (about 200–300 estimated in 1961), it is now considered rare in the Verde Valley and persists in only four or five areas statewide (Brown 1989). The species was recorded "near the dwellings" at Montezuma Castle (Sutton 1954). There are records for the Camp Verde region (Corman and Wise-Gervais 2005), but none recently from the MCNM complex.

Gambel's Quail *(Callipepla gambelii)*
Residency: yearlong
　　Dates present: January – December
　　Breeding status: known to breed in area
Abundance: abundant
Preferred habitat type: mixed desert upland shrub
Jackson (1941) listed this quail as a common yearlong resident with breeding noted from April through June. Sutton (1954) described it as "probably the most abundant bird in the Verde Valley," with young most evident in June or July. We found the species abundant throughout the Verde Valley watershed, including the Camp Verde area. We had more than 100 detections of this species during MCNM complex surveys, in every month of the year, within mesquite riparian, upland shrub, and grassland habitats. Occasionally we found them in riparian woodland. We also detected young frequently during summers from 1991 to 1997.

Pied-billed Grebe *(Podilymbus podiceps)*
Residency: yearlong
　　Dates present: January – December
　　Breeding status: known to breed in area
Abundance: uncommon
Preferred habitat type: open water; marsh
The Pied-billed Grebe is present throughout the Verde Valley, including Peck's Lake (Sutton 1954, Johnson and Sogge 1995). Almost all of our sightings were from open water and marsh at Montezuma Well, with 18 birds detected between 2 January and 19 April and between 30 August and 14 October. Young were observed from June through August, as the species is multi-brooded in Arizona.

Least Bittern *(Ixobychus exilis)*
Residency: yearlong
　　Dates present: January – December

Breeding status: known to breed in area
Abundance: rare
Preferred habitat type: marsh

The Least Bittern is present in the Verde Valley, such as at Peck's Lake (Sutton 1954, Corman and Wise-Gervais 2005). Almost all of our sightings were from marsh areas, with breeding detected at Tavaci Marsh near Tuzigoot National Monument (Johnson and Sogge 1995).

Great Blue Heron *(Ardea vivescens)*
Residency: yearlong
    Dates present: January – December
    Breeding status: known to breed in area
Abundance: uncommon
Preferred habitat type: open water

This species is considered uncommon, though readily seen in riparian zones throughout the Verde Valley watershed, including the Camp Verde area (Sutton 1954, Johnson and Sogge 1995). We documented 41 detections from 8 January through 19 October in riparian areas and open water, especially slow-moving areas of Beaver Creek. The Great Blue Heron is also found in scattered heronries along the Verde River corridor with young present April through July (Corman and Wise-Gervais 2005).

Snowy Egret *(Egretta thula)*
Residency: winter; migrant
    Dates present: October – March
    Breeding status: not known to breed in area
Abundance: rare
Preferred habitat type: open water

Snowy Egret is listed as uncommon to rare by Jackson (1941) and Sutton (1954). There are no records in the MCNM complex since 1953, and we did not detect this species during our surveys, although Snowy Egrets are uncommon-to-rare fall through spring migrants elsewhere in the Verde Valley (Johnson and Sogge 1995, Radd in litt.). Elsewhere in the state, the range of this species is expanding, with occurrences near open water, marshes, and riparian areas (Phillips et al. 1964).

Cattle Egret *(Bubulcus ibis)*
Residency: migrant
    Dates present: April – June
    Breeding status: not yet known to breed in area
Abundance: uncommon
Preferred habitat type: marsh

Cattle egrets began their expansion into North America in the 1950s, and the movement has been so rapid that they now breed in scattered locations as far north as extreme southern Canada (Telfair 1994). Since Cattle Egrets were first documented nesting near Yuma, Arizona, in 1991, their breeding range in the state has steadily expanded northward (Corman and Wise-Gervais 2005). This species was first reported at Camp Verde in June 1974 (Alden and Mills 1974) and then in Cottonwood in April 1984 (Stejskal and Witzeman 1984). The species has been observed very close to the MCNM complex at Lake Montezuma (April 1983, Witzeman 1983b). Given continued range expansion and increases in abundance in newly settled areas, this species will probably become quite common throughout the MCNM complex and Camp Verde region.

Green Heron • *(Butorides striatus)*
Residency: yearlong
    Dates present: January – December
    Breeding status: known to breed in area
Abundance: rare
Preferred habitat type: marsh

Sutton (1954) listed this heron species as an uncommon breeder in late 1930s. In the 1950s (Sutton 1954) recorded the species as "casual" during summer within the Verde Valley. The bird is secretive and usually observed near open water, marshes, and riparian areas (Phillips et al. 1964). The Green Heron is now a confirmed breeder at Page Springs and Tavasci Marsh (Radd in litt.), and probably elsewhere along the Verde River and its tributaries (Corman and Wise-Gervais 2005).

Black-crowned Night-Heron
(Nycticorax nycticorax)

Residency: yearlong
  Dates present: January – December
  Breeding status: known to breed in area
Abundance: rare
Preferred habitat type: marsh
Historically, the Black-crowned Night-Heron was an uncommon breeder in the area, with eggs collected in May a few kilometers below Camp Verde in the early 1900s. This heron breeds locally at Page Springs (Radd in litt.) and elsewhere in the Camp Verde region (Corman and Wise-Gervais 2005). It is not listed as a breeder within the MCNM complex by B. Jackson (1941), we did not observe this heron during our surveys, and no sightings have been recorded there since 1951 (Sutton 1954).

White-faced Ibis *(Plegadis chihi)*
Residency: winter; migrant
  Dates present: September – March
  Breeding status: not known to breed in area
Abundance: common
Preferred habitat type: marsh
The White-faced Ibis is now a fairly common migrant species in the Verde Valley, including Peck's Lake and Tavasci Marsh (Johnson and Sogge 1995, Radd in litt.). The species is more common now than in the 1950s when Sutton (1954) listed only one record from the Verde Valley (from 1951). Appropriate marsh habitat is rare within MCNM complex, and there has been only one recent record from 5 September 1991.

Wood Stork *(Mycteria americana)*
Residency: migrant
  Dates present: June- July
  Breeding status: not known to breed in area
Abundance: irregular-casual
Preferred habitat type: marsh
Jackson (1941) listed this species as a rare summer visitor (June through July, 1935–1940) in the Verde Valley, but there have been no subsequent records at the MCNM complex. Sutton (1954) lists the "Wood Ibis" as rare, but may be referring only to Jackson's records.

Black Vulture *(Coragyps atratus)*
Residency: migrant
  Dates present: not certain
  Breeding status: not known to breed in area
Abundance: rare
Preferred habitat type: mixed desert upland shrub
This Black Vulture spread from Mexico into southern Arizona in the early 1900s (Phillips et al. 1964). Sutton (1954) describes the Black Vulture as "extremely rare" for the Verde Valley, and reports that in 1947 "Scroeder" observed one flying over Montezuma Well. However, Sutton (1954) did not feel the 1947 observation was certain, and so considered the vulture "hypothetical." Monson and Phillips (1981) also state that the Verde Valley records were not considered authentic. The Black Vulture appears restricted to southern Arizona (Corman and Wise-Gervais 2005).

Turkey Vulture • *(Cathartes aura)*
Residency: breeder
  Dates present: March – September
  Breeding status: believed to breed in area
Abundance: common
Preferred habitat type: mixed desert upland shrub
The Turkey Vulture is an abundant migrant throughout the Verde Valley watershed including the Camp Verde area (Corman and Wise-Gervais 2005). This species can be seen on almost any visit to the Camp Verde region during summer, but usually in low numbers. We made 39 observations of this species at the MCNM complex between 23 March and 30 September. However, it is much more prevalent than our number of detections would suggest, because point counts are not effective for sampling large, soaring raptors.

Osprey • *(Pandion haliaetus)*
Residency: migrant
  Dates present: September – April
  Breeding status: not known to breed in area
Abundance: rare
Preferred habitat type: open water
Although rarely observed along open water

along Beaver Creek, this species may be more common than the detection rate would suggest. It was recorded along Beaver Creek several kilometers upstream of Montezuma Well on 23 October 1998 and at Lake Montezuma on 25 February 1980 and 18 September 1982 (Radd in litt.). There are two other records in the MCNM complex (19 April 1980 and 6 September 1991) At Clear Creek, Stejskal and Rosenberg (1990) reported a bird during the winter season (February 1990).

Bald Eagle *(Haliaeetus leucocephalus)*
Residency: yearlong
　Dates present: January – December
　Breeding status: known to breed in area
Abundance: uncommon
Preferred habitat type: riparian; open water
Historically, this species was extremely rare in winter (Sutton 1954), although it was reported to breed at Stoneman Lake in 1880s by Mearns (per Sutton 1954). Regional and national status of this endangered species has improved over the last 30 years, and now it is a common breeding resident in the Verde Valley upstream of Peck's Lake and downstream of Camp Verde (Glinski 1998). We had five sightings between 8 January and 5 March of Bald Eagles flying or perched in large trees along Beaver Creek.

Northern Harrier • *(Circus cyaneus)*
Residency: migrant; winter
　Dates present: August – March
　Breeding status: not known to breed in area
Abundance: uncommon
Preferred habitat type: mixed desert upland shrub
Historically, this raptor species was considered casually present in winter and spring. The first MCNM complex record was on September 17, sometime between 1937 and 1939 (Jackson 1941). Sutton (1954) reports two sightings of Northern Harrier at Montezuma Castle, but considers the sightings uncertain. Pasture and grassland habitat is limited in the MCNM complex, so harriers would not be expected to occur commonly. However, they probably occur occasionally as a "flyover" in upland habitats and the pasture at the Montezuma Well portion. Glinski (1998) reports them as rare to uncommon winter visitors within the Verde Valley.

Sharp-shinned Hawk • *(Accipiter striatus)*
Residency: winter; migrant
　Dates present: September – April
　Breeding status: not known to breed in area
Abundance: uncommon
Preferred habitat type: riparian; mixed desert upland shrub
Historically, this species was a casual visitor during fall through winter. A dead Sharp-shinned Hawk was recorded at Montezuma Castle in 1946 and five observations are listed at Montezuma Well by Sutton (1954). One specimen was collected on 15 September 1927. The Sharp-shinned Hawk breeds in riparian woodlands and forests, and at higher elevations elsewhere, but can be found in virtually any habitat during migration and winter (Glinski 1998).

Cooper's Hawk • *(Accipiter cooperii)*
Residency: yearlong
　Dates present: January – December
　Breeding status: known to breed in area
Abundance: uncommon
Preferred habitat type: riparian; mixed desert upland shrub
This hawk is an uncommon breeder, mainly in riparian zones, and a yearlong resident throughout the Verde Valley watershed, including the Camp Verde area (Corman and Wise-Gervais 2005). The species has bred at the MCNM complex since at least 1916, with multiple records of nesting in sycamores (summarized by Sutton 1954). We had 22 detections during our surveys (19 January and 24 April through 28 October), with most of those in riparian woodland, but we noted this species foraging throughout the MCNM complex and the entire Camp Verde region. We also found one active nest each year from 1991 to 1996 at the Montezuma Castle portion of the MCNM complex.

Northern Goshawk *(Accipiter gentilis)*
Residency: migrant
    Dates present: November – December; March – April
    Breeding status: not known to breed in area
Abundance: rare
Preferred habitat type: riparian

We had one sighting of two soaring Goshawks on 6 May 1997, probably migrating individuals. This hawk breeds in dense coniferous forests, but may be found in almost any habitat during post-breeding dispersal and migration (Wiens et al. 2006).

Common Black-Hawk • *(Buteogallus anthracinus)*
Residency: breeding
    Dates present: January – December
    Breeding status: known to breed in area
Abundance: uncommon
Preferred habitat type: riparian

This bird is an uncommon riparian breeder that now occurs 20 March through 14 October throughout the Verde Valley watershed, including the Camp Verde area (Corman and Wise-Gervais 2005). The species appears to have been previously absent from this region, as it was not reported at the MCNM complex by Jackson (1941), and then noted as "extremely rare" with only one record at Montezuma Well by Sutton (1954). Now locally more common, the Common Black Hawk is often seen soaring above riparian areas during the spring and early summer, with nesting pairs about every 1.6 kilometers apart along suitable portions of Wet Beaver Creek and the Verde River (Radd in litt.). We had 28 sightings during our surveys and found one nest each year (1991–1996) at the Montezuma Castle portion, and one nest at the Montezuma Well portion from 1995 to 1996.

Harris's Hawk • *(Parabuteo unicinctus)*
Residency: migrant
    Dates present: August – December
    Breeding status: not known to breed in area
Abundance: rare
Preferred habitat type: mixed desert upland shrub

There is one record of Harris's Hawk breeding during 1886 in the Verde Valley (Monson and Phillips 1981), but Jackson (1941) listed the species as casual, with sightings in March, April, and September. Sutton (1954) considered the species as "rare, hypothetical." The Camp Verde region is presently north of the normal breeding range (Corman and Wise-Gervais 2005), and there is no appropriate breeding habitat in the MCNM complex or the Verde Valley, as this species is a desert hawk typically associated with cactus-dominated habitats. Recent records in Cottonwood in January 1991 (Stejskal and Rosenberg 1991b) and near Clarkdale in August 1989 during fall migration (Rosenberg and Stejskal 1990), lend support to the credibility of Jackson's earlier sightings.

Swainson's Hawk • *(Buteo swainsoni)*
Residency: migrant
    Dates present: October – November; March – April
    Breeding status: not known to breed in area
Abundance: rare
Preferred habitat type: mixed desert upland shrub

Taylor and Jackson (1916) described the Swainson's Hawk as "quite common along Verde River below Camp Verde, where they were seen on several occasions...." This species has experienced substantial declines in the region during recent decades, as there is less grassland habitat around Camp Verde and no suitable areas within the MCNM complex. This hawk is now a rare spring and fall migrant bird in the Verde Valley, with only a few records per year, including one recent record on 24 September 1980 (Hyde in litt.).

Red-tailed Hawk *(Buteo jamaicensis)*
Residency: yearlong
    Dates present: January – December
    Breeding status: known to breed in area
Abundance: uncommon
Preferred habitat type: riparian; mixed desert upland shrub

Red-tailed Hawks are uncommon breeders

throughout the Verde Valley watershed, including the Camp Verde area, and although uncommon in terms of numbers, they can be seen soaring over the MCNM complex during most visits, especially in spring and early summer. We counted 69 individuals during our surveys, with birds observed in riparian trees and soaring over and foraging in virtually every terrestrial habitat. We also found active nests in the same riparian area each year at the Montezuma Castle portion from 1991 to 1993 and one at the Montezuma Well portion from 1991 to 1997.

Ferruginous Hawk *(Buteo regalis)*
Residency: migrant; winter
    Dates present: November – May
    Breeding status: not known to breed in area
Abundance: rare
Preferred habitat type: mixed desert upland shrub
Jackson (1941) and Sutton (1954) do not report this species within the MCNM complex, and Sutton (1954) lists only one record for the Verde Valley. This hawk is an irregular to casual migrant and wintering visitor in the Peck's Lake area (Johnson and Sogge 1995), with a few birds wintering in grassland areas from Cornville to Sedona (Glinski 1998, Radd in litt.). There is one substantiated record on 24 April 1980 in Camp Verde, but the species probably flies over the area more often than the single sighting suggests. This hawk hunts in open grassland and shrub habitats, vegetation types certainly abundant in the Camp Verde region.

Rough-legged Hawk *(Buteo lagopus)*
Residency: winter
    Dates present: November – February
    Breeding status: not known to breed in area
Abundance: rare
Preferred habitat type: riparian; mixed desert upland shrub
The Rough-legged Hawk was not reported by Jackson (1941), but Sutton (1954) noted three "hypothetical" observations at Montezuma Well. Radd (in litt.) noted the species as rare during the winter in the Verde Valley, as did Johnson and Sogge (1995).

Golden Eagle *(Aquila chrysaetos)*
Residency: winter
    Dates present: October – February
    Breeding status: not known to breed in area
Abundance: rare
Preferred habitat type: mixed desert upland shrub
Historically, the Golden Eagle was extremely rare within the Verde Valley, possibly due to human persecution, with only one record at Montezuma Well during summer 1953 (Sutton 1954). The species has probably increased in abundance since then, as it is currently a rare but regular breeder in parts of the Verde Valley (Glinski 1998, Corman and Wise-Gervais 2005), and is occasionally seen in areas between Camp Verde and Cottonwood. We detected four of these eagles between 1 October and 2 February on our surveys, within open grassland and shrub habitat associated with steep cliffs.

American Kestrel *(Falco sparverius)*
Residency: yearlong
    Dates present: January – December
    Breeding status: known to breed in area
Abundance: uncommon
Preferred habitat type: riparian; mixed desert upland shrub
There are many historical records of this species in the Camp Verde region, including breeding records (Taylor and Jackson 1916, Jackson 1941, Sutton 1954). Although uncommon in terms of numbers, we regularly observed American Kestrels (71 detections) perched on treetops in riparian areas and telephone lines in the uplands. We observed the birds foraging in all habitat types, but most frequently in more open areas. It breeds in cavities of riparian trees and possibly in cliffs, and we observed adults going in and out of tree cavities and juveniles near those cavities.

Merlin *(Falco columbarius)*
Residency: winter
    Dates present: November – February
    Breeding status: not known to breed in area

Abundance: rare
Preferred habitat type: mixed desert upland shrub
This is a rare to casual winter visitor to the Verde Valley, where it prefers open grasslands and shrub areas (Glinski 1998, Johnson and Sogge 1995). There are two records at Montezuma Well, one in 1949 (Sutton 1954) and one on 14 October 1986 (Thornburg in litt.). There was also one Merlin recorded at Montezuma Castle on 29 November 1997 (Radd in litt.).

Peregrine Falcon • *(Falco peregrinus)*
Residency: yearlong
   Dates present: January – December
   Breeding status: known to breed in area
Abundance: rare
Preferred habitat type: riparian; mixed desert upland shrub
A letter from J. Hickey (1939) describes nesting Peregrines in what appears to be the cliffs above the Castle, including the collection (by Mearns) of an egg, three juveniles, and two adults from 1884 to 1887. No historical records are noted by Jackson (1941) or Sutton (1954). The loss of this species was probably related to the range-wide decline through the 1970s as a result of the use of the pesticide DDT, though collection of eggs and nestlings within the MCNM complex may have been a factor locally (Hickey and Anderson 1968). The Peregrine is now a rare breeder and visitor within the Verde Valley (Johnson and Sogge 1995), with the closest known breeding territories in Sycamore Creek (above Clarkdale) and along the Verde River downstream of Camp Verde. This falcon is no longer breeding within MCNM complex. There is one recent record in the MCNM complex on 22 December 1994, while the most current previous record was 1953. If regional Peregrine populations continue to increase, this species may once again breed within the MCNM complex.

Virginia Rail *(Rallus limicola)*
Residency: yearlong
   Dates present: January – December
   Breeding status: known to breed in area
Abundance: uncommon
Preferred habitat type: marsh
Virginia Rails are fairly common yearlong in the Camp Verde region and along the Verde Valley watershed (Johnson and Sogge 1995), breeding at Tavasci Marsh. Historically, the species was a rare fall and winter visitor to Verde Valley, with records only from Montezuma Well (Sutton 1954). We found five individuals from 27 December to 6 March on our surveys, but this is a secretive species and it may occur at Montezuma Well more frequently than our sightings suggest.

Sora *(Porzana carolina)*
Residency: yearlong
   Dates present: January – December
   Breeding status: known to breed in area
Abundance: uncommon
Preferred habitat type: marsh
Historically, this species was a rare fall and winter visitor to Verde Valley, with records from Montezuma Well and Peck's Lake (Sutton 1954). Presently the species is fairly common yearlong in the Verde Valley (Johnson and Sogge 1995); it is a secretive species that may occur at Montezuma Well more frequently than sightings suggest. We found the Sora in marsh-associated habitat at Montezuma Well on 23 and 30 August 1991, and on 6 March 1999.

Common Moorhen *(Gallinula chloropus)*
Residency: yearlong
   Dates present: January – December
   Breeding status: known to breed in area
Abundance: rare
Preferred habitat type: marsh
Currently this Moorhen is uncommon yearlong within the Camp Verde region, nesting at Page Springs and Tavasci Marsh (Corman and Wise-Gervais 2005). Historically, the bird was noted as "extremely rare," with one record (31 October 1953) at Montezuma Well (Sutton 1954). We sighted two birds on 28 July 1993 in marsh habitat associated with Montezuma Well.

American Coot *(Fulica americana)*
Residency: yearlong
   Dates present: January – December
   Breeding status: known to breed in area
Abundance: common
Preferred habitat type: open water; marsh
Common historically (Sutton 1954) and currently (Johnson and Sogge 1995) in the Verde Valley, this species is a breeder and yearlong resident throughout the Verde Valley watershed, including the Camp Verde area. Although rare with regard to abundance, it is regularly seen at Montezuma Well, and we detected eight individuals between 12 January and 5 March and on 9 October.

Black-bellied Plover *(Pluvialis squatarola)*
Residency: migrant
   Dates present: October
   Breeding status: not known to breed in area
Abundance: irregular-casual
Preferred habitat type: riparian
There is one record of Black-bellied Plover at Montezuma Well on 4 October 1947 reported by Sutton (1954), but he considered the sighting "hypothetical." Monson and Phillips (1981) considered it an uncommon transient statewide.

Mountain Plover *(Charadrius montanus)*
The only record of this species in the MCNM complex is its inclusion in Jackson's (1941) report. In her report, the species name is followed by a question mark in the occurrence table, indicating that Jackson was unsure of the correct identification. The subsequent inclusion of this species in Frost's (1947) checklist appears to be based only on Jackson (1941). Other shorebirds do migrate through the area (Johnson and Sogge 1995), and thus this Mountain Plover record may be a misidentification.

Killdeer • *(Charadrius vociferus)*
Residency: yearlong
   Dates present: January – December
   Breeding status: known to breed in area
Abundance: uncommon
Preferred habitat type: open water; riparian
The Killdeer is an uncommon breeder and yearlong resident throughout the Verde Valley watershed, including the Camp Verde area (Corman and Wise-Gervais 2005). There are many records from at least 1916 (Taylor and Jackson 1916). Although uncommon in terms of numbers, this vociferous shorebird is regularly heard (and sometimes seen), especially during the spring and summer when it responds to human intrusions with loud alarm calls. It is found primarily on open sandy or rocky areas along Beaver Creek and sewage ponds. We detected 28 individuals during our surveys, and found one active nest at Montezuma Castle in 1995.

Willet • *(Catoptrophorus semipalmatus)*
Residency: migrant
   Dates present: November
   Breeding status: not known to breed in area
Abundance: rare
Preferred habitat type: open water; riparian
The Willet is a rare spring and fall migrant in the Camp Verde region (Johnson and Sogge 1995). There is one record at Montezuma Well in 1953 (Sutton 1954). We did not detect this species during our surveys.

Spotted Sandpiper • *(Actitus macularis)*
Residency: yearlong
   Dates present: January – December
   Breeding status: known to breed in area
Abundance: uncommon
Preferred habitat type: riparian
This is the most widely distributed breeding sandpiper throughout Arizona riparian areas (Corman and Wise-Gervais 2005). Historically, the bird was noted as "extremely rare" in the Verde region and was unrecorded by Jackson (1941). There were only two records noted by Sutton (1954), with those sightings occurring on open sandy and wetland areas along Beaver Creek. We counted five individuals from 17 February to 10 March and from 18 July to 11 September, and although it may be

observed more frequently than in the past, it is still not common within the MCNM complex. We suspect that this species may be breeding, based on adult "distraction behavior" noted along Wet Beaver Creek. In addition, Corman and Wise-Gervais (2005) note breeding along the Verde River and at Verde Hot Springs.

Baird's Sandpiper • *(Calidris bairdii)*
Residency: migrant
    Dates present: October – November
    Breeding status: not known to breed in area
Abundance: irregular-casual
Preferred habitat type: riparian
Baird's Sandpiper is a common to fairly common fall transient statewide (Monson and Phillips 1981). There is one record at Montezuma Well, on an unspecified date in 1948 (Sutton 1954).

Wilson's Snipe • *(Gallinago delicate)*
Residency: migrant
    Dates present: October – December
    Breeding status: not known to breed in area
Abundance: rare
Preferred habitat type: riparian; marsh
This bird was formerly known as a subspecies of the Common Snipe (Gallinago gallinago), inhabiting relatively short marsh and wetland vegetation, often along rivers and streams. Sutton (1954) reported two fall records, one at Montezuma Castle (1940) and one at Montezuma Well (1949). The species is an uncommon fall migrant through spring in the Verde Valley (Johnson and Sogge 1995, Radd in litt.).

Wilson's Phalarope • *(Phalaropus tricolor)*
Residency: migrant
    Dates present: October – November; February – April
    Breeding status: not known to breed in area
Abundance: uncommon
Preferred habitat type: open water; riparian
Wilson's Phalarope is a common to fairly common transient statewide (Monson and Phillips 1981), but an uncommon spring and fall migrant at Peck's Lake (Johnson and Sogge 1995). There is one record at Montezuma Well in September 1953 (Sutton 1954).

Rock Pigeon • *(Columba livia)*
Residency: yearlong
    Dates present: January – December
    Breeding status: known to breed in area
Abundance: common
Preferred habitat type: mixed desert upland shrub
The Rock Pigeon is a common breeder and yearlong resident throughout the Verde Valley watershed, including the Camp Verde area (Corman and Wise-Gervais 2005). There were no historical records in the MCNM complex through 1954 (Taylor and Jackson 1916, Jackson 1941, Sutton 1954), but the species appears to have increased with growing regional urbanization. There is one record of nesting in the cliffs above Montezuma Castle in the 1984 breeding season (Witzeman and Stejskal 1984). We had 13 detections of this species on our surveys, with detections occurring between 2 March to 8 November.

White-winged Dove • *(Zenaida asiatica)*
Residency: breeding
    Dates present: March – September
    Breeding status: known to breed in area
Abundance: common
Preferred habitat type: mixed desert upland shrub
The abundance and distribution of this dove appears to have fluctuated locally over the past 100 years, with a northward range expansion since the 1870s (Phillips et al. 1964). Not noted at all within the Verde Valley by Mearns in the late 1800s (Phillips 1942), but found to be common below Camp Verde by Taylor and Jackson (1916). Later, Jackson (1941) listed the bird as casual from May through September, with no record of breeding. Sutton (1954) describes it as common summer resident throughout the Verde Valley, and notes a nest in mesquites near the Castle residence. The statewide abundance and distribution of this species has been influenced by changes in agricultural practices, urbanization, and

water use (Brown 1989). There are no recent records from the MCNM complex, though it is an uncommon breeder in Clarkdale and on the Black Hills foothills, with single birds seen north to Oak Creek Canyon during the breeding season (Radd in litt.).

Mourning Dove • *(Zenaida macroura)*
Residency: breeding
    Dates present: March – September
    Breeding status: known to breed in area
Abundance: common
Preferred habitat type: mixed desert upland shrub
Presently, the Mourning Dove is a common riparian-upland breeder and yearlong resident throughout the Verde Valley watershed, including the Camp Verde area (Corman and Wise-Gervais 2005). The species was described as "numerous" at Montezuma Castle and "common" at Montezuma Well by Taylor and Jackson (1916), with notes of two active nests. It was noted as a breeding species by Jackson (1941). This was one of the birds most commonly observed during our surveys; we detected 371 individuals in trees and shrubs within upland and riparian areas, and found many active nests.

Common Ground Dove *(Columbina passerina)*
Residency: breeding
    Dates present: March – September
    Breeding status: known to breed in area
Abundance: rare
Preferred habitat type: mixed desert upland shrub
Sutton (1954) describes a 1913 sighting of the Common Ground Dove at Camp Verde, and Jackson (1916) recorded one at the mouth of Beaver Creek. There is one record at Montezuma Castle from May 1991.

Thick-billed Parrot
(Rhynchopsitta pachyrhyncha)
In the 1800s, this species ranged casually as far north as "Beaver Creek near Camp Verde," breeding in pine forests but moving through other habitats (Phillips et al. 1964). This rather general geographic description seems to be at or very close to the Camp Verde region and the MCNM complex, as are two specimens collected at "Fort Verde" in 1886 and 1887. Therefore, it is quite likely that the Thick-billed Parrot was present within the area to some degree in the past, but it no longer occurs in the region.

Yellow-billed Cuckoo •
(Coccyzus americanus)
Residency: breeding
    Dates present: May – September
    Breeding status: known to breed in area
Abundance: uncommon
Preferred habitat type: riparian
The cuckoo is an uncommon riparian breeder in appropriate habitat throughout the Verde Valley watershed, including the Camp Verde area (Holmes et al. 2008). Historically, Taylor and Jackson (1916) reported cuckoos as common in timber along Beaver Creek up to at least three kilometers above Montezuma Well. Jackson (1941) recorded them as casual from June through September, but did not consider them a breeder within the MCNM complex. Sutton (1954) described cuckoos as uncommon summer transients "usually seen only for a day or two, or at most for a week." This implies that they were not breeding within the MCNM complex at that time. The species is currently breeding at the MCNM complex, and appears to be more common now than from the 1930s to 1950s. The cuckoo is a species of special concern in Arizona, and a candidate species for federal listing. In the western United States, the cuckoo is a riparian-obligate that breeds only in mature gallery riparian woodlands. A secretive species, it was probably not well sampled with our general bird surveys and warrants a focused inventory. We did, however, detect 36 birds between 26 May and 6 September within dense, tall sycamore and cottonwood riparian habitat, and noted active nests throughout the summer on Wet Beaver Creek.

Greater Roadrunner •
(Geococcyx californianus)
Residency: yearlong
    Dates present: January – December

Breeding status: known to breed in area
Abundance: uncommon
Preferred habitat type: mixed desert upland shrub

Although large, the roadrunner is surprisingly elusive, but they may become tame and quite bold if they habituate to people. The bird is an uncommon upland breeder and yearlong resident throughout the Verde Valley watershed, including the Camp Verde area. Males maintain large territories, so there are probably not many resident individuals within the MCNM complex, but we did make 11 detections between 31 January and 24 September. We also found one active nest at the Montezuma Castle in 1995.

Groove-billed Ani •
(Crotophaga sulcirostris)
Residency: migrant
    Dates present: May; September – October
    Breeding status: not known to breed in area
Abundance: irregular-casual
Preferred habitat type: riparian

The Groove-billed Ani is "a straggler from Mexico" with two pre-1980 records in northern Arizona (Monson and Phillips 1981). An Ani was photographed at Rim Rock on 21 October 1982 during fall migration (Witzeman 1983a); other nearby sightings include one at the Walker Basin Trail (about six kilometers west of the MCNM complex) on 10 May 1998, and one at Lake Montezuma on 24 and 26 September 1998. These individuals probably moved through the region by following the riparian corridor, and may, therefore, have passed through Camp Verde and Montezuma Castle on the MCNM complex en route to Rim Rock or Lake Montezuma.

Owls. Being nocturnal, owls are not adequately surveyed with the morning point counts that we used. Usually quiet and inconspicuous during the day, owls are often not seen during general bird surveys, and are undoubtedly more common than the number of detections suggest. Several of the species listed below were detected during focused nocturnal owl surveys, conducted with tape playback, and would have been entirely missed during our daytime survey efforts.

Barn Owl • *(Tyto alba)*
Residency: yearlong
    Dates present: January – December
    Breeding status: known to breed in area
Abundance: rare
Preferred habitat type: mixed desert upland shrub

Historically widespread but uncommon, Barn Owls are now generally absent in the Camp Verde region. Sutton (1954) notes many specimens from Fort Verde in the late 1800s. Based on local reports, Taylor and Jackson (1916) describe the owl as "not uncommon" in the Verde Valley, extending up Beaver Creek at least to Montezuma Well. They described a Barn Owl nest in a dirt bank. Given its local abundance and ability to nest in cliffs and dirt banks, the Barn Owl may have nested within the MCNM complex. This is a species that has clearly declined locally and regionally, and since 1916 there has only been one record in the MCNM complex (April 1984). However, a few sightings have been recorded at Tavasci Marsh and along the adjacent Verde River (Radd in litt.).

Western Screech-Owl •
(Megascops kennicottii)
Residency: yearlong
    Dates present: January – December
    Breeding status: known to breed in area
Abundance: rare
Preferred habitat type: riparian

Taylor and Jackson (1916) recorded Screech Owls as common in the trees along Camp Verde region streams. Jackson (1941) considered the bird as casual. Sutton (1954) described a 1942 sighting of an adult and three young near Montezuma Castle, but noted that records were rare since 1942. We noted 12 detections during our surveys along riparian areas, and one probable nest in a particularly large cottonwood tree cavity.

Great Horned Owl *(Bubo virginianus)*
Residency: yearlong

Dates present: January – December
Breeding status: known to breed in area
Abundance: uncommon
Preferred habitat type: riparian; mixed desert upland shrub

This species is presently an uncommon breeder throughout the Verde Valley watershed, including the Camp Verde area (Corman and Wise-Gervais 2005). With its ability to a live in a wide range of habitats, the Great Horned Owl is one of the most widespread owls in North America. It was found frequently by Taylor and Jackson (1916), and noted as a regular breeder within the MCNM complex by Jackson (1941) and Sutton (1954). We recorded 25 individuals during our surveys between 31 January and 15 November.

Northern Pygmy-Owl • *(Glaucidium gnoma)*
Residency: migrant
Dates present: April
Breeding status: not known to breed in area
Abundance: rare
Preferred habitat type: riparian

This owl is a fairly common breeding resident of coniferous forests in central Arizona (Corman and Wise-Gervais 2005). They are occasional transients or winter visitors into nearby lowlands, particularly along forested drainages (Monson and Phillips 1981). There are many winter sightings of transients and winter visitors to the Camp Verde region and there is one record in April 1993 at the MCNM complex.

Elf Owl • *(Micrathene whitneyi)*
Residency: yearlong
Dates present: January – December
Breeding status: known to breed in area
Abundance: rare
Preferred habitat type: riparian

Generally, the Elf Owl is a sparse summer resident and uncommon breeder in central Arizona (Phillips et al. 1964). There are summer records of these owls along Oak Creek near Cornville (Stejskal and Rosenberg 1991a). The species is generally found along riparian corridors in the Camp Verde region, and we detected a pair using tape playback on 14 July 1992 during a nocturnal owl survey along Beaver Creek.

Spotted Owl *(Strix occidentalis)*
Residency: winter
Dates present: December
Breeding status: not known to breed in area
Abundance: rare; irregular-casual
Preferred habitat type: mixed desert upland shrub; riparian

Spotted Owls breed primarily in higher-elevation conifer forests, but they do occur rarely during post-breeding dispersal in the lowlands (Willey and van Riper 2007). Monson and Phillips (1981) indicated that dispersing owls are "not far from mountains" in winter. One Spotted Owl was observed in Page Springs on 4 November 1985 (V. Gilmore in litt.). Frost's (1947) MCNM complex checklist includes an entry for this owl, but there is no other supporting information for occurrence in the complex, and Sutton (1954) stated that this species needed further confirmation before accepting the record.

Short-eared Owl *(Asio flammeus)*
Residency: winter
Dates present: November – May
Breeding status: not known to breed in area
Abundance: rare
Preferred habitat type: mixed desert upland shrub

This open shrub and grassland species is generally a rare winter visitor to "open areas" statewide (Phillips et al. 1964). One Short-eared Owl was detected in May 1994 and another in May 1995 during our surveys, at night along the main road to the Castle.

Northern Saw-whet Owl *(Aegolius acadicus)*
This owl is listed in Frost's (1947) checklist, but considered by Sutton (1954) as a questionable sighting. There are no additional data or documentation of this species occurring at the MCNM complex, but Corman and Wise-Gervais (2005) show breeding records within the mountain ranges

that border the Camp Verde region.

Caprimulgids. *Nighthawks and poorwills are active primarily at dusk. Our bird surveys, based on morning counts, are not an effective method for sampling such crepuscular, aerial species. Therefore, the lack or scarcity of current records may be a function of the avian sampling techniques used, rather than an absence of species within this bird group.*

Lesser Nighthawk ● *(Chordeiles acutipennis)*
Residency: breeding
    Dates present: May – July
    Breeding status: known to breed in area
Abundance: uncommon
Preferred habitat type: mixed desert upland shrub
This species was first recorded at the MCNM complex as a "Texas Nighthawk" by Taylor and Jackson (1916), with four to six birds regularly flying nightly over Montezuma Well. Jackson (1941) lists this species as "fairly common" to "common" in summer. Sutton (1954) also describes it as a common summer resident. The bird is common in summer and is thought to breed in the Verde Valley watershed, including the Camp Verde area (Corman and Wise-Gervais 2005). The Lesser Nighthawk is also an uncommon to common summer resident in Sonoran habitat zones statewide (Phillips et al. 1964).

Common Nighthawk ● *(Chordeiles minor)*
Residency: breeding
    Dates present: May – July
    Breeding status: known to breed in area
Abundance: rare
Preferred habitat type: mixed desert upland shrub
Corman and Wise-Gervais (2005) and Phillips et al. (1964) described this species as a common summer resident in northern Arizona and the Upper Sonoran Zones in central Arizona, and there have been several accepted records from the Verde Valley. There is no mention in historical literature of this species occurring at the MCNM complex, but we did detect one on 19 June 1991. Easily misidentified as a Lesser Nighthawk or overlooked due to its crepuscular habits, it may be more common than our single sighting suggests.

Common Poorwill ● *(Phalaenoptilus nuttallii)*
Residency: breeding
    Dates present: May – July
    Breeding status: thought to breed in area
Abundance: uncommon
Preferred habitat type: mixed desert upland shrub
Taylor and Jackson (1916) noted the Common Poorwill as uncommon in upland and riparian habitats throughout the Verde Valley, but it was not recorded within the MCNM complex by Jackson (1941). However, Sutton (1954) describes a dead poorwill found near Montezuma Castle in 1942 and lists the species as a rare transient. This rare summer resident is a suspected breeder in the Camp Verde region (Corman and Wise-Gervais 2005). During our surveys we detected 10 individuals between 15 March and 24 September. Phillips et al. (1964) described it as a common summer resident in Sonoran habitat zones in central Arizona. It may be locally more common now than early in this century, or it might simply have been overlooked due to its crepuscular habits.

Swifts. *Our surveys were not focused on aerial foragers and, thus, probably seriously undercounted all swifts. However, it provides a relative index of abundance and gives some idea of swift species arrival and departure dates.*

Vaux's Swift ● *(Chaetura vauxi)*
Residency: migrant
    Dates present: March
    Breeding status: not known to breed in area
Abundance: rare
Preferred habitat type: riparian
A fairly common but irregular migrant in central Arizona (Phillips et al. 1964), this species may occur more frequently in the MCNM complex, but could be overlooked or misidentified if flying with the more common White-throated Swift. A sighting

on 27 March 1982 by Dr. Hyde at the Well picnic area is the only known record at the MCNM complex.

White-throated Swift • *(Aeronautes saxatalis)*
Residency: breeding
    Dates present: March – October
    Breeding status: known to breed in area
Abundance: common
Preferred habitat type: riparian

This swift is an aerial forager that nests in cliffs near riparian areas and is a common breeder throughout the Verde Valley watershed, including the Camp Verde area (Corman and Wise-Gervais 2005). Taylor and Jackson (1916) noted "perhaps a hundred" at Montezuma Well, while Sutton (1954) describes it as an uncommon summer resident. We found the species present from 6 March to 14 October, counting over 40 individuals that were occupying appropriate nesting cliffs throughout the breeding season.

Blue-throated Hummingbird •
(Lampornis clemenciae)
Residency: unknown
    Dates present: April – September
    Breeding status: unknown
Abundance: irregular-casual
Preferred habitat type: riparian; mixed desert upland shrub

Jackson (1941) described this hummingbird as present April through September, and breeding from April through May within the MCNM complex. However, there are no subsequent records (Sutton 1954), although Monson and Phillips (1981) report "unverified" sightings in Oak Creek. The Camp Verde region is considerably outside the normal breeding range of this species, and Blue-throated Hummingbirds do not tend to wander from their southeastern Arizona stronghold (Corman and Wise-Gervais 2005). The similar-sized Magnificent Hummingbird (Eugenes fulgens) is a casual visitor to the Mogollon Rim and nearby Oak Creek (see next species account), so we are inclined to believe Jackson's Blue-throated sightings may have been a misidentification.

Magnificent Hummingbird • *(Eugenes fulgens)*
Residency: breeding
    Dates present: March – August
    Breeding status: believed to breed in area
Abundance: rare
Preferred habitat type: riparian; mixed desert upland shrub

This hummingbird species has recently expanded into the Camp Verde region, most likely the result of increasing numbers of hummingbird feeders. We observed it between 25 March and 30 August.

Anna's Hummingbird • *(Calypte anna)*
Residency: migrant
    Dates present: June – August
    Breeding status: not known to breed in area
Abundance: rare
Preferred habitat type: riparian; mixed desert upland shrub

Although the Verde Valley is outside the general range for Anna's Hummingbird, there are records of males near Sedona (March through May 1987; Stejskal and Witzeman 1987) and Oak Creek Canyon (January 1985; Witzeman and Stejskal 1986). This species' breeding range in Arizona is quickly expanding, and now likely includes much of the riparian habitat in the Camp Verde region, at least in limited numbers (Corman and Wise-Gervais 2005). We had two detections during our surveys on 6 June 1990 and 23 August 1991.

Costa's Hummingbird • *(Calypte costae)*
Residency: migrant
    Dates present: May
    Breeding status: not known to breed in area
Abundance: rare
Preferred habitat type: riparian; mixed desert upland shrub

Costa's Hummingbird was noted by Taylor and Jackson (1916) near the mouth of Beaver Creek. Phillips (1942) reported one or two specimens taken in the 1880s and suggested this speices was a casual visitor at Camp Verde. It is now an uncommon spring migrant in the Verde Valley (Radd in litt.). We observed one individual on 8 May 1991.

Broad-tailed Hummingbird •
(Selasphorus platycercus)
Residency: migrant
    Dates present: March – May
    Breeding status: not known to breed in area
Abundance: rare
Preferred habitat type: riparian; mixed desert upland shrub
This hummingbird is a fairly common breeder at higher elevations above the Mogollon Rim, but an uncommon (Radd in litt.) to rare (Johnson and Sogge 1995) spring and fall migrant elsewhere in the Camp Verde region. We had four observations of this species between 23 March and 27 April.

Rufous Hummingbird • *(Selasphorus rufus)*
Residency: migrant
    Dates present: June – September
    Breeding status: not known to breed in area
Abundance: uncommon
Preferred habitat type: riparian; mixed desert upland shrub
The Rufous Hummingbird is an uncommon spring and fall migrant in central Arizona (Phillips et al. 1964) and elsewhere in the Verde Valley (Johnson and Sogge 1995). We observed four individuals between 10 June and 23 August.

Belted Kingfisher • *(Ceryle alcyon)*
Residency: yearlong
    Dates present: January – December
    Breeding status: known to breed in area
Abundance: uncommon
Preferred habitat type: riparian
Now a species of concern in Arizona, the Belted Kingfisher was noted as casual yearlong in the MCNM complex by Jackson (1941) and as a common resident along streams and ponds in the Verde Valley by Sutton (1954). There are no early records of nesting and recent nesting records are rare, with only a few reported in the Verde Valley and surrounding area (Witzeman and Stejskal 1987, Rosenberg and Stejskal 1989). Radd (in litt.) reported this species breeding at Tapco in 1997 and Page Springs in 1998. Breeding populations within the Verde River appear to be steadily increasing (Corman and Wise-Gervais 2005). During our surveys, we detected 60 individuals on open water along Beaver Creek riparian areas. We also found one active nest on Wet Beaver Creek in 1996.

Lewis's Woodpecker • *(Melanerpes lewis)*
Residency: migrant; winter
    Dates present: December – April
    Breeding status: not known to breed in area
Abundance: irregular-casual
Preferred habitat type: riparian; mixed desert upland shrub
Sutton (1954) describes this woodpecker species as "extremely rare," known only from several spring sightings at Montezuma Well. The species is an irregular winter visitor in the Camp Verde region and throughout the Verde Valley (Johnson and Sogge 1995).

Red-headed Woodpecker •
(Melanerpes erythrocephalus)
Residency: migrant
    Dates present: August
    Breeding status: not known to breed in area
Abundance: irregular-casual
Preferred habitat type: riparian
Generally distributed east of the Rockies, there is an extralimital record in Arizona from June 1894 (Phillips et al. 1964). During our surveys, we observed a single subadult foraging in large cottonwood trees at the Montezuma Well picnic area. This record was submitted to American Birds and accepted by the Arizona Bird Records Committee.

Gila Woodpecker • *(Melanerpes uropygialis)*
Residency: yearlong
    Dates present: January – December
    Breeding status: known to breed in area
Abundance: common
Preferred habitat type: riparian
Sutton (1954) notes the absence of this

species in Mearns' studies at Camp Verde during the 1880s, although it became "fairly abundant" in the valley and MCNM complex by 1916 (Taylor and Jackson 1916). Now frequently seen and heard in riparian habitat and at the picnic area at Montezuma Well, this woodpecker has become a common riparian breeder and yearlong resident throughout the Verde Valley watershed, including the Camp Verde area (Corman and Wise-Gervais 2005). We detected over 283 individuals on our surveys and found several active nests in gallery sycamore and cottonwood trees along Wet Beaver Creek and the lower Verde River.

Red-naped Sapsucker •
(Sphyrapicus nuchalis)
Residency: migrant
    Dates present: September – November
    Breeding status: not known to breed in area
Abundance: irregular-casual
Preferred habitat type: riparian
Jackson (1941) and Sutton (1954) list this species as casual in the MCNM complex during winter. This is a relatively secretive and easily overlooked woodpecker species and may, therefore, be a more frequent visitor to the MCNM complex and Verde Valley than our data suggest. We detected four individuals between 24 September and 8 November.

Williamson's Sapsucker •
(Sphyrapicus thyroideus)
Residency: migrant
    Dates present: October – November
    Breeding status: not known to breed in area
Abundance: irregular-casual
Preferred habitat type: riparian
Sutton (1954) reported a 1938 sighting at Montezuma Castle. This woodpecker is infrequently encountered in the Camp Verde region. One bird was seen at Camp Verde in October 1993 (Rosenberg et al. 1994a) and at Mingus Mountain during fall 1996 (Radd in litt.).

Ladder-backed Woodpecker •
(Picoides scalaris)
Residency: yearlong
    Dates present: January – December
    Breeding status: known to breed in area
Abundance: uncommon
Preferred habitat type: riparian
The Ladder-backed Woodpecker is an uncommon breeder and yearlong resident in the Verde Valley watershed, including the Camp Verde area (Sogge and Johnson 1998). We detected 102 birds during our surveys within the MCNM complex, principally in riparian and mesquite bosque habitats. We also noted active nests and breeding evidence from recently fledged young.

Northern Flicker • *(Colaptes auratus)*
Residency: yearlong
    Dates present: January – December
    Breeding status: known to breed in area
Abundance: uncommon
Preferred habitat type: riparian
All records that we could find of this species pertain to the "Red-shafted" group (*C. a. cafer*). This woodpecker was reported as a resident breeder by Jackson (1941) and Sutton (1954) in the MCNM complex. Although uncommon in terms of numbers, this species is often seen and heard in riparian areas and at the Montezuma Well picnic area. We detected 72 individuals during our surveys. In the Camp Verde region and the Verde Valley, it is most common during winter, with few breeding locally (Radd in litt.). We noted yearlong occupancy in appropriate riparian habitat and territorial behavior throughout the breeding period.

Northern Beardless-Tyrannulet •
(Camptostoma imberbe)
Sutton (1954) reported this tyrannulet as rare in the Verde Valley, and "known from Montezuma Castle, Montezuma Well, and Sedona." There is no other evidence of sighting records, and Phillips et al. (1964) did not list any winter records near the Verde Valley.

Olive-sided Flycatcher • *(Contopus cooperi)*
Residency: migrant
    Dates present: May
    Breeding status: not known to breed in area
Abundance: rare
Preferred habitat type: riparian

This flycatcher was seen perched on a tall sycamore along Beaver Creek in the western section of Montezuma Castle on 8 May 1991. It is a rare migrant in the Camp Verde region and throughout the Verde River valley (Radd in litt.).

Western Wood-Pewee •
(Contopus sordidulus)
Residency: breeding; migrant
    Dates present: June – September
    Breeding status: known to breed in area
Abundance: uncommon
Preferred habitat type: riparian

The Western Wood-Pewee is an uncommon breeder throughout the Verde Valley watershed, including the Camp Verde region (Corman and Wise-Gervais 2005). Phillips et al. (1964) reported verified breeding birds in the Verde Valley, primarily in oak and pine woodlands (Radd in litt.). Taylor and Jackson (1916), Jackson (1941) and Sutton (1954) describe MCNM complex summer sightings of this species, but with no note of breeding. During our surveys we detected 37 pewees between 21 June and 6 September, in cottonwood and sycamore riparian habitat. We noted breeding in the MCNM complex and made regular detections throughout the summer of singing-territorial behavior. We also found that northbound migrants were common into mid-June.

Willow Flycatcher • *(Empidonax traillii)*
Residency: breeding; migrant
    Dates present: May – September
    Breeding status: known to breed in area
Abundance: rare
Preferred habitat type: riparian

The Southwestern Willow Flycatcher (*E. t. extimus*) is a federally listed endangered species. This bird is a rare breeder in isolated riparian zones within the Verde Valley watershed, including the Camp Verde region (Corman and Wise-Gervais 2005). It breeds in small numbers at one site along the Verde River near Camp Verde. However, there are no records of breeding within the MCNM complex, probably because there is no appropriate habitat. We detected one singing male in a very small patch of sycamore trees at the Montezuma Well outflow on 15 May 1996. Migrants are known to sing, and this bird may have been a migrant.

Gray Flycatcher • *(Empidonax wrightii)*
Residency: migrant
    Dates present: May – November
    Breeding status: not known to breed in area
Abundance: rare
Preferred habitat type: mixed desert upland shrub

The Gray Flycatcher is a common breeder in pinyon-juniper habitat in central Arizona, breeding along the periphery of the Verde Valley. It probably occurs "casually" to "rarely" as a spring and fall migrant at the MCNM complex and in the Camp Verde region. One was collected on 10 May 1925, but recent observations are lacking.

Western Flycatcher (Pacific-slope or Cordilleran) •
(Empidonax difficilis or E. occidentalis)
Residency: migrant
    Dates present: April – September
    Breeding status: known to breed in area
Abundance: uncommon
Preferred habitat type: riparian

The two recently separated flycatcher species are difficult to differentiate in the field and until recently were considered a single species. One individual was detected on 24 April 1993, and at the time of detection this bird was noted only as Western Flycatcher. Both Pacific-slope and Cordilleran Flycatchers have been reported in the Verde Valley (Phillips et al. 1964), although the Cordilleran may be more common. This flycatcher breeds on Mingus Mountain (Radd in litt.). Pacific-slope Flycatchers are rare migrants in the Verde Valley (Radd in litt.).

Black Phoebe • *(Sayornis nigricans)*
Residency: yearlong
    Dates present: January – December
    Breeding status: known to breed in area
Abundance: common
Preferred habitat type: riparian

This phoebe is a common riparian breeder and yearlong resident throughout the entire Verde Valley watershed (Corman and Wise-Gervais 2005). The Black Phoebe builds mud nests under overhanging rocks, cliffs, and bridges, usually near water, and we often observed it foraging along the Verde, Beaver Creek, and at Montezuma Well. Sutton (1954) reported that one or two Black Phoebes nested at Montezuma Well, and one specimen was collected 2 June 1893. We made more than 90 detections of this species in riparian habitats, and found several active nests from 1991 through 1995.

Say's Phoebe • *(Sayornis saya)*
Residency: yearlong
    Dates present: January – December
    Breeding status: known to breed in area
Abundance: uncommon
Preferred habitat type: riparian

Although uncommon in terms of numbers, Say's Phoebes are very conspicuous and can usually be seen on spring or summer visits to MCNM complex upland habitats. They are yearlong residents throughout the Verde Valley watershed, including the Camp Verde area (Corman and Wise-Gervais 2005), and become even more common in fall and winter in the Verde Valley (Radd in litt.). We counted 63 birds during our MCNM complex surveys and found active nests.

Vermilion Flycatcher • *(Pyrocephalus rubinus)*
Residency: breeding
    Dates present: March – September
    Breeding status: known to breed in area
Abundance: rare
Preferred habitat type: riparian

Taylor and Jackson (1916) found this species to be fairly common, and Sutton (1941) reported this flycatcher as often seen around Montezuma Well. Both Jackson (1941) and Sutton (1954) list it as an uncommon but regular breeder in the MCNM complex. We recorded only 12 individuals between 10 March and 28 September, and two nests in riparian habitat, during our surveys. The Vermilion Flycatcher is now an uncommon breeder in the Verde Valley watershed (Corman and Wise-Gervais 2005). Based on the historical records, it appears that the Vermilion Flycatcher was previously more common in the Camp Verde region.

Ash-throated Flycatcher •
(Myiarchus cinerascens)
Residency: breeding
    Dates present: March – September
    Breeding status: known to breed in area
Abundance: common
Preferred habitat type: mixed desert upland shrub

Reported as common by Taylor and Jackson (1916), Jackson (1941), and Sutton (1954), the Ash-throated Flycatcher is still a common spring and summer breeder. Between 12 March and 25 September mainly in upland, but also in mesquite riparian habitats, we detected 237 individuals and found several nests. This is one of the most frequently seen birds at the MCNM complex, primarily because of its large size and conspicuous behavior.

Brown-crested Flycatcher •
(Myiarchus tyrannulus)
Residency: breeding
    Dates present: April – July
    Breeding status: known to breed in area
Abundance: common
Preferred habitat type: riparian

This flycatcher is a breeder in western Arizona and south of the Mogollon Rim. It has experienced population reductions due to the loss of large riparian trees, in which it finds cavities for nesting (Corman and Wise-Gervais 2005). Its presence in the MCNM complex was not reported by Jackson (1941) or Sutton (1954), but we recorded 156 individuals during our surveys. The presence of the flycatcher in the MCNM complex is an indication of the present quality of the riparian habitat. We inferred nesting of the Brown-crested Flycatcher in gallery cottonwood and sycamore riparian habitats between 24 April and 18 July by our observations of adult presence throughout summer and adults with recently fledged young near possible nest cavities

Cassin's Kingbird • *(Tyrannus vociferans)*
Residency: breeding
    Dates present: April – August
    Breeding status: known to breed in area
Abundance: uncommon
Preferred habitat type: riparian; mixed desert upland shrub

This kingbird is presently an uncommon (but sometimes locally common) breeder throughout the Verde Valley watershed, including the Camp Verde area (Corman and Wise-Gervais 2005). Jackson (1941) and Sutton (1954) also list the Cassin's Kingbird as an uncommon breeder within the MCNM complex. Sutton (1941) reports adults and young at Montezuma Castle and Montezuma Well. We found 44 individuals and active nests, primarily in riparian, but also some upland areas between 16 April and 30 August.

Western Kingbird • *(Tyrannus verticalis)*
Residency: breeding
    Dates present: April – August
    Breeding status: known to breed in area
Abundance: uncommon
Preferred habitat type: riparian; mixed desert upland shrub

Jackson (1941) and Sutton (1954) list the Western Kingbird as a fairly common breeder within the MCNM complex. We detected 42 individuals of this species between 16 April and 30 August during our surveys, and observed adults with recently fledged young. Taylor and Jackson (1916) reported this as one of the most abundant birds at all of their camps in the Verde Valley.

Loggerhead Shrike • *(Lanius ludovicianus)*
Residency: yearlong
    Dates Present: January – December
    Breeding status: known to breed in area
Abundance: uncommon
Preferred habitat type: mixed desert upland shrub

The yearlong presence of Loggerhead Shrikes indicates local nesting in the Camp Verde region, but no evidence exists of nesting within MCNM complex. Sutton (1954) reported the shrikes as "rarely seen" yearlong residents, and made no mention of nesting within the Verde Valley. This bird is currently an uncommon yearlong resident; we detected 17 individuals during our surveys within the Verde Valley.

Bell's Vireo • *(Vireo bellii)*
Residency: breeding
    Dates present: April – August
    Breeding status: known to breed in area
Abundance: uncommon
Preferred habitat type: mixed desert upland shrub; riparian

Although found to be abundant in the area by Taylor and Jackson (1916), the number of Bell's Vireos appears to have declined by the mid-1900s. This species was not reported in the MCNM complex by Jackson (1941) and was listed as uncommon by Sutton (1954). One specimen was collected on 8 August 1916. Between 18 April and 3 August we detected 25 of these vireos during our 1994–1996 surveys, and also noted that it is a frequent host of Brown-headed Cowbird parasitism within the MCNM complex.

Gray Vireo • *(Vireo vicinior)*
Residency: migrant
    Dates present: April – August
    Breeding status: not known to breed in area
Abundance: rare
Preferred habitat type: riparian; mixed desert upland shrub

Taylor and Jackson (1916) reported this species as fairly common in oak brush "in the vicinity of" Montezuma Well. Three specimens were collected by Taylor at Montezuma Well in early August 1916. We found this species to be a rare, spring-through-fall visitor, detecting eight birds in riparian and upland habitats between 30 April and 6 August.

Plumbeous Vireo • *(Vireo plumbeus)*
Residency: migrant
　Dates present: October – January; April – May
　Breeding status: not known to breed in area
Abundance: rare
Preferred habitat type: riparian; mixed desert upland shrub

Sutton (1954) described this vireo as a rare summer visitor (May through August) in the Verde Valley, with no mention of records specific to the MCNM complex. One specimen was collected by W. P. Taylor at Montezuma Well on 8 August 1916. Quiet and easily overlooked when not vocalizing, and often misidentified as a Ruby-crowned Kinglet, it may occur more frequently in fall through spring than our six records suggest.

Hutton's Vireo • *(Vireo huttoni)*
Residency: migrant
　Dates present: March
　Breeding status: not known to breed in area
Abundance: irregular-casual
Preferred habitat type: riparian

We detected a single bird of this species; it was seen foraging and calling in the tall cottonwoods and junipers along Beaver Creek south of the Montezuma Well picnic area. Another Hutton's Vireo was observed on 8 March 1998 upstream of Montezuma Well at the Wet Beaver Creek campground (Radd in litt.). There seems to have been a northward expansion in Arizona of this species (Corman and Wise-Gervais 2005), and it may become a future breeder in the Camp Verde region.

Warbling Vireo • *(Vireo gilvus)*
Residency: breeding; migrant
　Dates present: May – September
　Breeding status: known to breed in area
Abundance: rare
Preferred habitat type: riparian

Jackson (1941) reported this vireo as a casual fall migrant, but neither Jackson (1938) or Sutton (1954) provide any evidence of breeding in the MCNM complex. Statewide, this bird is a common summer resident in riparian habitat with willows, maples, and box-elder (Phillips et al. 1964), but is primarily a migrant in the Verde Valley (Radd in litt.). We did record six of these birds between 16 May and 5 September, and found one active nest in 1996. We had one detection on 28 June, which is considered late for a migrant, and may have been a wandering non-breeder.

Steller's Jay • *(Cyanocitta stelleri)*
Residency: migrant
　Dates present: November – December
　Breeding status: not known to breed in area
Abundance: irregular-casual
Preferred habitat type: riparian; mixed desert upland shrub

Sutton (1954) noted that this jay strays casually into the Verde Valley in winter. We did not detect any during our surveys.

Western Scrub-Jay •
(Aphelocoma californica)
Residency: yearlong
　Dates present: January – December
　Breeding status: known to breed in area
Abundance: rare
Preferred habitat type: mixed desert upland shrub

One Western Scrub-Jay was reported (as "Woodhouse's Jay") by Taylor and Jackson (1916), and the species was described as a casual visitor in the MCNM complex from November through February (Jackson 1941). Sutton (1954) stated that this jay was reported feeding daily at Montezuma Castle bird feeders during the early 1940s. This may have been a temporary situation, as subsequent records in the MCNM complex are few. Scrub-Jays are common in dense pinyon pine and juniper habitats in the Verde Valley. This habitat does not occur within the MCNM complex, but Scrub-Jays are occasionally seen in the MCNM complex, usually in the highest elevations where sparse pinyon and juniper occur. Although our 18 detections occurred throughout most of the year, we found no evidence of breeding. The species is known to breed within the Camp Verde region (Corman and Wise-Gervais 2005).

Mexican Jay • *(Aphelocoma ultramarina)*
Residency: migrant
  Dates present: December
  Breeding status: not known to breed in area
Abundance: irregular-casual
Preferred habitat type: mixed desert upland shrub
Recorded (as "Arizona Jay") by Jackson (1938) during a Christmas Bird Count at the MCNM complex, this jay was described as casual from August through February (Jackson 1941). Phillips et al. (1964) show no records in the Verde Valley, and state that most Mexican Jay sightings in central Arizona are actually misidentified Scrub Jays. However, given B. Jackson's familiarity with Scrub-Jays, we believe her identification of Mexican Jay was probably accurate.

Pinyon Jay • *(Gymnorhinus cyanocephalus)*
Residency: migrant
  Dates present: August – October
  Breeding status: not known to breed in area
Abundance: rare
Preferred habitat type: mixed desert upland shrub
This jay breeds in pinyon-juniper habitats, but can be found in a variety of shrub and woodland habitats during migration and winter; thus, it is considered an irregular fall and winter visitor to the Camp Verde region. We had three detections of this species between 30 August and 28 October during our MCNM complex surveys.

American Crow •
(Corvus brachyrhynchos)
Residency: migrant
  Dates present: September – April
  Breeding status: not known to breed in area
Abundance: uncommon
Preferred habitat type: mixed desert upland shrub; riparian
Sutton (1954) listed the American Crow as an occasional winter visitor to the Verde Valley, reporting flocks of about 30 and 80 in "fields at Montezuma Well" in 1947 and 1948, respectively. This species is local and usually uncommon in Arizona (Phillips et al. 1964), especially at lower elevations. Although we did not detect this species during our surveys, one sighting was made at Wet Beaver Creek ranger station on 2 May 1998 (Radd in litt.), and three other detections have been reported: 2 January 1991(Thornburg in litt.), 2 September 1991, and 27 April 1993.

Common Raven • *(Corvus corax)*
Residency: yearlong
  Dates present: January – December
  Breeding status: known to breed in area
Abundance: common
Preferred habitat type: mixed desert upland shrub
This raven is a common breeder and yearlong resident throughout the Verde Valley watershed, including the Camp Verde region (Corman and Wise-Gervais 2005). It is one of the most frequently seen birds (though not in large numbers), often soaring high above the MCNM complex. We detected 131 individuals during our surveys and found two nests each year from 1992 to 1997 near the Castle. Another nest was found in 1992 in the cliffs at the west border of Montezuma Castle.

Horned Lark • *(Eremophila alpestris)*
Residency: winter
  Dates present: February
  Breeding status: not known to breed in area
Abundance: rare
Preferred habitat type: mixed desert upland shrub
Historically, the Horned Lark was a rare winter visitor (Sutton 1954), with records at and near Montezuma Well. This species prefers open dirt and rocky areas, and short grasslands which are not common in the MCNM complex. We noted one bird, observed on 17 February 1997 at the Montezuma Well picnic area.

Swallows. *Our surveys were not focused on aerial foragers, and we undoubtedly under-*

counted all swallows. However, our surveys do provide a relative index of abundance and some idea of arrival and departure dates.

Purple Martin • *(Progne subis)*
Residency: migrant
    Dates present: May
    Breeding status: not known to breed in area
Abundance: irregular-casual
Preferred habitat type: mixed desert upland shrub
One of these martins was collected at "Beaver Creek near Fort Verde" in May 1888. Taylor and Jackson (1916) described "a few" Purple Martins regularly seen at Montezuma Well. However, the species is not listed by Jackson (1941) and was reported as "extremely rare" by Sutton (1954). We know of only one recent record from May 1978. This suggests that the Purple Martin, which experienced a local population decline in the early 1900s, has not recovered in the Camp Verde region.

Violet-green Swallow •
(Tachycineta thalassina)
Residency: breeding
    Dates present: March – August
    Breeding status: known to breed in area
Abundance: common
Preferred habitat type: riparian; mixed desert upland shrub
This swallow is a common breeder mainly in riparian zones throughout the Verde Valley watershed, including the Camp Verde region (Corman and Wise-Gervais 2005). We detected 92 individuals during our surveys between 5 March and 4 August. This is a tree- and cliff-nesting species, and breeding status was confirmed with active nests and from observations of adults with recently fledged young in and around riparian trees throughout the summer. Flocks can usually be seen flying low over Beaver Creek or high over the MCNM complex from March through August.

Northern Rough-winged Swallow •
(Stelgidopteryx serripennis)
Residency: breeding
    Dates present: March – July
    Breeding status: known to breed in area
Abundance: common
Preferred habitat type: riparian
This swallow, a bank-nesting species, is a common breeder throughout the Verde Valley watershed, including the Camp Verde area (Corman and Wise-Gervais 2005). It was listed by Jackson (1941) and Sutton (1954) as a fairly common to common breeder within the MCNM complex. We had 26 detections between 6 March and 18 July and observed adults hovering outside of small burrow openings in the dirt banks along Beaver Creek on the west side of the Montezuma Castle portion in June 1997.

Bank Swallow • *(Riparia riparia)*
Residency: migrant
    Dates present: April – June
    Breeding status: not known to breed in area
Abundance: rare
Preferred habitat type: mixed desert upland shrub
Historically, this swallow has been rare in the Camp Verde region, and there are no recent records at Montezuma Well. Taylor and Jackson (1916) reported this swallow at the mouth of Beaver Creek, while Jackson (1941) did not report the species, and Sutton (1954) mentioned reports at Montezuma Well but considered this swallow "hypothetical." Phillips et al. (1964) described it as "fairly common to rare" throughout the state. Thus, the exact historical status within the MCNM complex is uncertain. Although we did not detect this species during our surveys, it is now considered a rare migrant in the Verde Valley (Radd in litt.).

Cliff Swallow • *(Petrochelidon pyrrhonota)*
Residency: breeding
    Dates present: March – July
    Breeding status: known to breed in area
Abundance: common
Preferred habitat type: open water; mixed desert upland shrub
Recorded as a breeder within the MCNM complex at least as far back as 1916 (Taylor and

Jackson 1916), Cliff Swallows build gourd-shaped nests of mud under overhanging cliffs, bridges, and eaves of buildings. Each year more than 50 nests are built in the cliffs at and near Montezuma Castle, and adults can easily be seen flying in and near riparian areas along Beaver Creek. We detected 54 birds between 13 March and 19 July, and observed many active nests on the cliffs each year, especially near Montezuma Castle.

Mountain Chickadee • *(Poecile gambeli)*
Residency: migrant; winter
    Dates present: December – February
    Breeding status: not known to breed in area
Abundance: irregular-casual
Preferred habitat type: mixed desert upland shrub
Mountain Chickadees breed and reside in the higher elevations above the Verde Valley, but occasionally appear in the Verde Valley in late fall and winter (Sutton 1954, Stejskal and Witzeman 1985). There are two recent records at the MCNM complex, one from 22 December 1987 (Thornburg in litt.) and one made during our surveys on 17 February 1997.

Bridled Titmouse •
(Baeolophus wollweberi)
Residency: yearlong
    Dates present: January – December
    Breeding status: known to breed in area
Abundance: common
Preferred habitat type: riparian
Taylor and Jackson (1916) found this species to be common, and Jackson (1941) and Sutton (1954) also list it as a fairly common to common resident breeder within the MCNM complex. We found this species present on almost all of our surveys, detecting 123 individuals principally in sycamore and cottonwood riparian habitats. We also recorded breeding, with at least two active nests found each year from 1991 through 1997. It is considered a common riparian breeder and yearlong resident throughout the Verde Valley watershed, including the Camp Verde region (Corman and Wise-Gervais 2005).

Juniper Titmouse • *(Baeolophus ridgwai)*
Residency: yearlong
    Dates present: January – December
    Breeding status: known to breed in area
Abundance: uncommon
Preferred habitat type: mixed desert upland shrub
The Juniper Titmouse uses primarily upland shrub and trees, but also riparian areas during the winter. Although this titmouse was reported near the MCNM complex by Taylor and Jackson (1916), it was not included in Jackson's (1941) checklist. Sutton (1954) recorded it as present in the northern part of the Verde Valley, but it was unknown at the MCNM complex after 1916. We had 25 detections of this species during our surveys; thus it appears that the Juniper Titmouse is more common now than during the mid-1900s. Given our detection dates and available habitat, it is possible that they breed in the MCNM complex, but we were unable to locate any active nests. Corman and Wise-Gervais (2005) found this titmouse a common breeder and yearlong resident throughout the Verde Valley watershed, including the Camp Verde region.

Verdin • *(Auriparus flaviceps)*
Residency: yearlong
    Dates present: January – December
    Breeding status: known to breed in area
Abundance: uncommon
Preferred habitat type: mixed desert upland shrub
The Verdin is an uncommon breeder and yearlong resident throughout the Verde Valley watershed, including the Camp Verde region (Corman and Wise-Gervais 2005). Jackson (1941) recorded the Verdin being present only in October and November. Sutton (1954) described it as uncommon with records only from October through June, reporting sightings of adults and immature birds at Montezuma Castle. We detected the species primarily in the transition zone between riparian and upland, particularly in mesquite habitat, recording 42 birds during the summer, in addition to finding several active nests.

Bushtit • *(Psaltriparus minimus)*
Residency: yearlong
    Dates present: January – December
    Breeding status: known to breed in area
Abundance: common
Preferred habitat type: mixed desert upland shrub

A common resident of Upper Sonoran woodlands throughout Arizona (Monson and Phillips 1981), the Bushtit is also common in the Camp Verde region. We detected 64 individuals and nests, primarily in MCNM complex upland habitats, but also in riparian locations during the winter.

White-breasted Nuthatch • *(Sitta carolinensis)*
Residency: winter; migrant
    Dates present: February – March; August – October
    Breeding status: not known to breed in area
Abundance: irregular-casual
Preferred habitat type: riparian

Sutton (1954) described the White-breasted Nuthatch as a rare winter visitor reported from Montezuma Well. We detected 10 birds from 17 February to 13 March and from 9 August to 19 October, with most sightings being in large sycamores and cottonwoods, including the Well picnic area.

Pygmy Nuthatch • *(Sitta pygmaea)*
Residency: migrant
    Dates present: July
    Breeding status: not known to breed in area
Abundance: rare
Preferred habitat type: riparian

This species has been reported to be a fairly common to common resident in pines statewide, and casual at lower elevations (Phillips et al. 1964). We recorded one observation on 28 July 1993.

Brown Creeper • *(Certhia americana)*
Residency: winter
    Dates present: January – March
    Breeding status: not known to breed in area
Abundance: rare
Preferred habitat type: riparian

Sutton (1954) reported this species as a rare winter visitor (though more common in some years), with sightings throughout the Verde region and at Montezuma Well. We had six detections of this species in the MCNM complex between 19 January and 20 March.

Cactus Wren
(Campylorhynchus brunneicapillus)
Residency: breeding
    Dates present: March – May
    Breeding status: known to breed in area
Abundance: rare
Preferred habitat type: mixed desert upland shrub

This wren species was historically absent from the upper Verde River valley (Phillips et al. 1964). Sutton (1954) reported it as "extremely rare" in the Verde Valley, noting just one record from Montezuma Well from 18 May 1947. Sighting records began to increase in the 1960s (Monson and Phillips 1981), and there is one recent record from 14 March 1985. In the Verde Valley area, the Cactus Wren is at the northern edge of its range and is a very local and rare breeder in portions where appropriate habitat exists (such as the slopes above Beaver Creek). The bird is generally found at slightly higher elevations than those at the MCNM complex (Radd in litt.).

Rock Wren • *(Salpinctes obsoletus)*
Residency: yearlong
    Dates present: January – December
    Breeding status: known to breed in area
Abundance: common
Preferred habitat type: mixed desert upland shrub

A very vocal and readily detected species in areas with cliffs and large rocks, this wren is a common breeder and yearlong resident throughout the Verde Valley and the Camp Verde area. We detected 138 Rock Wrens during our MCNM complex surveys, primarily in rock and cliff areas in the uplands. Rock Wren nests are well concealed deep inside cracks in cliffs or jumbles of boulders, and we found only one active nest, but the species is much more prevalent as a breeder than the small number of located nests would suggest.

Canyon Wren • *(Catherpes mexicanus)*
Residency: yearlong
    Dates present: January – December
    Breeding status: known to breed in area
Abundance: common
Preferred habitat type: mixed desert upland shrub

This wren is a fairly common, permanent resident of the Camp Verde region. We found 79 birds in our surveys, principally in cliffs and canyons. Canyon Wrens can be heard on almost any day, spring through fall, in appropriate habitat within the MCNM complex. Although we did not find any nests, breeding status is implied from the presence of singing individuals in appropriate breeding habitat throughout the summer.

Bewick's Wren • *(Thryomanes bewickii)*
Residency: yearlong
    Dates present: January – December
    Breeding status: known to breed in area
Abundance: common
Preferred habitat type: riparian

The species is a common riparian-upland breeder throughout the Verde Valley watershed, including the Camp Verde area (Corman and Wise-Gervais 2005). Bewick's Wren is a very vocal species that is readily heard in and near riparian areas. We detected 550 of these wrens during our surveys, with more birds found during the breeding period. During the breeding season they were found primarily in riparian habitat, but some use was detected in the adjacent upland habitats in the post-breeding season and winter. The difference in abundance between breeding and nonbreeding season may actually be a function of detection, rather than actual population size differences. We found many nests and saw recently fledged birds in the MCNM complex.

House Wren • *(Troglodytes aedon)*
Residency: migrant; winter
    Dates present: February – May; August – October
    Breeding status: not known to breed in area
Abundance: rare
Preferred habitat type: riparian

This wren is a casual migrant and rare winter visitor and is uncommon fall through winter in the Verde Valley (Radd in litt.). Jackson (1941) reported this as a casual species from April through July. Sutton (1954) noted sightings from both Montezuma Castle and Montezuma Well. We detected six birds during our surveys from 19 February to 8 May; and from 23 August to 17 October.

Marsh Wren • *(Cistothorus palustris)*
Residency: breeder
    Dates present: January
    Breeding status: not known to breed in area
Abundance: rare
Preferred habitat type: marsh

Though uncommon throughout most of Arizona (Phillips et al. 1964), Marsh Wrens are fairly common to common fall through spring in wetland habitats in the Verde Valley (Radd in litt.). Corman and Wise-Gervais (2005) list this species as a rare marsh breeder in the Verde Valley watershed. Marsh Wrens were not reported by Jackson (1941) or Sutton (1954) at the MCNM complex, but we had one sighting on 20 January 1992 at Montezuma Well. This species probably visits the MCNM complex more often than our data suggest, although suitable habitat is limited.

Ruby-crowned Kinglet • *(Regulus calendula)*
Residency: migrant; winter
    Dates present: January – June; October – December
    Breeding status: not known to breed in area
Abundance: common
Preferred habitat type: riparian

This is one of the most readily detectable winter birds in the Verde Valley and MCNM complex. Although Jackson (1941) and Sutton (1954) reported it present yearlong, we did not detect the species outside of the migratory and winter seasons. We counted 90 individuals from 4 January to 4 June and from 9 October to 13 December in our surveys, primarily in riparian habitats and upland habitats with trees.

Blue-gray Gnatcatcher • *(Polioptila caerulea)*
Residency: yearlong
    Dates present: January – December
    Breeding status: known to breed in area
Abundance: uncommon
Preferred habitat type: mixed desert upland shrub
This species is an uncommon breeder and yearlong resident throughout the Verde Valley watershed, including the Camp Verde region, occurring primarily in mesquite bosque and adjacent shrubby uplands (Corman and Wise-Gervais 2005). Taylor and Jackson (1916) reported hearing this gnatcatcher at Montezuma Well. Sutton (1954) notes only one other record at Montezuma Castle (in 1938 by B. Jackson) and makes no reference to breeding within the MCNM complex. We found the species to be casual during winter and fall, uncommon in the spring, and a summer breeder, detecting 69 individuals in our surveys. We also found active nests that had been parasitized by Brown-headed Cowbirds. It appears that Blue-gray Gnatcatchers may be a more frequent breeder now than in the mid-1900s.

Western Bluebird • *(Sialia mexicana)*
Residency: winter; migrant
    Dates present: January – April; October – December
    Breeding status: not known to breed in area
Abundance: rare
Preferred habitat type: mixed desert upland shrub
Usually detected in small flocks, often flying over the Camp Verde region, the Western Bluebird is fairly common in winter and uncommon during spring and fall. It may be seen or heard in any habitat within the MCNM complex, though most frequently in areas with trees. On our surveys we found 27 bluebirds in the riparian corridors and juniper uplands from 8 January to 21 April and from 14 October to 5 December.

Mountain Bluebird • *(Sialia currucoides)*
Residency: winter; migrant
    Dates present: December – March
    Breeding status: not known to breed in area
Abundance: uncommon
Preferred habitat type: mixed desert upland shrub
This bluebird is an irregular to uncommon winter visitor to the Camp Verde region. Historically, it was observed occasionally at Montezuma Well during winter (Sutton 1954). We counted six individuals during our surveys from 5 December to 26 March, primarily in riparian woodlands and uplands. The Mountain Bluebird will use more sparse upland habitat than the Western Bluebird typically occupies.

Townsend's Solitaire • *(Myadestes townsendi)*
Residency: winter
    Dates present: December – April
    Breeding status: not known to breed in area
Abundance: uncommon
Preferred habitat type: riparian; mixed desert upland shrub
Jackson (1941) reported Townsend's Solitaires as casual to uncommon in February and March. Sutton (1954) described them as common winter visitors to the Verde Valley, but made no specific mention of records within the MCNM complex. This species is found primarily in riparian woodlands and juniper uplands. We made 12 detections of this species between 5 December and 16 April. They are now rare to uncommon fall through spring elsewhere in the Verde Valley (Johnson and Sogge 1995, Radd in litt.).

Swainson's Thrush • *(Catharus ustulatus)*
This thrush is a very uncommon spring migrant in central Arizona, with only one verified record (May 1887) for the Verde Valley (Phillips et al. 1964). Frost's (1947) checklist includes an entry for this bird, but there is no other supporting information, and Sutton (1954) did not list this species at the MCNM complex. Swainson's Thrush is easily confused with the Hermit Thrush, which is a much more common migrant and winter visitor.

Hermit Thrush • *(Catharus guttatus)*
Residency: winter

Dates present: October – March
Breeding status: not known to breed in area
Abundance: rare
Preferred habitat type: riparian

This bird is an uncommon species throughout the Verde Valley during the winter (Radd in litt.), found almost exclusively in riparian woodland with dense understory vegetation. The species is described as casual to rare in the MCNM complex during winter by Jackson (1941) and Sutton (1954). A secretive species that can easily go undetected, it may also be a regular but uncommon fall and spring migrant, as we detected five birds between 14 October and 1 March.

American Robin • *(Turdus migratorius)*
Residency: yearlong
Dates present: January – December
Breeding status: known to breed in area
Abundance: common
Preferred habitat type: riparian

This riparian woodland breeder uses juniper uplands and grasslands in migration and during winter. We noted the species as a common fall through spring visitor, and a fairly common summer breeder, detecting 94 robins during our MCNM complex surveys and finding a number of active nests. The American Robin is a common breeder and yearlong resident throughout the Verde Valley watershed, including the Camp Verde area (Corman and Wise-Gervais 2005).

Northern Mockingbird • *(Mimus polyglottus)*
Residency: breeding
Dates present: April – September
Breeding status: known to breed in area
Abundance: common
Preferred habitat type: mixed desert upland shrub

The Northern Mockingbird is a very common upland breeder throughout the Verde Valley watershed, including the Camp Verde area (Corman and Wise-Gervais 2005). Taylor and Jackson (1916) noted that mockingbirds were "seen daily," and Jackson (1941) reported them as fairly common breeders within the MCNM complex. In addition to finding many nests during our studies, we detected 149 individuals between 14 April and 24 September.

Sage Thrasher • *(Oreoscoptes montanus)*
Residency: migrant
Dates present: August – December
Breeding status: not known to breed in area
Abundance: rare
Preferred habitat type: mixed desert upland shrub

Sutton (1954) reported this thrasher as an uncommon spring and fall migrant in the Verde Valley, usually found in small flocks of 10 to 12 birds using upland mesquite. No records specific to the MCNM complex are listed in Jackson (1941) or Sutton (1954). On our surveys, we detected 10 individuals, mostly between 23 August and 18 December, with one record 16 March. We detected most of our birds in upland shrub, often foraging on juniper berries.

Brown Thrasher • *(Toxostoma rufum)*
Residency: migrant
Dates present: May
Breeding status: not known to breed in area
Abundance: irregular-casual
Preferred habitat type: mixed desert upland shrub

The Brown Thrasher is a rare spring vagrant in Arizona (Stejskal and Witzeman 1987), found only during the spring migration period. The one existing record from the Verde Valley area is from May 1988 (Witzeman and Stejskal 1988).

Crissal Thrasher • *(Toxostoma crissale)*
Residency: yearlong
Dates present: January – December
Breeding status: known to breed in area
Abundance: uncommon
Preferred habitat type: mixed desert upland shrub

The Crissal Thrasher is an upland breeder and yearlong resident throughout the Verde Valley watershed, including the Camp Verde area (Corman and Wise-Gervais 2005). One bird was collected on 2 April 1886 at Montezuma Well. This is a secretive and difficult-to-find species except during the breeding season, when it can be very vocal.

We found 21 birds during our surveys and several nests within the MCNM complex's mesquite bosque and shrubby uplands.

Le Conte's Thrasher • *(Toxostoma lecontei)*
Frost (1947) includes this species in his Montezuma Castle bird checklist, but there are no other supporting data. All thrashers are secretive and often difficult to view clearly, and the LeConte's Thrasher could be easily confused with the resident Crissal Thrasher. Moreover, Monson and Phillips (1981) state that there are no accepted records from the Verde Valley even though Rosenberg et al. (1980) described a sight record at Tuzigoot National Monument in May 1980. Given the rarity of this species in the region, its non-migratory habitats, strict habitat selection, and the potential for misidentification, we believe that this thrasher needs additional documentation in the Camp Verde region.

American Pipit • *(Anthus rubescens)*
Residency: winter
Dates present: December – March
Breeding status: not known to breed in area
Abundance: rare
Preferred habitat type: riparian
Historically, the American Pipit has been uncommon in winter (Sutton 1954), with flocks being noted only at Montezuma Well. We found five of these birds during our surveys between 26 December and 21 March and one summer record on 29 June. This casual winter visitor favors open streamside areas.

Cedar Waxwing • *(Bombycilla cedrorum)*
Residency: migrant
Dates present: November; March – April
Breeding status: not known to breed in area
Abundance: rare
Preferred habitat type: riparian
This species is a rare fall and uncommon spring migrant in the Verde Valley (Johnson and Sogge 1995, Radd in litt.). Sutton (1954) describes these waxwings as "extremely rare," and known only from Montezuma Castle where they were observed eating "sycamore balls and hackberries." We had three detections: 29 April 1992, 8 November 1997, and 6 March 1999. The species typically occurs irregularly and in highly variable numbers from fall to spring in lowland riparian woodlands.

Phainopepla • *(Phainopepla nitens)*
Residency: breeding
Dates present: March – August
Breeding status: known to breed in area
Abundance: common
Preferred habitat type: mixed desert upland shrub
The Phainopepla is a common upland-riparian breeder throughout the Verde Valley watershed, including the Camp Verde area (Corman and Wise-Gervais 2005). The Phainopepla is one of the most common upland nesting species in the MCNM complex. Between 25 March and 9 August we recorded 155 Phainopepla during our survey efforts. The birds were detected in riparian woodlands (especially mesquite) and uplands with trees and tall shrubs, and many active nests were found. Phillips et al. (1964) noted that the Phainopepla winters in most of the lower Sonoran habitat zones statewide except the Verde Valley. However, "very small numbers" do winter locally in the Verde Valley (Radd in litt.).

European Starling *(Sturnus vulgaris)*
Residency: yearlong
Dates present: January – December
Breeding status: known to breed in area
Abundance: abundant
Preferred habitat type: riparian
Starlings were introduced to North America in 1890 and 1891, when about 100 birds were released in New York's Central Park (Cabe 1993). They spread rapidly and were present in the western United States by 1940. The species was first recorded in Arizona in 1946 and in the MCNM complex in 1949 (Sutton 1954). This is likely very close to its actual establishment date here and in the Verde Valley, because it was not listed in many ornithological records prior to the early

1940s. The species has become an abundant breeder and yearlong resident throughout the Verde Valley watershed, including the Camp Verde area (Corman and Wise-Gervais 2005). During our avian surveys we detected 149 starlings, predominantly in riparian habitats. Throughout the United States, starlings have become an unwanted "pest" species because they compete with native species for nesting cavities. We have observed starlings chasing American Kestrels and Northern Flickers away from potential nest cavities at the Montezuma Well picnic area. We also observed adults with nest material entering holes in large sycamores and cottonwoods (particularly at Montezuma Well), and juveniles were seen near nest holes.

Golden-winged Warbler •
(Vermivora chrysoptera)
Residency: migrant
    Dates present: September
    Breeding status: not known to breed in area
Abundance: irregular-casual
Preferred habitat type: riparian

The Golden-winged Warbler is a casual transient in Arizona (Monson and Phillips 1981). There is one record on 1 September 1988 at the MCNM complex (Rosenberg and Stejskal 1989) during the fall migration period.

Orange-crowned Warbler • *(Vermivora celata)*
Residency: migrant
    Dates present: April – May; September – October
    Breeding status: not known to breed in area
Abundance: uncommon
Preferred habitat type: riparian

This small warbler is a fairly common spring and fall migrant in the Verde Valley (Johnson and Sogge 1995). Reported by Sutton (1954) to be "extremely rare," there was one 1953 record from Montezuma Castle. During our surveys, we detected nine individuals in riparian woodlands and adjacent uplands with trees and tall shrubs from 4 April to 22 May and from 20 September to 19 October.

Nashville Warbler • *(Vermivora ruficapilla)*
Residency: migrant
    Dates present: April – May; September – October
    Breeding status: not known to breed in area
Abundance: uncommon
Preferred habitat type: riparian

The Nashville Warbler is an uncommon spring and fall migrant in the Verde Valley (Johnson and Sogge 1995). We had only one record on 2 September 1991, from riparian woodlands in the MCNM complex.

Virginia's Warbler • *(Vermivora virginiae)*
Residency: migrant
    Dates present: March – June
    Breeding status: not known to breed in area
Abundance: rare
Preferred habitat type: riparian

This species is a common summer resident in brush habitats at high elevations in Arizona, and a fairly common transient during migration (Phillips et al. 1964). Between 27 March and 28 June we recorded six of this species in riparian woodlands within the MCNM complex.

Lucy's Warbler • *(Vermivora luciae)*
Residency: breeding
    Dates present: March – August
    Breeding status: known to breed in area
Abundance: common
Preferred habitat type: riparian

This warbler was reported in the MCNM complex as "plentiful" by Taylor and Jackson (1916) and common by Jackson (1941) and Sutton (1954). We found 325 Lucy's Warblers during our surveys in riparian woodlands and mesquite bosque, from 16 March to 6 August. Even though the adults are very vocal during the breeding season, this species is a probably a more common nester than implied by the single nest found in the MCNM complex, since it is a cavity nester and often escapes detection. We only discovered one nest, and that was parasitized by Brown-headed Cowbirds. The species is a common riparian-upland breeder throughout the Verde Valley watershed, including the Camp Verde area (Corman and Wise-Gervais 2005).

Yellow Warbler • *(Dendroica petechia)*
Residency: breeding
    Dates present: April – September
    Breeding status: known to breed in area
Abundance: common
Preferred habitat type: riparian

The Yellow Warbler is a common riparian breeder throughout the Verde Valley watershed, including the Camp Verde area (Corman and Wise-Gervais 2005). There were three specimens collected by Taylor at Montezuma Well in early August 1916. Historically, it was also reported as a common breeder by Taylor and Jackson (1916), Jackson (1941) and Sutton (1954); and Taylor and Jackson (1916) report a pair of these warblers feeding a young cowbird. We counted 154 birds of this species between 3 April and 6 September during our surveys at the MCNM complex. We also found two Yellow Warbler nests in riparian woodlands, as well as juveniles, and in one instance verified it as a cowbird host in the MCNM complex. This species often nests high above ground, so nests may go undetected and may be a more common nester than suggested by the two nests that we found.

Yellow-rumped Warbler •
(Dendroica coronata)
Residency: migrant; winter
    Dates present: October – May
    Breeding status: not known to breed in area
Abundance: uncommon
Preferred habitat type: riparian

Johnson and Sogge (1995) reported the Yellow-rumped Warbler as an uncommon spring and fall migrant in the Verde Valley. We also found this species to be an uncommon winter visitor, detecting 71 birds between 9 October and 28 May, principally in riparian woodlands.

Black-throated Blue Warbler •
(Dendroica caerulescens)
Residency: migrant
    Dates present: October
    Breeding status: not known to breed in area
Abundance: irregular-casual
Preferred habitat type: riparian

The Black-throated Blue Warbler is a casual transient statewide in Arizona (Phillips et al. 1964). There is only one record at Montezuma Castle, taken on 6 October 1975 (Witzeman et al. 1976). We did not detect this species during our surveys.

Black-throated Gray Warbler • *(Dendroica nigrescens)*
Residency: migrant
    Dates present: March – April
    Breeding status: not known to breed in area
Abundance: rare
Preferred habitat type: riparian

This warbler breeds in pinyon and juniper forests along the Mogollon Rim (Phillips et al. 1964), and is an uncommon to fairly common spring and fall migrant statewide (Phillips et al. 1964), as well as in some areas of the Verde Valley (Johnson and Sogge 1995). The species was reported by Jackson (1941) only as a casual spring migrant. Sutton (1954) stated that the only records from Montezuma Castle and Montezuma Well were in May. We found the bird rare, being present only during March and April.

Townsend's Warbler • *(Dendroica townsendi)*
Residency: migrant
    Dates present: January; March
    Breeding status: not known to breed in area
Abundance: rare
Preferred habitat type: riparian

Townsend's Warbler is a fairly common transient in lower elevations throughout Arizona (Phillips et al. 1964). We had only two records, on 31 March 1984 and 2 January 1993, from our MCNM complex surveys.

Grace's Warbler • *(Dendroica graciae)*
Residency: migrant
    Dates present: May; September
    Breeding status: not known to breed in area
Abundance: rare

Preferred habitat type: riparian

This warbler is a casual Sonoran Zone migrant in Arizona (Phillips et al. 1964) and a breeder on Mingus Mountain (Radd in litt.). Jackson (1941) reports this warbler as casual in May and September, but there have been no records at the MCNM complex since then and we did not detect this species during our surveys.

Northern Waterthrush •
(Seiurus noveboracensis)
Residency: migrant
   Dates present: September
   Breeding status: not known to breed in area
Abundance: rare
Referred habitat type: riparian

This waterthrush is an uncommon streamside transient statewide (Phillips et al. 1964) and a rare spring and fall migrant in dense riparian areas in the Camp Verde region (Johnson and Sogge 1995). We had only one record of a Northern Waterthrush during our surveys at the MCNM complex, on 2 September 1991.

MacGillivray's Warbler • *(Oporornis tolmiei)*
Residency: migrant
   Dates present: May; August – September
   Breeding status: not known to breed in area
Abundance: rare
Preferred habitat type: riparian

We found the MacGillivray's Warbler to be a casual spring and uncommon fall migrant, detecting 10 individuals during surveys on 8 May and between 23 August and 25 September. All birds were found in riparian woodlands, particularly those with wet, dense understory vegetation. Sutton (1954) reports this species as casual in summer, with sightings at Montezuma Castle and Montezuma Well.

Common Yellowthroat • *(Geothlypis trichas)*
Residency: breeding
   Dates present: April – July
   Breeding status: known to breed in area
Abundance: irregular-casual
Preferred habitat type: riparian

The Common Yellowthroat breeds in dense riparian areas throughout the Verde Valley (Johnson and Sogge 1995, Corman and Wise-Gervais 2005). It was described by Sutton (1954) as a rare spring and fall migrant. We counted eight individuals between 16 April and 27 June during our surveys, all in marsh and other wetlands, often near riparian woodlands. Although we did not find any nests, breeding was inferred in the MCNM complex from the presence of singing adults throughout the breeding season.

Hooded Warbler • *(Wilsonia citrina)*
Residency: migrant
   Dates present: April
   Breeding status: not known to breed in area
Abundance: irregular-casual
Preferred habitat type: riparian

A very rare migrant, only one Hooded Warbler, a male, has been reported in the area. It was detected on 19 April 1991 foraging in the tall cottonwoods just west of the Montezuma Well picnic area. This record was accepted by the Arizona Bird Records Committee, and published by Rosenberg and Stejskal (1991).

Wilson's Warbler • *(Wilsonia pusilla)*
Residency: migrant
   Dates present: April – September
   Breeding status: not known to breed in area
Abundance: uncommon
Preferred habitat type: riparian

The Wilson's Warbler is found fairly frequently as a migrant in lowland riparian areas in the Camp Verde region through the first and second week of June, more rarely later in summer (Radd in litt.). We detected seven migrant birds between 16 April and 24 September in riparian woodlands. It was described by Sutton (1954) as an uncommon summer resident in the Verde Valley, but no records of breeding were noted within the MCNM complex.

Yellow-breasted Chat • *(Icteria virens)*
Residency: breeding
   Dates present: May – August
   Breeding status: known to breed in area

Abundance: common
Preferred habitat type: riparian; mixed desert upland shrub

The Yellow-breasted Chat is a common breeder statewide in riparian and mesquite bosque habitats (Phillips et al. 1964). The species is also a common riparian breeder throughout the Verde Valley watershed, including the Camp Verde region (Corman and Wise-Gervais 2005). There were 50 chats detected on our surveys, with birds present between 8 May and 23 August. We also found several nests, and all were parasitized by the Brown-headed Cowbird.

Hepatic Tanager • *(Piranga flava)*
Residency: migrant
   Dates present: April – May; August
   Breeding status: not known to breed in area
Abundance: irregular-casual
Preferred habitat type: mixed desert upland shrub

The Hepatic Tanager is a common summer resident in oak, pinyon pine, and other pine habitats in central Arizona, with several local records noted in Phillips et al. (1964) and breeding season detections at Mingus Mountain in 1998 (Radd in litt.). Jackson (1941) lists this species as casual to uncommon from April through August at the MCNM complex, but it is quite possible that these sightings are misidentifications of Summer Tanagers. There is some probability of detecting this species moving through the MCNM complex, as we made two Hepatic Tanager detections on 16 May 1992 and 20 May 1994.

Summer Tanager • *(Piranga rubra)*
Residency: breeding
   Dates present: April – September
   Breeding status: known to breed in area
Abundance: common
Preferred habitat type: riparian; mixed desert upland shrub

The Summer Tanager is a common riparian breeder throughout the Verde Valley watershed, including the Camp Verde area (Corman and Wise-Gervais 2005). It was seen "almost daily" near Montezuma Well by Taylor and Jackson (1916), and listed as a common summer resident by Sutton (1954). We found the species uncommon in spring, but more abundant during the summer as a breeder and as a fall migrant, detecting 111 individuals between 23 April and 25 September on our surveys. We found nests in MCNM complex riparian habitat and adjacent mesquite bosque, and recorded recently fledged juveniles with adults.

Western Tanager • *(Piranga ludoviciana)*
Residency: migrant
   Dates present: May – September
   Breeding status: not known to breed in area
Abundance: uncommon
Preferred habitat type: riparian; mixed desert upland shrub

We found the Western Tanager to be an uncommon spring transient and a casual summer and fall visitor, detecting 19 birds between 8 May and 25 September. We found birds in riparian woodlands, mesquite bosque, and the uplands. The birds regularly occur as migrants in June and July in the Verde River area, and are among the earliest southbound migrants.

Northern Cardinal • *(Carinalis cardinalis)*
Residency: yearlong
   Dates present: January – December
   Breeding status: known to breed in area
Abundance: common
Preferred habitat type: riparian

The Northern Cardinal is a common breeder throughout the Verde Valley watershed, including the Camp Verde area (Corman and Wise-Gervais 2005). We found this species fairly abundant during spring and the summer breeding period, but uncommon in fall and winter. We detected 69 individuals during our surveys, but the species is more quiet and secretive, and therefore harder to detect, during the fall and winter. They nest primarily in riparian woodlands with dense understory; we found several nests and have verified this species as a cowbird host in the MCNM complex.

Black-headed Grosbeak •
(Pheucticus melanocephalus)
Residency: migrant
    Dates present: May; August – September
    Breeding status: known to breed in area
Abundance: uncommon
Preferred habitat type: riparian

The Black-headed Grosbeak is an uncommon but regular migrant throughout Arizona (Phillips et al. 1964). Sutton (1954) reported this bird as a rare spring and summer visitor to the MCNM complex. We found the bird to be a rare spring and fall migrant, and a casual summer visitor and breeder, recording only three (19 May, 9 August, and 20 September) during our surveys. But the species probably occurs more frequently in the MCNM complex than our number of detections imply. We did find one active nest in riparian woodland during 1995.

Blue Grosbeak • *(Passerina caerulea)*
Residency: breeding; migrant
    Dates present: May – September
    Breeding status: known to breed in area
Abundance: uncommon
Preferred habitat type: riparian; mixed desert upland shrub

The Blue Grosbeak is a fairly common breeder in willow and cottonwood associations of Sonoran habitat zones (Monson and Phillips 1981). The species is also a common riparian breeder throughout the Verde Valley watershed, including the Camp Verde region (Corman and Wise-Gervais 2005). We found the species an uncommon spring and summer breeder and a rare fall migrant, detecting 125 birds between 16 May and 13 September during our surveys. We discovered several active nests, principally in riparian woodlands with shrubby vegetation, and documented this species as a Brown-headed Cowbird host in the MCNM complex.

Lazuli Bunting • *(Passerina amoena)*
Residency: breeding; migrant
    Dates present: April – September
    Breeding status: known to breed in area
Abundance: uncommon
Preferred habitat type: riparian; mixed desert upland shrub

We found this bunting to be an uncommon spring and summer breeder, and rare in the fall, occupying primarily riparian woodlands and shrub. Between 30 April and 25 September we detected 31 individuals on our surveys, found nests, and banded a female with a brood patch. Taylor and Jackson (1916) heard this species "nearly every day" at Montezuma Well, so it may have bred within the MCNM complex at that time, although there was no confirmation.

Indigo Bunting • *(Passerina cyanea)*
Residency: breeding; migrant
    Dates present: May – August
    Breeding status: known to breed in area
Abundance: uncommon
Preferred habitat type: riparian; mixed desert upland shrub

Found within larger and more continuous stands of deciduous trees, the Indigo Bunting is an uncommon breeder in the Verde Valley watershed, including the Camp Verde area (Corman and Wise-Gervais 2005). There are breeding records at Oak Creek and in Prescott noted by Phillips et al. (1964). We found the species to be a rare spring and summer visitor to the MCNM complex, detecting 25 individuals between 11 May and 23 August on our surveys. Given these records, and the regular sightings of singing individuals throughout the breeding season, we consider this a probable breeding species within the MCNM complex.

Green-tailed Towhee • *(Piplio chlorurus)*
Residency: migrant
    Dates present: April; August – September
    Breeding status: not known to breed in area
Abundance: irregular-casual
Preferred habitat type: riparian

Sutton (1954) reported this species as an uncommon winter transient in the Verde Valley. We also found this towhee to be a casual fall and spring migrant, with three sightings, on 6 September 1988, 30 August 1991, and 22 April 1997. In April 1997, we captured and banded a Green-tailed Towhee at Montezuma Castle. This shy, secretive species is easily overlooked during winter and migration, and may be more common than sightings indicate.

Spotted Towhee • *(Pipilo maculatus)*
Residency: yearlong
   Dates present: February – August
   Breeding status: known to breed in area
Abundance: uncommon
Preferred habitat type: riparian

The Spotted Towhee occupies very dense habitats, such as riparian woodlands with a dense understory, and is more often heard singing or scratching on the ground than actually seen. The species is a common riparian breeder throughout the Verde Valley watershed, including the Camp Verde region (Corman and Wise-Gervais 2005). We found the species to be an uncommon migrant and summer breeder, detecting 23 individuals from 17 February through 9 August. Breeding is inferred in the MCNM complex from the presence of adults throughout breeding season.

Canyon Towhee • *(Pipilo fuscus)*
Residency: yearlong
   Dates present: January – December
   Breeding status: known to breed in area
Abundance: uncommon
Preferred habitat type: mixed desert upland shrub

Sutton (1954) reported the Canyon Towhee as a common yearlong resident in the Verde Valley, where it prefers arid grasslands with arroyos and scattered mesquite and barberry (Radd in litt.). During our surveys we found this towhee to be a rare to uncommon resident and breeder, detecting 15 individuals throughout the year in upland shrubs and trees. We also found the species nesting in the MCNM complex. Taylor and Jackson (1916) described this towhee as relatively common near Montezuma Well, although it was later listed in the MCNM complex by Jackson (1941) as a casual spring and fall migrant. This casts uncertainty as to the actual historical status of the shy and secretive species that can be easily confused with Abert's Towhee (see the following species description), even though we have now demonstrated it to be a resident of the MCNM complex.

Abert's Towhee • *(Pipilo aberti)*
Residency: yearlong
   Dates present: January – December
   Breeding status: known to breed in area
Abundance: common
Preferred habitat type: riparian; mixed desert upland shrub

We found Abert's Towhee to be a common permanent resident and breeder, detecting 109 birds during our MCNM complex surveys. This species is present throughout the year, and we found several active nests in riparian habitat with dense understory. Like other towhees, this species secretive and shy, most often detected when heard calling and ground scratching. Corman and Wise-Gervais (2005) describe it as a common riparian breeder throughout the Verde Valley watershed, including the Camp Verde region.

Rufous-crowned Sparrow •
(Aimophila ruficeps)
Residency: breeding; migrant
   Dates present: February – May; September – November
   Breeding status: not known to breed in area
Abundance: irregular-casual
Preferred habitat type: mixed desert upland shrub

Sutton (1954) reported this sparrow as rare, with several observations at Montezuma Castle and Montezuma Well between 1944 and 1954. The species is very local and otherwise uncommon in the Verde Valley (Radd in litt.). We found the species to be a casual visitor fall through spring, detecting nine birds from 25 February to 8 May and from 24 September to 28 November during our surveys. All individuals were detected in brushy upland slopes, often with rocky outcroppings and boulders.

American Tree Sparrow • *(Spizella arborea)*
Residency: migrant
    Dates present: March; January
    Breeding status: not known to breed in area
Abundance: irregular-casual
Preferred habitat type: mixed desert upland shrub
Sutton (1954) describes this species as "extremely rare" in the Verde Valley. Jackson (1941) reports casual sightings in April, July, and November, and a suspected breeding record within the MCNM complex; however, the breeding record was not accepted by A. R. Phillips (Phillips et al. 1964), which may call the other Jackson identifications into question. More recently, four were observed in the Chino Valley (just north of the Verde Valley) in January of 1983 (Witzeman 1983c), and we detected one American Tree Sparrow at the MCNM complex in March 1997.

Chipping Sparrow • *(Spizella passerina)*
Residency: migrant
    Dates present: March – November
    Breeding status: not known to breed in area
Abundance: uncommon
Preferred habitat type: mixed desert upland shrub; riparian
This bird was described as a "rather common" summer resident in open wooded parts of the Upper Sonoran Zone in central Arizona (Monson and Phillips 1981). In the Camp Verde region we found this sparrow a fairly common fall migrant, but uncommon to rare in spring and summer. We detected 62 birds during our surveys between 13 March and 8 November, mostly in riparian and upland habitats.

Brewer's Sparrow • *(Spizella breweri)*
Residency: breeding; migrant
    Dates present: April; August – September
    Breeding status: known to breed in area
Abundance: uncommon
Preferred habitat type: mixed desert upland shrub
The Brewer's Sparrow is a common spring and fall migrant in the Verde Valley (Johnson and Sogge 1995), with verified winter records noted in Phillips et al. (1964). A breeding "colony" in the 1880s was described by Phillips et al. (1964) "from Camp Verde to Fossil Creek," so local breeding is possible. Jackson (1941) reported collecting a specimen at Montezuma Castle on 27 April 1940, and seeing one in May that looked as if it was carrying nesting material. Sutton (1954) reported one other record at Montezuma Well in 1953. We detected seven birds in upland habitats, on 17 April and between 30 August and 24 September. However, no breeding has ever been confirmed in the MCNM complex, and we were unable to find any nests of this species.

Black-chinned Sparrow •
(Spizella atrogularis)
Residency: migrant
    Dates present: March – May
    Breeding status: not known to breed in area
Abundance: irregular-casual
Preferred habitat type: mixed desert upland shrub
This sparrow currently breeds at mid-elevations on Mingus Mountain (Radd in litt.) and in brushy, scattered juniper slopes northeast of Camp Verde (Corman and Wise-Gervais 2005). Sutton (1954) described this sparrow as "extremely rare" in the Verde Valley, with only two records in 1947 and 1952 from Montezuma Well.

Vesper Sparrow • *(Pooecetes gramineus)*
Residency: winter; migrant
    Dates present: November – February
    Breeding status: not known to breed in area
Abundance: rare
Preferred habitat type: mixed desert upland shrub
The Vesper Sparrow has been seen fall through spring in the Verde Valley (Radd in litt.). Sutton (1954) reported this species as rare and recorded at Montezuma Castle. We did not detect any of this species during our surveys, and there are no recent records from the MCNM complex.

Lark Sparrow • *(Chondestes grammacus)*
Residency: breeding; migrant

Dates present: May – September
Breeding status: known to breed in area
Abundance: uncommon
Preferred habitat type: mixed desert upland shrub; riparian

This species prefers open grassland habitats with scattered trees and shrubs and is an uncommon breeder throughout the Verde Valley watershed, including the Camp Verde region (Corman and Wise-Gervais 2005). At the MCNM complex, we found the Lark Sparrow to be a casual spring and fall migrant, and a rare to uncommon summer breeder. We did detect 14 individuals between 16 May and 20 September on our surveys and found only one active nest.

Black-throated Sparrow •
(Amphispiza bilineata)
Residency: breeding
Dates present: February – November
Breeding status: known to breed in area
Abundance: common
Preferred habitat type: mixed desert upland shrub

The Black-throated Sparrow is a common upland breeder throughout the Verde Valley watershed, including the Camp Verde region (Corman and Wise-Gervais 2005). It is the most abundant upland breeding species within the MCNM complex, and a frequent host to cowbird nest parasitism (Johnson and van Riper 2004). It was listed as common breeder by Taylor and Jackson (1916), Jackson (1941) and Sutton (1954). We detected 649 individuals during our surveys between 29 January and 12 December and found many nests.

Sage Sparrow • *(Amphispiza belli)*
Residency: migrant
Dates present: January – March
Breeding status: not known to breed in area
Abundance: rare
Preferred habitat type: mixed desert upland shrub

The Sage Sparrow nests in northeastern Arizona and is an uncommon spring and fall migrant and a rare winter visitor in the Verde Valley (Johnson and Sogge 1995, Radd in litt.). In the MCNM complex, historically, this sparrow has been casual in fall and winter, with only two records at Montezuma Well (Sutton 1954). We detected three birds in upland shrub and scattered trees between 31 January and 12 March.

Lark Bunting • *(Calamospiza melanocorys)*
Residency: migrant
Dates present: December; June
Breeding status: not known to breed in area
Abundance: irregular-casual
Preferred habitat type: mixed desert upland shrub

Lark Buntings are regular migrants in the Verde Valley (Phillips et al. 1964). Six specimens were collected at Fort Verde in the 1880s, so winter occurrence is possible; however, the only record for this species in the MCNM complex is a June 1935 notation in the monthly park reports (Superintendents of MOCA 1932). The MCNM complex is outside of the normal breeding range of this bunting, and the 1935 notation may have been a misidentification.

Song Sparrow • *(Melospiza melodia)*
Residency: yearlong
Dates present: January – December
Breeding status: known to breed in area
Abundance: rare
Preferred habitat type: riparian

The Song Sparrow is a common breeding species in some dense riparian habitats in the Verde Valley (Johnson and Sogge 1995) and across Arizona (Phillips et al. 1964). It was reported as uncommon in the MCNM complex by Taylor and Jackson (1916) and as a casual winter visitor by Jackson (1941). It is still relatively uncommon in the MCNM complex (we only detected 13 individuals during our surveys), perhaps due to the limited amount of dense understory vegetation in the riparian zone.

Lincoln's Sparrow • *(Melospiza lincolnii)*
Residency: migrant
Dates present: February – April; September – October
Breeding status: not known to breed in area
Abundance: uncommon

Preferred habitat type: riparian; mixed desert upland shrub

Shy and secretive during winter and migration, Lincoln's Sparrow is sometimes misidentified as a Song Sparrow. This species is fairly common in fall through spring in other areas of the Verde Valley (Johnson and Sogge 1995, Radd in litt.). In the MCNM complex we found it to be a rare spring and fall migrant, detecting nine individuals from 17 February to 21 April and from 28 September through 14 October.

White-throated Sparrow •
(Zonotrichia albicollis)
Residency: migrant
    Dates present: January – June
    Breeding status: not known to breed in area
Abundance: uncommon
Preferred habitat type: mixed desert upland shrub

This species looks very much like the White-crowned Sparrow, and is usually seen in mixed sparrow flocks. It can be found in any habitat, usually near trees, shrubs, and grassland, and we detected three birds between 19 January and 4 June. Sutton (1954) described a single record during the winter of 1953–1954 at Montezuma Well. Although probably not very common in the MCNM complex, this species likely occurs more often than is detected because it is overlooked or misidentified in its association with other sparrows.

White-crowned Sparrow •
(Zonotrichia leucophrys)
Residency: winter; migrant
    Dates present: January – May; September – December
    Breeding status: not known to breed in area
Abundance: abundant
Preferred habitat type: mixed desert upland shrub

We have found this sparrow an abundant spring and fall migrant and winter visitor throughout the Verde Valley watershed, including the Camp Verde region. It is found in all habitats with trees and shrubs in pasture-grassland, and is frequently observed wintering with other sparrow species. We detected 172 individuals during our surveys, with detections from 4 January to 26 May and from 24 September to 13 December. During the winter we almost always found this species in flocks ranging from 10 to over 150 individuals. The White-crowned Sparrow can be commonly found during winter in the pasture-grassland near the Montezuma Well picnic area.

Dark-eyed Junco • *(Junco hyemalis)*
Residency: winter; migrant
    Dates present: January – April; October – December
    Breeding status: not known to breed in area
Abundance: common
Preferred habitat type: mixed desert upland shrub

An easily detected wintering species, the Dark-eyed Junco usually occurs in flocks of 5 to 25 individuals in riparian habitat (especially mesquite bosque) and uplands with trees and shrubs. It is a common winter resident throughout the Verde Valley watershed, including the Camp Verde area. Sutton (1954) noted five subspecies present in the Verde Valley; "Slate-colored" (rare, but several records at Montezuma Well and Montezuma Castle), "Oregon" (common), "Pink-sided" (fairly common), "Gray-headed" (casual, but recorded at Montezuma Well), and "Red-backed" (rare and sporadic). We found it to be a common winter and fall resident in the MCNM complex and uncommon in the spring, noting 69 individuals from 4 January to 20 April and from 14 October to 13 December during our surveys,

Red-winged Blackbird •
(Agelaius phoeniceus)
Residency: breeder
    Dates present: February – July
    Breeding status: known to breed in area
Abundance: common
Preferred habitat type: marsh

The Red-winged Blackbird is a common riparian breeder in marshland habitats throughout the Verde Valley watershed, including the Camp Verde region (Corman and Wise-Gervais 2005). At the MCNM complex we found the species to be an irregular winter, spring, and summer transient, and counted 15 individuals between 17 February and 26 July during our surveys. The species was usually observed in flocks, either flying over, perched amid riparian woodlands, or near the marsh at Montezuma Well.

Western Meadowlark • *(Sturnella neglecta)*
Residency: yearlong
    Dates present: January – December
    Breeding status: known to breed in area
Abundance: common
Preferred habitat type: mixed desert upland shrub
The Western Meadowlark is found in grasslands and open shrub uplands and is a common breeder throughout the Verde Valley watershed, including the Camp Verde region (Corman and Wise-Gervais 2005). It was reported by Sutton (1954) as a common resident breeder in the fields at Montezuma Well. We found the species to be an uncommon fall through spring visitor, detecting 15 individuals between 19 October and 26 March on our surveys. We found no evidence of current breeding within the MCNM complex, even though meadowlarks are very common in the fields and open-housing areas immediately outside the MCNM complex boundaries. Meadowlarks may no longer breed in the MCNM complex due to habitat changes in the field adjacent to the Montezuma Well picnic area. This field is now fallow, and supports large patches of thick, brushy vegetation which is not suitable breeding habitat for meadowlarks.

Yellow-headed Blackbird •
(Xanthocephalus xanthocephalus)
Residency: migrant
    Dates present: April; November
    Breeding status: not known to breed in area
Abundance: rare
Preferred habitat type: marsh
This blackbird is an uncommon spring and fall migrant in wetland areas in the lower elevations of the Verde Valley (Johnson and Sogge 1995). Sutton (1954) reported this species as rare, stating they were "known from" Montezuma Well. We detected only one individual on 30 April 1992 in marsh habitat within the MCNM complex.

Brewer's Blackbird •
(Euphagus cyanocephalus)
Residency: migrant
    Dates present: May – September
    Breeding status: not known to breed in area
Abundance: uncommon
Preferred habitat type: riparian; marsh
Brewer's Blackbirds winter in some portions of the Verde Valley (Johnson and Sogge 1995). It is a rare spring and fall migrant to the MCNM complex, usually seen in flocks flying over the MCNM complex or perched in riparian trees. Between 1991 and 1997 we detected three migrant blackbirds on 8 May, 16 May, and 20 September.

Great-tailed Grackle • *(Qusicalus mexicanus)*
Residency: yearlong
    Dates present: January – December
    Breeding status: known to breed in area
Abundance: common
Preferred habitat type: riparian; mixed desert upland shrub
A relatively recent arrival to the Verde Valley, the Great-tailed Grackle is now a common breeder throughout the Verde Valley watershed, including the Camp Verde region (Corman and Wise-Gervais 2005). Jackson (1941) and Sutton (1954) mention them as recent arrivals to the MCNM complex. We had two records on 9 April 1992 and 26 July 1994. Even though there are few sightings reported in the MCNM complex, grackles are fairly common and widespread elsewhere in the Verde Valley (Johnson and Sogge 1995).

Bronzed Cowbird • *(Molothrus aeneus)*
Residency: yearlong
    Dates present: January – December
    Breeding status: known to breed in area
Abundance: uncommon
Preferred habitat type: riparian; mixed desert

upland shrub

The first records for this species in Arizona were from 1909, and a range expansion was first noted in 1951 (Lowther 1995). There have been several recent sightings in the riparian habitats of the Verde Valley, including a male performing a breeding display along Beaver Creek on 28 April 1990 (Thornburg in litt.). We observed a female at the Tuzigoot Bridge in June 1995. Bronzed Cowbirds have also recently been reported breeding in Jerome and Clarkdale (Radd in litt.). Given the northward expansion of this species, it will almost inevitably be detected within the MCNM complex. Although a brood parasite like its congener the Brown-headed Cowbird, it appears that the Bronzed Cowbird may be less of a conservation and management issue because it tends to parasitize fewer host species, and may have a preference for orioles (Lowther 1995).

Brown-headed Cowbird • *(Molothrus ater)*
Residency: breeding
    Dates present: April – September
    Breeding status: known to breed in area
Abundance: common
Preferred habitat type: riparian; mixed desert upland shrub

Colonizing the Camp Verde region in the late 1800s, this species is now a common breeder throughout the region (Corman and Wise-Gervais 2005). Specimens were first collected at Fort Verde in the 1880s, and Taylor and Jackson (1916) recorded a cowbird chick being fed by adult Yellow Warblers in the MCNM complex. However, Jackson (1941) reported only two sightings, which suggests that cowbirds may have been uncommon until that time. Their numbers and distribution probably increased as irrigated agriculture and urban development spread throughout the valley. In the mid-1900s, Sutton (1954) described cowbirds as present from April through October, and "common in fields with cattle." We found this cowbird to be a very common spring and summer resident and breeder, detecting 393 individuals between 3 April and 5 September during our surveys. Detections of this species ranged widely over the MCNM complex, with birds found primarily in riparian woodlands and uplands with taller trees and shrubs. We recorded evidence of Brown-headed Cowbirds parasitizing nests of at least six bird species in the MCNM complex (Sogge and Johnson 1998, Johnson and van Riper 2004).

Hooded Oriole • *(Icterus cucullatus)*
Residency: breeding
    Dates present: April – September
    Breeding status: known to breed in area
Abundance: uncommon
Preferred habitat type: riparian

Taylor and Jackson (1916) saw this oriole "frequently in timber at all points visited," including Montezuma Well. This oriole is described as an uncommon riparian breeder in the Verde Valley watershed, including the Camp Verde area by Holmes et al. (2008). We found it uncommon in spring and more common as a summer breeder, counting 63 individuals of this species between 9 April and 2 September during our surveys, and finding two active nests in riparian woodlands.

Bullock's Oriole • *(Icterus bullockii)*
Residency: breeding
    Dates present: April – September
    Breeding status: known to breed in area
Abundance: uncommon
Preferred habitat type: riparian

The species is a common riparian breeder throughout the Verde Valley watershed, including the Camp Verde region (Corman and Wise-Gervais 2005). We found the species uncommon in the spring and a summer breeder, detecting 39 birds on our surveys between 9 April and 23 August. Nesting evidence in MCNM complex riparian woodlands came from the presence of adults during the breeding season and observing recently fledged juvenile birds.

Scott's Oriole • *(Icterus parisorum)*
Residency: migrant
    Dates present: March
    Breeding status: known to breed in area
Abundance: rare
Preferred habitat type: riparian
Scott's Oriole is a riparian breeder throughout the Verde Valley watershed, including the Camp Verde region (Corman and Wise-Gervais 2005). Sutton (1954) reported two records at Montezuma Well. Although a common summer resident of oak and yucca habitats in central Arizona (Phillips et al. 1964), there have been no sightings in the MCNM complex since 1954.

Purple Finch • *(Carpodacus purpureus)*
Residency: winter
    Dates present: September
    Breeding status: not known to breed in area
Abundance: irregular-casual
Preferred habitat type: mixed desert upland shrub
Duvall (1945) and Phillips et al. (1964) stated that this finch occasionally migrates through and winters in the Verde Valley. We found that six winter specimens had been collected in 1886 at Fort Verde. Witzeman and Stejskal (1985) reported that "small numbers" were recorded at Montezuma Castle in the winter of 1984, and we found the Purple Finch to be casual in fall and winter, adding one more recent record during our surveys, on 24 September 1991.

Cassin's Finch • *(Carposacus cassinii)*
Residency: winter
    Dates present: January – March
    Breeding status: not known to breed in area
Abundance: irregular-casual
Preferred habitat type: mixed desert upland shrub
Phillips et al. (1964) described Cassin's Finch as an irregular winter visitor to the Verde Valley. We found the species to be a casual winter transient, occurring in small flocks (four to eight birds). During our formal MCNM complex surveys we observed six individuals between 19 January and 1 March, one feeding on the seeds of "sycamores balls" along Beaver Creek.

House Finch *(Carpodacus mexicanus)*
Residency: yearlong
    Dates present: January – December
    Breeding status: known to breed in area
Abundance: abundant
Preferred habitat type: mixed desert upland shrub
Now the most common passerine species in North America, the House Finch is also a common species in the Verde Valley and the MCNM complex, and has been since at least the early 1900s (Taylor and Jackson 1916, Jackson 1941, Sutton 1954). We found 507 individuals during our surveys, with birds using riparian areas, particularly during the nonbreeding season. There are many descriptions of nests within the MCNM complex, especially near residential areas, and we found 10 active nests in upland shrubs and trees and along cliffs.

Red Crossbill • *(Loxia curvirostra)*
Residency: migrant
    Dates present: January; October, December
    Breeding status: not known to breed in area
Abundance: irregular-casual
Preferred habitat type: mixed desert upland shrub
Although typically a high-elevation species, crossbills are infrequently noted in the Verde Valley and Sedona areas (Rosenberg et al. 1994b), and Monson and Phillips (1981) describe them as "sparse and irregular" in winter in the lowlands. Sutton (1954) reported crossbills as "extremely rare," noting one record of four females at Montezuma Well in 1953. Witzeman and Stejskal (1985) described three crossbills "near Montezuma Well" from 29 December 1984 to 10 January 1985. We have records of two detections of migrant individuals during October 1953 and January 1985.

Pine Siskin • *(Carduelis pinus)*
Residency: winter;migrant
    Dates present: January – March
    Breeding status: not known to breed in area
Abundance: irregular-casual
Preferred habitat type: mixed desert upland shrub
The Pine Siskin breeds in the mountains surrounding the Camp Verde region, and is

a common winter resident throughout the Verde Valley watershed. There have been recent sightings during all seasons within the Verde Valley, more commonly in spring and fall (Radd in litt.). Jackson (1941) noted this as a casual species, with sightings at the MCNM complex in February and September. Sutton (1954) considered it a rare winter transient. We detected eight birds between 12 January and 17 March on our MCNM complex surveys, and classify this species as an irregular winter visitor.

Lesser Goldfinch • *(Carduelis psaltria)*
Residency: yearlong
  Dates present: January – December
  Breeding status: known to breed in area
Abundance: uncommon
Preferred habitat type: mixed desert upland shrub
The Lesser Goldfinch is a fairly common breeder in Sonoran Zone riparian habitats statewide (Monson and Phillips 1981). We found the species to be an uncommon summer breeder, and rare to uncommon for the remainder of the year. Most of the 54 individuals that we detected were found in riparian woodlands and adjacent uplands in summer, as well as fields and shrubby areas during winter. This bird breeds as late as September and October in Verde Valley (Radd in litt.), so our formal summer surveys (ending in August) may have missed some nesting attempts, but we did find one active nest in July.

American Goldfinch • *(Carduelis tristis)*
Residency: winter; migrant
  Dates present: February; June; November – December
  Breeding status: not known to breed in area
Abundance: rare
Preferred habitat type: riparian
The American Goldfinch is a casual winter and spring transient in the Verde Valley watershed, including the Camp Verde region, with late spring migrants being occasionally noted into early June (Monson and Phillips 1981). Jackson (1941) recorded this species as casual from December through April, but Sutton (1954) appears to have believed these were misidentified Lesser ("Arkansas") Goldfinches, as he reported "no positive records" of American Goldfinches since the late 1800s. On our surveys we detected six migrant birds on an irregular basis (February, June, November, and December) in riparian woodlands.

Evening Grosbeak •
(Coccothraustes vespertinus)
Residency: winter; migrant
  Dates present: January; May
  Breeding status: not known to breed in area
Abundance: uncommon
Preferred habitat type: riparian
Sutton (1954) reported this species as rare. The Evening Grosbeak is currently an uncommon fall through spring visitor to portions of the Camp Verde region and the Verde Valley (Johnson and Sogge 1995). There have been winter "invasions" of Evening Grosbeaks reported along Wet Beaver Creek in 1982 (Witzeman 1982) and 1984 (Witzeman and Stejskal 1985). The most recent sightings within the MCNM complex are recorded from May 1982 and January 1985.

House Sparrow • *(Passer domesticus)*
Residency: yearlong
  Dates present: January – December
  Breeding status: known to breed in area
Abundance: abundant
Preferred habitat type: mixed desert upland shrub
This species was introduced into eastern North America from Europe in 1850 and 1851 (Lowther and Cink 1992) when about 100 birds were released in Brooklyn, New York, with subsequent releases in San Francisco (1871 and 1872) and Salt Lake City (1873 and 1874). By 1916 the sparrow was present and abundant within portions of the Verde Valley including the MCNM complex (Taylor and Jackson 1916). The House Sparrow is now an abundant breeder and yearlong resident in housing developments throughout the Verde Valley watershed, including the Camp Verde region (Corman and Wise-Gervais 2005).

In the MCNM complex, Jackson (1941) made her first sighting in October 1938 or 1939, so they may not have been abundant in the MCNM complex during her tenure (1937–1941). Sutton (1954) described it as rare at Montezuma Castle but "regular" at Montezuma Well, where it bred in eaves of park housing units. Currently, this species is rare in the MCNM complex, still found in association with human housing or urban development. There are no recent breeding records, with the last two known nests active in 9 April 1980 and 14 June 1991.

CONCLUSIONS

The fact that well over 200 bird species have been reported at the MCNM complex in this historical assessment, is a testament to the tremendous avian species diversity of the Verde Valley in central Arizona. The wide variety of birds found in the MCNM complex is largely due to the diversity of habitats within and around the Verde Valley. We have identified close to 100 species of birds that are known or suspected to breed (currently or historically) within just the MCNM complex boundaries.

By comparing current population numbers with earlier historical records, we found that a number of species seem to have either increased or decreased over the past several decades throughout the Camp Verde region. Concomitantly, we found certain avian species that appear to have experienced no major changes in population numbers. Due to the small size of the Camp Verde region, and especially the MCNM complex (340 ha), events and influences outside the region's boundaries have had a significant influence on changing numbers of birds. In fact, we found that most avian species whose populations have changed across the past several decades have also done so over the entire southwestern United States. There are, however, a number of species that seem to be directly influenced by changing ecosystems and human-related actions.

A number of waterfowl species that we found during our surveys that had increased within the MCNM complex over the past century, have also increased in abundance in other areas of North America. For example, the Canada Goose has increased to the point that in some locations (e.g., golf courses) it is now considered a pest. The Wood Duck has increased as more nesting boxes have been placed throughout watersheds, and as riparian trees have aged and a greater number of nesting cavities have become available. We documented increases in the Great-blue Heron and White-faced Ibis, two species that have increased generally thoughout the Southwest, largely due to increased agricultural acreage and increased irrigation ditches that carry water to those fields.

With the cessation of use of the pesticide DDT in the 1960s (Hickey and Anderson 1968), coupled with decreases in shooting mortality, many raptorial species have also experienced significant increases in numbers (e.g., Glinski 1998). These increases have resulted in several raptor species becoming much more common not only throughout North America, but also within the MCNM complex and Camp Verde region. Bald Eagles and Osprey now nest throughout Arizona and are often sighted along the Verde River and Beaver Creek searching for prey, and may soon return to the MCNM complex. The Peregrine Falcon is becoming increasingly more common in the area, and will undoubtedly become a MCNM complex nesting species, as it was in the historical past. The expansion of the Black Hawk as a nester in the MCNM complex could be related to climate changes, or to maturation of the riparian woodland forest throughout the Verde River valley, or both.

Growth of the human population over the past several decades throughout the Camp Verde region has also been to the advantage of some bird species. The use of sugar feeders in residential areas hay have enabled some hummingbird species to expand into this region. Seed feeders have also enhanced many bird species. Additional sewage treatment facilities have increased numbers of the Common Yellowthroat and several blackbird species, and in the MCNM complex this has become an extremely important habitat. Those species that capitalize on human refuse outside the MCNM complex (e.g., Great-tailed Grackle, Common Raven) have also increased in numbers.

It also appears that a changing environment has allowed several passerine species to expand into the Verde region. The Verdin is now a common nesting species in the MCNM complex, and the Hutton's Vireo, the Blue-gray Gnatcatcher, and Spotted Towhee have all increased in numbers.

Other species have declined in numbers across the Camp Verde region. Several species of waterfowl, such as the Northern Pintails, have decreased throughout western North America and are correspondingly now rarely encountered in the MCNM complex. The same holds true for the Wood Stork. Barn Owl numbers have drastically declined in much of North America, principally due to the use of rodenticides (Marti 1992), and that also holds true in the MCNM complex and Camp Verde region. Reasons for the declines of species such as the Yellow-billed Cuckoo, Vermilion Flycatcher, and Vesper Sparrow, are less clear.

Several species that formerly were found in low numbers no longer occur within the MCNM complex and Camp Verde region. The Thick-billed Parrot has been extirpated, while the Northern Beardless-Tyrannulet and Groove-billed Ani are examples of species that were initially misidentified or whose ranges have contracted out of the region. Although most oriole species have not experienced changes in number, Scott's Oriole is an exception with decreasing numbers.

Changes in human use and management of the landscape have directly influenced species in the MCNM complex and within the Camp Verde region. The great reduction in agriculture and loss of grain crops has virtually eliminated the pheasant from the MCNM complex and region. At one time, the Camp Verde region had one of the highest pheasant densities in the state of Arizona. On a finer scale, the changing management of the pasture at Montezuma Well has resulted in the elimination of the Western Meadow Lark from the MCNM complex. Changes in riparian habitat, along with associated shrub understory, have certainly negatively influenced the distributional patterns of Willow Flycatchers and Yellow-billed Cuckoos.

In summary, over the past century there have been many changes in avian community structure at MCNM complex and within the Camp Verde region. A suite of species have had increases in numbers of individuals, while other species have had reductions in total numbers of birds, and a few have been lost from the region. This historical assessment and summary of changes in avian community composition should provide a foundation upon which future MCNM complex avian research and management actions can be based.

ACKNOWLEDGMENTS

We thank the many people who assisted with field work on this project, including Sarah Allen, Beth Bardwell, Jeri DeYoung, Laura Ellison, David Felley, John Grahame, Kathy Hiett, Patti Hodgetts, Jennifer Owen, Eben Paxton, and Helen Yard. Our special thanks go to Paul Super for his skills and energy. Charles Drost provided excellent advice and administrative assistance throughout the project. We are indebted to the late Virginia Gilmore for sharing her knowledge of the local bird communities and for providing valuable observations from her files and those of the Northern Arizona Audubon Society and the late Dr. Sidney Hyde. Ashley Thornburg and Murrelet Halterman also contributed useful bird-observation data from the MCNM complex. Roger Radd, Troy Corman, Sandra van Riper, and Kimball Garrett reviewed earlier drafts of this species list and provided many useful comments. This work was supported by funds from the National Park Service and the U.S. Geological Survey Southwest Biological Science Center.

LITERATURE CITED

Alden, S., and S. Mills. 1974. Nesting Season: Southwest Region. American Birds 28:934.

American Ornithologists' Union. 1998. 7th Ed. Check-list of North American Birds. Washington, D.C.

Brown, D. E. 1989. Arizona Game Birds. University of Arizona Press, Tucson.

Cabe, P. R. 1993. European Starling (*Sturnus vulgaris*). In The Birds of North America, no. 48, edited by A. Poole and F. Gill. Academy of Natural Sciences, Philadelphia, and American Ornithologists' Union, Washington, D.C.

Corman, T. E., and Wise-Gervais, editors. 2005. Arizona Breeding Bird Atlas. University of New Mexico Press, Albuquerque.

Duvall, A. J. 1945. Variation in *Carpodacus purpureus* and *Carpodacus cassinii*. Condor 47:202–205.

Ellison, L. E. and C. van Riper III. 1998. A comparison of small-mammal communities in a desert riparian floodplain. Journal of Mammalogy 79: 972–985.

Frost, C. B. 1947. A Check List of Birds Recorded in the Montezuma Castle Area. Unpublished report (request from Montezuma Castle National Monument, P.O. Box 219, Camp Verde, AZ 86322).

Glinski, R. L. 1998. Raptors of Arizona. University of Arizona Press, Tucson.

Hickey, J. J. 1939. Correspondence to B. Jackson. Unpublished letter (request from Montezuma Castle National Monument, P.O. Box 219, Camp Verde, AZ 86322).

Hickey J. J., and D. W. Anderson. 1968. Chlorinated hydrocarbons and eggshell changes in raptorial and fish-eating birds. Science 162:271–273.

Holmes, J. A., M. J. Johnson, and C. Calvo. 2008. Yellow billed cuckoo distribution, habitat use and breeding ecology in the Verde Watershed of Arizona, 2003–2004. Final Report, Arizona Game and Fish Heritage Program, Phoenix.

Jackson, B. 1938. Christmas bird census for 1938. Unpublished report (request from Montezuma Castle National Monument, P.O. Box 219, Camp Verde, Arizona 86322).

Jackson, B. 1939. Birds at Montezuma Castle National Monument. Unpublished report (request from Montezuma Castle National Monument, P.O. Box 219, Camp Verde, Arizona 86322).

Jackson, B. 1941. Birds of Montezuma Castle. Southwestern Monuments Association: Southwestern Monuments Special Report, no. 28 (request from Montezuma Castle National Monument, P.O. Box 219, Camp Verde, Arizona 86322).

Johnson, M. J., and M. K. Sogge. 1995. A Checklist of Birds of Tuzigoot National Monument and Vicinity. Southwest Parks and Monuments Association, Tucson, Arizona.

Johnson, M. J., and C. van Riper, III. 2004. Brown-headed Cowbird parasitism of the Black-throated Sparrow in central Arizona. Journal of Field Ornithology 75:303–311.

Lowther, P. E. 1995. Bronzed Cowbird (*Molothrus aeneus*). In The Birds of North America, no. 144, edited by A. Poole and F. Gill. Academy of Natural Sciences, Philadelphia, and American Ornithologists' Union, Washington, D.C.

Lowther, P. E., and C. L. Cink. 1992. House Sparrow (*Passer domesticus*). In The Birds of North America, no. 12, edited by A. Poole and F. Gill. Academy of Natural Sciences, Philadelphia, and American Ornithologists' Union, Washington, D.C.

Marti, C. D. 1992. Barn Owl (*Tyto alba*). In The Birds of North America, no. 1., edited by A. Poole, P. Stettenheim, and F. Gill. Academy of Natural Sciences, Philadelphia, and American Ornithologists' Union. Washington, D. C.

Monson, G., and A. R. Phillips. 1981. Annotated Checklist of the Birds of Arizona, 2nd ed. University of Arizona Press, Tucson.

O'Brien, C., D. W. Blinn, and C. van Riper III. 2003. Waterfowl, acanthocephalans, and amphipods at Montezuma Well, Arizona – What is the role of parasites? DOI, USGS FS-125-03.

Phillips, A. R. 1942. Correspondence to B. Jackson. Unpublished letter (request from Montezuma Castle National Monument, P.O. Box 219, Camp Verde, AZ 86322).

Phillips, A. R., J. Marshall, and G. Monson. 1964. The Birds of Arizona. University of Arizona Press, Tucson.

Ralph, C. J., G. R. Geupel, P. Pyle, T. E. Martin, and D. F. DeSante. 1993. Handbook of field methods for monitoring landbirds. Pacific Southwest Research Station, Forest Service, U.S. Department of Agriculture. Gen. Tech. Rep. PSW-GTR-144. Albany, California.

Reynolds, R. T., J. M. Scott, and R. A. Nussbaum. 1980. A variable circular-plot method for estimating bird numbers. Condor 82:309–313.

Rosenberg, G. H., and D. Stejskal. 1989. Fall Migration: Southwest Region. American Birds 43:145–148.

Rosenberg, G. H., and D. Stejskal. 1990. Fall Migration: Southwest Region. American Birds 44:136–136.

Rosenberg, G. H., and D. Stejskal. 1991. Spring Season: Southwest Region. American Birds 45:479.

Rosenberg, G. H., D. Stejskal, and C. D. Benesh. 1994a. Fall Season: Southwest Region. American Birds 48:136.

Rosenberg, G. H., D. Stejskal, and C. D. Benesh. 1994b. Winter Season: Southwest Region. American Birds 48:235.

Rosenberg, K. V., G. H. Rosenberg, and J.P. Hubbard. 1980. Spring Migration: Southwest Region. American Birds 34:805.

Schmidt, C. A., C. A. Drost, and W. L. Halvorson. 2006. Vascular Plant and Vertebrate Inventory of Montezuma Castle National Monument. USGS Open-File Report 2006–1163. U.S. Geological Survey, Southwest Biological Science Center, Sonoran Desert Research Station, University of Arizona, Tucson.

Sogge, M. K., and M. J. Johnson. 1998. Bird Community Ecology at Montezuma Castle National Park: Avian Inventory 1991–1994. Unpublished report. U.S. Geological Survey, Colorado Plateau Field Station, P.O. Box 5614, Flagstaff, AZ 86011.

Stejskal, D., C. D. Benesh, and G. H. Rosenberg. 1995. Winter Season: Southwest Region. American Birds 49:175.

Stejskal, D., and G. H. Rosenberg. 1990. Winter Season: Southwest Region. American Birds 44:302.

Stejskal, D., and G. H. Rosenberg. 1991a. Summer Season: Southwest Region. American Birds 45:1145.

Stejskal, D., and G. H. Rosenberg. 1991b. Winter Season: Southwest Region. American Birds 45:300.

Stejskal, D., and J. Witzeman. 1984. Spring Migration: Southwest Region. American Birds 38:942–944.

Stejskal, D., and J. Witzeman. 1985. Fall Migration: Southwest Region. American Birds 39:87.

Stejskal, D., and J. Witzeman. 1987. Spring Migration: Southwest Region. American Birds 41:471–472.

Stejskal, D., and J. Witzeman. 1988. Fall Migration: Southwest Region. American Birds 42:113.

Superintendents of MOCA. 1932. Monthly Park Reports. Unpublished (request from Montezuma Castle National Monument, P.O. Box 219, Camp Verde, AZ 86322).

Sutton, M. 1954. Birds of the Verde Valley: An Interpretive Treatment. Unpublished report (order from National Park Service, Western Archaeological Conservation Center Archives and Library, Tucson, Arizona).

Taylor, W. P., and H. H. T. Jackson. 1916. Biological Survey Reports, Verde Valley (Birds and Mammals). Unpublished report (request from Montezuma Castle National Monument, P.O. Box 219, Camp Verde, AZ 86322).

Telfair, R. C., II. 1994. Cattle Egret (*Bubulcus ibis*). In The Birds of North America, no. 113, edited by A. Poole and F. Gill, Academy of Natural Sciences, Philadelphia, and American Ornithologists' Union, Washington, D.C.

Verner, J. 1985. Assessment of counting techniques. In Current Ornithology, vol. 2, edited by R. F. Johnston., pp. 247–302. Plenum Press, New York.

Wiens, J. D., R. T. Reynolds, and B. R. Noon. 2006. Juvenile movement and natal dispersal of Northern Goshawks in Arizona. Condor 108:253–269.

Willey, D. A., and C. van Riper, III. 2007. Home range characteristics of Mexican Spotted Owls in the canyonlands of Utah. Journal of Raptor Research. 41:10–15.

Witzeman, J. 1982. Spring Migration: Southwest Region. American Birds 36:881–882.

Witzeman, J. 1983a. Fall Migration: Southwest Region. American Birds 37:209.

Witzeman, J. 1983b. Spring Migration: Southwest Region. American Birds 37:898–899.

Witzeman, J. 1983c. Winter Season: Southwest Region. American Birds 37:326.

Witzeman, J., and D. Stejskal. 1984. Nesting Season: Southwest Region. American Birds 38:1048.

Witzeman, J., and D. Stejskal. 1985. Winter Season: Southwest Region. American Birds 39:197.

Witzeman, J., and D. Stejskal. 1986. Winter Season: Southwest Region. American Birds 40:311.

Witzeman, J., and D. Stejskal. 1987. Nesting Season: Southwest Region. American Birds 41:1471.

Witzeman, J., and D. Stejskal. 1988. Winter Season: Southwest Region. American Birds 42:303.

Witzeman, J., J. P. Hubbard, and K. Kaufman. 1976. Fall Migration: Southwest Region. American Birds 30:109.

Large Mammal Conservation and Management

MULE DEER ANTLER GROWTH AND HUNTING MANAGEMENT ON THE NORTH KAIBAB, ARIZONA

Brian F. Wakeling

ABSTRACT

Mule deer (*Odocoileus hemionus*) management must be biologically sustainable and socially acceptable. Social expectations for hunter harvest and hunt quality often place greater restrictions on deer management than do biological limitations. The mule deer herd that inhabits the North Kaibab, Arizona (Game Management Unit 12A) is managed under Arizona Game and Fish Commission alternative deer-management guidelines that were designed to provide lower hunter densities, higher hunt success during late season hunts, and a greater opportunity to harvest an older age-class deer. I compared antler spread, antler points, and cementum age of mule deer bucks harvested on the North Kaibab from measurements taken at a mandatory hunter check station at Jacob Lake, Arizona. Mule deer antler points and antler spread increased with age to 5 years, after which antler points and spread did not substantially increase. Through a public process, the Arizona Game and Fish Commission established alternative deer management guidelines that included permit adjustments for late season hunts to ensure 55–75% of animals harvested were ≥3 years of age and 20–30% of the animals harvested were ≥5 years of age.

INTRODUCTION

Setting hunting seasons and allowable harvest is an important aspect of wildlife management under the authority of state wildlife management agencies. Seasons and harvest are routinely based on biological data and social desires. For instance, research indicates that mule deer buck to doe ratios must drop below 4–7:100 before reproductive capability of the herd is limited biologically (White et al. 2001, Bishop et al. 2005). Harvest, therefore, should not reduce the number of bucks beyond the point at which reproduction is reduced to maintain or grow deer populations, and deer populations should be surveyed routinely to monitor their relative status. Hunters express interest in a wide variety of hunt types (Manfredo et al. 2004), yet vocal proponents are often interested in higher buck to doe ratios with fewer hunters (Bishop et al. 2005, Wakeling and Watkins in press). In other words, social restraints are generally greater than are biological restraints on hunting.

Because the hunting public desires differing types of opportunities, the Arizona Game and Fish Commission (Commission) has adopted two philosophically different sets of management guidelines by which mule deer hunts are developed: standard and alternative management. The Commission is the five-member, policy-setting body for the Arizona Game and Fish Department (Department). Currently, all general-season opportunity is regulated through a lottery draw system. Units managed according to standard management guidelines adjust permits when post-season mule deer buck-to-doe ratios are beyond 10–20:100, fawn to doe ratios are beyond 40–50:100, measured hunt success is beyond 15–20%, or the population trend is not stable. Game Management Unit (Unit) 12A (East and West) encompasses the North Kaibab Plateau, Arizona, and is managed according to alternative deer management guidelines. Alternative deer management guidelines are designed to provide higher hunt success, lower hunter density, or a greater opportunity to harvest an older age-class buck. Alternative deer management guidelines include managing for buck to doe ratios up to 30:100, fawn to doe ratios up to 60:100, and provide for late season hunts that occur near the onset of the breeding season with low hunter densities.

Management of mule deer on the North Kaibab has often been controversial (Russo 1964, Swank 1998, Wakeling 2007). When several hunters expressed interest in introducing more restrictive guidelines during 2000 to ensure the quality of future harvests, the Department examined data and used a public input process before proposing amendments to the alternative management guidelines to the Commission. Hunters suggested that existing alternative guidelines, which included managing for a late season harvest that comprised 55–75% ≥3-year-old deer with no guidance regarding older age class deer, did not always provide the trophy-quality deer the public expected. Russo (1964) had examined 3,124 mule deer and found the preponderance of deer with ≥70 cm outside antler spread were >3 years of age. Because Russo (1964) relied on aging mule deer based on tooth eruption and wear patterns (Dimmick and Pelton 1996), he could examine relationships of antler growth only among broad age classes (e.g., 3–5 years). Because the Department had been collecting age data from harvested deer in some years during the late 1990s using cross sections of incisors and counts of cementum annuli (Dimmick and Pelton 1996), more precise age data was available to compare with antler growth measurements, which could then be used to evaluate existing hunt guidelines. Antler spread seemed to reach a maximum at five years of age, and the Commission ultimately adopted a guideline in 2001 that 20–30% of the late season harvest should comprise deer that were ≥5 years of age. My objective was to evaluate the effectiveness of the alternative deer management guidelines adopted by the Commission in 2001.

METHODS

Mule deer harvested by hunters in Units 12AW and 12AE are legally required to be checked at the Jacob Lake hunter check station. Deer age was determined by tooth eruption patterns for deer up to three years of age or by removal of a middle incisor for sectioning and counting of cementum annuli for deer >3 years of age (Dimmick and Pelton 1996). Antler points were enumerated for each antler, and the outside spread of the antlers was measured and recorded.

To determine what constituted potential antler growth in Unit 12A, the Department examined mean antler spread and mean antler points/side by age. The assumption in this exercise is that a deer with wider antlers and more points constitute a more desirable harvest for hunters than do deer with narrow antlers and few points.

Public comment was sought on changes to the alternative management guidelines as well as any other proposed changes to hunt guidelines for fall 2001 through spring 2002. Eleven public meetings with more than 200 attendees were held in Flagstaff, Fredonia, Kingman, Payson, Pinetop, Phoenix, Prescott, Safford, Sierra Vista,

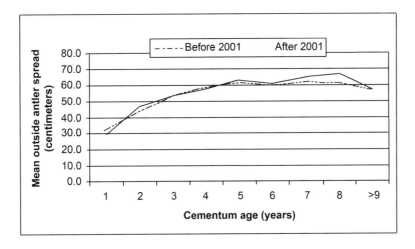

Figure 1. Mean outside antler spread (centimeters) for each age class of harvested mule deer (1995–2005) from Unit 12A, Arizona, before implementation and after implementation of amended alternative deer-management guidelines in 2001.

Tucson, and Yuma, Arizona, during January and February 2001. During the same time frame, written comments could be submitted for consideration, and the Department received more than 150. Comments were compiled and shared with the Commission at the April 2001 meeting, where deer season hunt recommendations were adopted for fall 2001. Very few of the comments from the public meetings or written correspondence specifically addressed the amendments to the alternative deer management guidelines.

Alternative management guidelines for deer were amended in April 2001 to include two additional parameters: (1) 55–75% of the harvest in late season hunts would comprise deer ≥3 years of age and (2) 20–30% of the harvest in the late season hunts would comprise deer ≥5 years of age. Consequently, I compared mean antler data by age class before and after guideline change implementation graphically. I also examined age data from harvests before and after implementation of amended guidelines, graphically and using chi square contingency tables (Zar 1984) to determine if differences in composition of the harvest could be detected. In addition, I examined the approved permit allocations before and after implementation of the amended guidelines.

RESULTS

I analyzed measurements from 1,794 male mule deer harvested in Unit 12A during 1995, 1996, 1998, 2003, 2004, and 2005 (908 before 2000, 846 after 2002). Maximum mean antler width and points/side were effectively achieved by the time a deer reached five years of age (Figures 1, 2). Mean antler points/side by age and mean outside spread by age had similar distributions both before and after implementation of amended alternative deer-management guidelines on the North Kaibab (Figures 1, 2). Because so few deer are harvested in older age classes (Figure 3), a single eight-year-old deer with multiple antler points in 2006 resulted in a peak in that age class (Figure 2).

Minor shifts have been observed in the composition of the harvest since implementation of the amended alternative guidelines in 2001. During the late hunts before 2001, 67% of the harvest comprised deer ≥3 years of age and 18% of the harvest

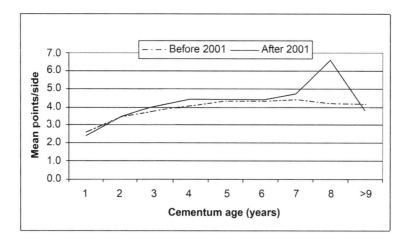

Figure 2. Mean antler points/side for each age class of harvested mule deer (1995–2005) from Unit 12A, Arizona, before implementation and after implementation of amended alternative deer-management guidelines in 2001.

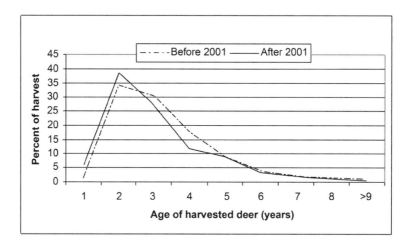

Figure 3. Age classes of mule deer as a percentage of harvest from all hunts (1995–2005) in Unit 12A, Arizona, (not only late hunts) before and after implementation of amended alternative management guidelines in 2001.

comprised deer ≥5 years of age. Since 2001, the late hunts comprised 71% of the harvest from deer ≥3 years of age and 28% from deer ≥5 years of age. In the late hunts, proportionally fewer younger deer, more deer >3 years of age ($\chi^2_c = 13.367$, 1 df, $P < 0.007$), and more deer >5 years of age ($\chi^2_c = 17.87$, 1 df, $P < 0.001$) have been harvested since 2001 (Figure 4).

Permit levels were somewhat greater in the five years prior to implementation of amended alternative deer-management guidelines than in the five years after implementation. During early hunts,

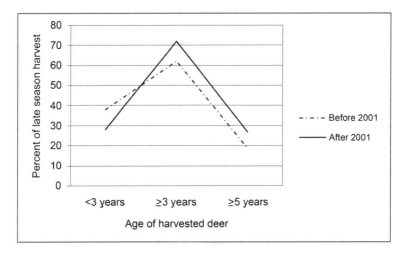

Figure 4. Percent of deer by age class harvested during the late season (1995–2005) in Unit 12A, Arizona, before and after implementation of the amended alternative management guidelines in 2001.

approved Unit 12A permits ranged from 1,100 to 1,400 before implementation, whereas approved early hunt permits ranged from 425 to 1,200 after implementation. Approved late hunt permits ranged from 200 to 300 before implementation and 150 to 225 afterward. Hunter success remained high (>50%) during all late season hunts across all years.

DISCUSSION

Using a public process, the Department was able to determine what a critical segment of the hunting customers desired in a region of the state. By coupling social desires with biological data, the Department was able to propose amended guidelines that were largely supported by the public and attainable without placing unreasonable restrictions on hunting opportunity. Relatively minor adjustments to the number of recommended permits during late season hunts (total permits in these hunts are generally between 200 and 250) allow the Department to maintain hunt quality in late hunts, while still providing hunters with greater numbers of permits in early hunts in Unit 12A.

Maximum mean outside antler spread of 64 cm is achieved by 5 years of age, although a small increase seems to occur during the seventh year. In fall 2007, a deer that measured >100 cm outside spread was inspected at the check station, although no data is yet available on the cementum age of the deer. A substantial proportion of the antler growth occurs by three years of age. This growth pattern is generally in agreement with that found by Russo (1964) and Anderson and Medin (1969). According to hunt guidelines, permit levels were influenced by population level and buck-to-doe ratios during early hunts, whereas late hunt permit levels were more influenced by harvest composition and population level.

Managing deer hunts with these metrics requires the acquisition of harvested deer teeth for sectioning and cementum-aging to determine precise ages; using tooth eruption would require broader categories in hunt guidelines. To acquire incisors, substantial effort through check stations or field checks is required to obtain a statistically valid sample. In practice, managing deer herds conservatively, hence further from the biological limitations, is often more costly to monitor because of public interest and scrutiny, although this costly effort is less critical or necessary in monitoring the biological status of the deer herd. Because the herds we monitored on the Kaibab are managed more conservatively, less revenue for the State and wildlife management is derived from them than could be realized on a more intensively managed herd, and fewer hunters are afforded the chance to participate in hunting in conservatively managed areas.

Arizona's mule deer management is largely successful in Unit 12A with few hunters voicing complaints about the quality of deer being harvested during late hunts. Other issues remain contentious among hunters, land management agencies, and the Department, such as forage monitoring, female harvests, and population estimates (see Wakeling 2007). Amendments to alternative management guidelines have yielded increased proportions of older age-class animals in the harvest, which should reflect larger-antlered deer, while hunter opportunity has not been reduced on early hunts.

LITERATURE CITED

Anderson, A. E., and D. E. Medin. 1969. Antler morphometry in a Colorado mule deer population. Journal of Wildlife Management 33:520–533.

Bishop, C. J., G. C. White, D. J. Freddy, and B. E. Watkins. 2005. Effect of limited antlered harvest on mule deer sex and age ratios. Wildlife Society Bulletin 33:662–668.

Dimmick, R. W., and M. R. Pelton. 1996. Criteria of sex and age. In Research and Management Techniques for Wildlife and Habitats. 5th edition, edited by T. A. Bookhout, pp. 169–214. The Wildlife Society, Bethesda, Maryland.

Manfredo, M. J., P. J. Fix, T. L. Teel, J. Smeltzer, and R. Kahn. 2004. Assessing demand for big-game hunting opportunities: Applying the multiple-satisfaction concept. Wildlife Society Bulletin 32:1147–1155.

Russo, J. P. 1964. The Kaibab North deer herd: Its history, problems, and management. Wildlife Bulletin 7. Arizona Game and Fish Department, Phoenix, Arizona.

Swank, W. G. 1998. History of the Kaibab deer herd, beginning in 1968. In Proceedings of the 1997 Deer-Elk Workshop, Rio Rico, Arizona, edited by J. C. deVos, Jr., pp. 14–23. Arizona Game and Fish Department, Phoenix, Arizona.

Wakeling, B. F. 2007. Wildlife management decisions and Type I and II errors. Western States and Provinces Deer and Elk Workshop Proceedings 6:35–42.

Wakeling, B. F., and B. E. Watkins. In press. Mule deer hunting: Public attitudes and agency management. Western States and Provinces Deer-Elk Workshop 7.

White, G. C., D. J. Freddy, R. B. Gill, and J. H. Ellenberger. 2001. Effect of adult sex ratio on mule deer and elk productivity in Colorado. Journal of Wildlife Management 65:543–551.

Zar, J. H. 1984. Biostatistical Analysis. 2nd Edition. Prentice-Hall, Inc., Englewood Cliffs, New Jersey.

FEMALE ELK HABITAT USE AFTER THE RODEO-CHEDISKI FIRE IN NORTHEAST ARIZONA

Kirby Bristow and Stan Cunningham

ABSTRACT

Between 18 June and 7 July 2002, the Rodeo-Chediski fire burned 184,096 ha of U. S. Forest Service, state, private, and White Mountain Apache Reservation land along the Mogollon Rim of Arizona. Before 1880, Arizona's ponderosa pine (*Pinus ponderosa*) forest communities, and the elk (*Cervus elaphus*) that inhabited them, were subjected to large-scale (>5,000 ha) episodic fires that occurred about every 2 to 10 years. Aggressive suppression of wildfires, historical livestock grazing practices, and timber management practices have rendered ponderosa pine forests in Arizona densely timbered. The over-accumulation of fuels, coupled with persistent drought conditions, have recently resulted in several stand-replacing fires in the western United States, a phenomenon which could be considered ecologically abnormal. Understanding impacts of these stand-replacing fires (such as the Rodeo-Chediski fire) on elk habitat, and how elk use areas recovering from fire could provide insights for improving forest and fire management to protect and enhance wildlife habitat. Beginning three years after containment, we investigated habitat selection and modeled habitat-use by female elk ($n = 11$) within the boundary of the Rodeo-Chediski fire. Female elk selected ponderosa pine habitats with 40–60% canopy cover that were classified as subjected to heavy to extreme burn intensity. Favorable precipitation in years following the fire, increased light transmission to the forest floor, and enhanced soil nutrient condition likely enhanced vigorous growth of forbs and shrubs that improved forage conditions and attracted elk. Forest treatments and prescribed fire designed to reduce canopy cover to 40–60% in a mosaic pattern, while reducing the impacts of roads and vehicle traffic, would likely improve habitat conditions for elk in ponderosa pine communities.

INTRODUCTION

Before 1880, Arizona ponderosa pine (*Pinus ponderosa*) forest communities, and the elk (*Cervus elaphus*) that inhabit them, evolved with large-scale (>5,000 ha) episodic fires that occurred about once every 2 to 10 years (Swetnam and Betancourt 1990). These naturally occurring fires are important for native ungulates because forage conditions after a burn are often improved by increased availability of forbs and grasses (Thill et al. 1990, Kucera and Mayer 1999). Furthermore, rapidly growing young or resprouting browse is usually more nutritious for ungulates than older browse.

Aggressive suppression of wildfires, historical livestock grazing practices, and timber management practices have rendered many Arizona ponderosa pine forests densely stocked (>3,000 stems/ha), with closed canopies that preclude sunlight from reaching the forest floor (Mast 2003). The resulting accumulation of litter increases fuels for wildfire and inhibits growth of grasses and forbs. Trees, shrubs, forbs, and grasses compete for limited nutrients, and this competition becomes more intense when extensive beds of organic litter cover the ground (Covington and Moore 1994, Kolb et al. 1994). Forests containing overly dense, small trees become further stressed by drought, enabling pathogens and insects to reach levels high enough to kill trees, further increasing fuels for wildfire (Mast 2003).

The constant accumulation of fuels eventually results in stand-replacing fires, fires that could be considered ecologically abnormal (Mast 2003). Such large-scale fires, exacerbated by fire suppression, could negatively affect elk populations, either directly via high fire mortality, or indirectly by decreasing the amount of cover necessary to avoid predators (Singer et al. 1997).

The Rodeo fire started 18 June 2002 and burned 50,000 ha in three days. The Chediski fire started 20 June 2002 in close proximity to the Rodeo fire, burning 4,400 ha during the first day. The fires merged on 23 June 2002 and had burned 132,536 ha by that date. On 7 July 2002, the largest fire in Arizona history was declared contained. Below the Mogollon Rim, the Rodeo-Chediski fire had burned 110,534 ha on the White Mountain Apache Reservation and 4,308 ha on the Tonto National Forest. Above the Mogollon Rim, the fire burned 65,776 ha on the Apache-Sitgreaves National Forest, 3,469 ha on private lands, and 9.2 ha on Arizona Game and Fish Department (AGFD) land, totaling 184,096 ha burned. High winds above the Mogollon Rim pushed the fire quickly and erratically, resulting in many unburned islands and leaving a mosaic of vegetation stands on the Apache Sitgreaves National Forest (Figure 1). Twenty-seven percent of the 70,084 ha burned on U.S. Forest Service (USFS) lands was considered high severity burn and 26% was considered moderate according to a rapid post-fire assessment conducted by the USFS Burned Area Emergency Rehabilitation (BAER) team. In contrast, 7% of the land was unburned and 40% was considered a low severity burn (Figure 1).

Habitat for large ungulates often benefits from fire because of improved forage quantity, quality, structure, and composition (Dills 1970, Hobbs and Spowart 1984, Carlson et al. 1993, Main and Richardson 2002). Fire generally results in increased diversity of forage species which may improve selective foraging opportunities (Riggs et al. 1996). Early successional grass and shrub leaves are usually more palatable than older, more decadent ones because as plants age, their concentrations of digestible fiber, minerals, and proteins decrease (Wilms et al. 1981). Fires of low to high intensity help to remove accumulated litter which can result in increased growth of herbaceous vegetation (Hulbert 1988). Additionally, increased nutrients deposited in ash, increased light intensity, reduced competition for water, and warmer soil temperatures can promote the growth of surviving plants (Pearson et al. 1972).

Several studies have documented short-term (i.e., six months to two years) increases in nutritional quality (Meneely and Schemnitz 1981, Hobbs and Spowart 1984, Wood 1988) and quantity (Dills 1970, Carlson et al. 1993) of ungulate forage after fires, as well as increased use of prescribed burn habitat by elk (Singer et al. 1989). However, stand-replacing fires can have varied effects on wildlife. After a stand-replacing Yellowstone fire, Singer et al. (1989) hypothesized rapid elk population growth. Later, they found that the elk population decreased post-fire, with increases in predation because of the lack of cover and production of weak, underweight calves as drought conditions continued (Singer et al. 1997). Notably, Arizona large predator communities differ from those in Yellowstone.

The Rodeo-Chediski fire may have im-

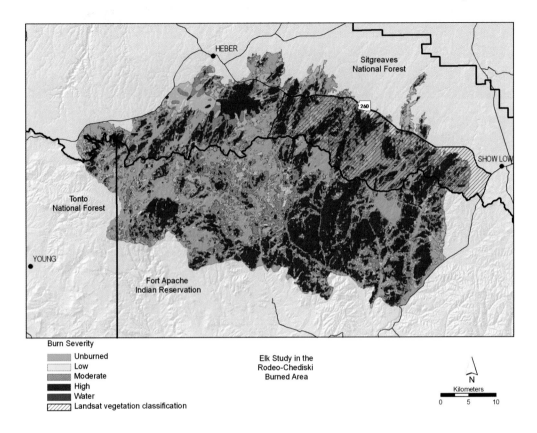

Figure 1. Study location showing burn severity estimates within the perimeter of the Rodeo-Chediski fire, Apache-Sitgreaves National Forest, east-central Arizona, 2005–2007.

proved habitat conditions for elk along the Mogollon Rim, certainly, the number, size, and shape of different overstory patches, along with the amount of edge habitat, increased within the fire perimeter. Furthermore, it is reasonable to expect that forage abundance (including some key species) increased with a reduction in overstory in areas of suitable soil conditions. However, we do not know whether the amount of high (27%) and moderate (26%) severity burn reduced thermal or hiding cover to unacceptable levels. To answer those questions and provide general insights about effects of wildfire on elk habitat after 3 to 5 years, we conducted a research project with the following objectives:

(1) investigate female elk use of current-condition forest and areas exposed to different burn intensities;
(2) model landscape-scale habitat selection by female elk across a varied habitat including burned and unburned areas; and
(3) evaluate forest stand structure to identify silvicultural treatments that meet elk habitat needs.

STUDY AREA

We conducted this project in east-central Arizona on the Apache-Sitgreaves National Forest (Figure 1), where the Rodeo-Chediski fire burned 65,776 ha in June 2002. The center of the burn perimeter is located 171 km northeast of Phoenix, Arizona, and is

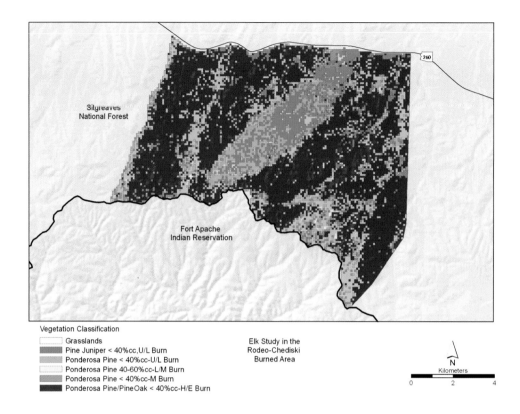

Figure 2. Map of study area showing habitat type classifications (U/L = unburned to lightly burned, L/M = light to medium burn, M = medium burn, H/E = heavy to extreme burn) for areas within the annual home range of female elk (n = 9), Apache-Sitgreaves National Forest, east-central Arizona, 2005–2007.

bordered to the north primarily by U. S. Highway 260. The towns of Payson to the west, Heber-Overgaard to the north, and Show Low to the east lie on or near the study border. Elevations range from 1,519 m to 2,356 m. Mean annual precipitation, measured at the Heber Ranger Station along the northern boundary of the fire, was 43.9 cm with 98.8 cm of snowfall. Mean temperature ranged from 0.4° C in January to 20.2° C in July.

We captured elk in the burned sites (N110°22'30", W34°15'00") between FS 300 where it intersects with U. S. Highway 260 to the west and Juniper Ridge to the east. The northern boundary was U. S. Highway 260 and the southern boundary was the White Mountain Apache Tribal lands. Final area of study was determined by where elk traveled within the boundaries of the fire.

Pre-fire habitat in this area comprised primarily ponderosa pine forest, with smaller patches of meadow-like openings, oak (*Quercus gambellii*), and pinyon (*Pinus edulis*)-juniper (*Juniperus deppeana*). Despite the high intensity of the fire below the Mogollon Rim, the high speed of the fire above the Mogollon Rim created a diverse mosaic of forest conditions. High burn intensity occurred on 27% of the area, 26% of the area received moderate intensity fire, 40% of the area at low intensity burn, and 7% of the area was unburned. The fire mosaic on the forest at the landscape scale (e.g., 4,000 ha) ranged from areas totally dominated by high intensity burn to mosaics of relatively small patches of unburned and high, moderate, and low intensity burned forest. The large size of the Rodeo-Chediski fire provided the opportunity to assess stand-replacing wildfire effects on elk across a forest landscape.

METHODS
Capture and Monitoring

Between July 2005 and February 2006, we captured female elk using clover traps baited with mineral blocks and alfalfa. Each animal was fitted with a Telonics (Mesa, AZ) global positioning system (GPS) radiocollar and unique ear tags. Each GPS radiocollar was programmed to attempt location acquisition every five hours throughout the monitoring period. Radiocollars sampled ambient temperature at location fixes, and all sampled data were stored on internal memory. Motion sensors with VHF (very high frequency) beacons allowed us to aerially monitor signals once monthly to determine if radiocollared animals were still alive. Spread Spectrum GPS radiocollars that transmitted stored location data on a monthly basis allowed us to monitor movements of some marked animals throughout the study. Radiocollars were programmed to fall off after 400 or 745 days for retrieval and data downloading.

We calculated seasonal (winter 15 Dec–15 Mar, spring 16 Mar–14 Jun, summer 15 Jun–13 Sep, and autumn 13 Sep–14 Dec) and annual minimum convex polygons (MCP) to estimate home ranges of individual female elk (Hayne 1949). We also calculated a master MCP outlining the annual home range of all elk to delineate the boundaries of the study area. We calculated the number of elk locations within each annual and seasonal home range and generated a similar number of random sites within that home range to establish availability of each habitat variable.

Habitat Classification

The BAER team created a burn severity geographic information system (GIS) cover (map) for all land within the perimeter of the Rodeo-Chediski fire. The map was derived from Landsat satellite imagery acquired 7 July 2002, from U. S. Geological Survey Data Center, Sioux Falls, South Dakota. Using spectral classification modeling, each 900 m^2 area was assigned a burn severity class: high, moderate, low, and unburned. The map was field-verified on the USFS portion of the burn. We were interested in elk use of habitats after recovery from fire, and because the BAER burn severity map was developed immediately following the fire with limited ground verification, we chose to create an additional habitat classification map.

We used remote sensing to delineate post-fire vegetation conditions because of the size of the project area and the scope of the analysis. The initial step in any image classification process is to determine the number and types of categories. There are two major levels of classification categories: informational classes and spectral classes. Informational classes are those categories of interest in the classification. We selected three levels of informational classes: vegetation (ponderosa pine, pine-oak, pinyon-juniper, and open grassland), canopy cover (CC, defined by USFS canopy cover data), and burn intensity (unburned, light, moderate, high). Spectral classes are groups of pixel brightness values that are uniform with respect to the informational classes across several spectral bands of the imagery. Groupings of homogenous pixels were identified using both point and spatial classifiers. We used supervised classification for habitat analysis. This approach uses samples of known identity to classify pixels of unknown identity, where 1) samples of known identity are those pixels located within training areas, 2) training areas are clearly identifiable areas of the imagery that have known properties on the landscape,

3) spectral properties of the imagery typify properties of the informational class and are homogeneous, and 4) training areas are clearly identifiable both on imagery and landscape.

We selected 15 three-hectare training areas for each possible combination of informational class (vegetation, canopy cover, burn intensity), for a total of 1,350 training areas. To ensure training areas represented individual informational classes, each was located ≥200 m from any boundary between informational classes. We recorded the location and extents of each training area using a GPS point averaging option to insure one-meter accuracy. We reviewed all training areas based on their informational class to ensure separation of spectral signatures. When necessary, additional training areas were collected to ensure separation of spectral signatures.

We used two types of imagery to identify habitat class: Landsat 7 thematic mapper and MODIS ASTER. We performed spectral classification analysis using ERDAS Imagine and ENVI image processing software. To classify the project area we used a maximum likelihood classifier at the 95% level. The resulting habitat classification was assessed for accuracy at the $\alpha = 0.05$ level using 778 random sites. We formatted the output of the classification as ArcGIS compatible GeoTiff raster layers of post-fire habitat availability. Additional GIS layers were accessed to provide information on elevation, percent slope, and slope aspect for the study area.

Habitat Selection

To establish availability of habitat characteristics, we generated 9,389 random sites across the study area using ArcGIS. We overlaid all elk locations and random sites on available GIS habitat characteristic covers and recorded values for each of the following habitat variables: BAER burn severity, elevation, slope, aspect, habitat class (vegetation-canopy-burn intensity), and habitat patch size. To predict annual and seasonal habitat use of female elk, we developed logistic regression models (Hosmer and Lemeshow 1989) using habitat variables that were not correlated ($\alpha<0.1$). We developed probable models *a priori*, and calculated Akaike's Information Criterion (AIC) to select the most parsimonious model (Burnham and Anderson 1992) using 0.5 as the cutpoint for classification of use and random sites.

To determine habitat selection by female elk, we compared the Bonferroni 90% simultaneous confidence interval (Byers et al. 1984) of the percentage of locations in each BAER burn severity and habitat class to the expected frequency distribution, based on random sites. We calculated Jacob's D (Jacob 1974) values (-1.0 to 1.0) to examine the extent of selection (positive values) or avoidance (negative values), where the closer the value is to 1.0 or –1.0, the stronger the relationship. We used the same methods to determine landscape habitat selection among seasons and time of locations (night = two hours after sunset to two hours before sunrise, crepuscular = two hours before sunrise and two hours after sunset, and day = two hours after sunrise to two hours before sunset).

RESULTS

Capture and Monitoring

Between August 2005 and February 2006, we captured 14 female elk. We deployed 4 Spread Spectrum radiocollars for about 340 days each and 10 store-on-board radiocollars for about 480 days each. We were unable to recover one radiocollar, two of the elk were legally harvested by hunters, and one radiocollar failed to obtain any location fixes during deployment. For habitat selection we used only those locations that overlapped the areas where we had classified habitat with remote sensing (study area). We downloaded 20,543 locations from the 13 recovered radiocollars and although 20,070

Table 1. Ranking of logistic regression models[1] with K parameters for female elk (n = 9) annual, winter, spring, summer, and autumn habitat use within the perimeter of the Rodeo-Chediski fire, Apache-Sitgreaves National Forest, east-central Arizona, 2005-2007.

Model		K	-2 log likelihood	elk correctly classified (%)	randoms correctly classified (%)	AIC	Delta AIC
Annual							
1	Global	5	25938.36	59.2	56.1	25948.36	0.00
2	Habitat-patch	2	26138.86	55.9	58.7	26140.86	192.50
3	Migratory	3	26192.72	60.3	50.3	26198.72	250.36
4	Habitat type	1	26305.02	66.7	42.1	26307.02	358.66
5	Landform	3	26328.40	57.0	50.1	26334.40	386.04
Winter							
1	Global	5	7338.28	65.1	54.7	7348.28	0.00
2	Migratory	3	7430.31	63.4	53.6	7436.31	88.03
3	Habitat-patch	2	7497.97	64.9	53.8	7501.97	153.69
4	Landform	3	7542.33	61.3	52.5	7548.33	200.05
5	Habitat type	1	7570.33	69.6	43.1	7572.33	224.05
Spring							
1	Global	5	9312.26	62.1	52.2	9322.26	0.00
2	Landform	3	9439.10	63.1	50.0	9445.10	112.84
3	Migratory	3	9432.07	61.1	50.6	9438.07	115.81
4	Habitat-patch	2	9463.05	70.8	41.4	9467.05	144.79
5	Habitat type	1	9530.28	76.8	29.5	9532.28	210.02
Summer							
1	Global	5	6091.78	68.3	47.1	6101.78	0.00
2	Migratory	3	6126.12	73.1	40.1	6132.12	30.34
3	Habitat type	1	6151.12	73.7	38.5	6153.12	51.34
4	Habitat-patch	2	6145.20	73.7	38.5	6149.20	47.42
5	Landform	3	6225.00	56.3	52.3	6231.00	129.22
Autumn							
1	Global	5	2672.71	59.7	49.1	2682.71	0.00
2	Migratory	3	2686.19	59.8	46.9	2692.19	9.48
3	Landform	3	2690.25	59.8	49.9	2696.25	13.54
4	Habitat type	1	2701.13	84.1	22.5	2702.13	19.42
5	Habitat-patch	2	2699.94	80.7	26.0	2703.94	21.23

[1] P-values for all models were <0.001 (n = 9,555 for annual, n = 2,798 for winter, n = 3,467 for spring, n = 2,289 for summer, and n = 1,001 for autumn), and degrees of freedom were equal to the number of variables included in the model. Models are presented in order of parsimony. Models contained the following variables:
Global model = Habitat type, elevation, percent slope, slope aspect, and patch size
Habitat-patch size model = Habitat type and patch size
Migratory model = Habitat type, elevation, and slope
Habitat type model = Habitat type only
Landform model = Elevation, slope, and slope aspect

Table 2. Annual, winter, spring, summer, autumn, nighttime (two hours after sunset to two hours before sunrise), crepuscular (two hours before sunrise to two hours after sunset), and daytime (two hours after sunrise to two hours before sunset) use of habitat type (vegetation, canopy cover, burn intensity) by female elk (n = 9) compared to random locations within the perimeter of the Rodeo-Chediski fire, Apache-Sitgreaves National Forest, east-central Arizona, 2005–2007.

Habitat type	No. of locations observed	No. of locations expected	Bonferroni 90% CI	Jacobs' D^1
Annual				
Grasslands, 0% canopy, unburned – extreme burn intensity	73	113	57–96	-0.20
Pinyon-juniper, <40% canopy, unburned – light burn intensity	782	1,078	716–850	-0.18
Pipo^2 – Quga^3, 40-60% canopy, heavy – extreme burn intensity	6,376	5,512	6,268–6,478	0.19
Pipo – Quga, <40% canopy, unburned – light burn intensity	568	935	506–621	-0.27
Pipo – Quga, <40% canopy, medium burn intensity	461	610	411–506	-0.15
Pipo – Quga, 40-60% canopy, light – medium burn intensity	1,295	1,307	1,223–1,376	
Winter				
Grasslands, 0% canopy, unburned – extreme burn intensity	19	34	8–31	-0.27
Pinyon-juniper, <40% canopy, unburned – light burn intensity	199	342	168–229	-0.29
Pipo – Quga, 40-60% canopy, heavy – extreme burn intensity	1,947	1,593	1,891–2,003	0.27
Pipo – Quga, <40% canopy, unburned – light burn intensity	153	283	126–176	-0.32
Pipo – Quga, <40% canopy, medium burn intensity	147	173	120–176	
Pipo – Quga, 40-60% canopy, light – medium burn intensity	333	373	294–373	
Spring				
Grasslands, 0% canopy, unburned – extreme burn intensity	34	59	21–49	-0.26
Pinyon-juniper, <40% canopy, unburned – light burn intensity	361	416	319–402	-0.08
Pipo – Quga, 40-60% canopy, heavy – extreme burn intensity	2,182	1,979	2,115–2,247	0.12
Pipo – Quga, <40% canopy, unburned – light burn intensity	246	326	211–281	-0.15
Pipo – Quga, <40% canopy, medium burn intensity	164	222	135–191	-0.16
Pipo – Quga, 40-60% canopy, light – medium burn intensity	480	465	430–527	
Summer				
Grasslands, 0% canopy, unburned – extreme burn intensity	14	11	5–21	
Pinyon-juniper, <40% canopy, unburned – light burn intensity	113	188	87–137	-0.27
Pipo – Quga, 40-60% canopy, heavy – extreme burn intensity	1,673	1,397	1,623–1,724	0.27
Pipo – Quga, <40% canopy, unburned – light burn intensity	76	228	55–96	-0.53
Pipo – Quga, <40% canopy, medium burn intensity	106	133	82–128	-0.12

Habitat/burn				
Pipo – Quga, 40-60% canopy, light – medium burn intensity	307	332	268–346	
Autumn				
Grasslands, 0% canopy, unburned – extreme burn intensity	6	11	3–11	
Pinyon-juniper, <40% canopy, unburned – light burn intensity	109	134	86–132	-0.12
Pipo – Quga, 40-60% canopy, heavy – extreme burn intensity	574	544	537–609	
Pipo – Quga, <40% canopy, unburned – light burn intensity	93	97	72–114	
Pipo – Quga, <40% canopy, medium burn intensity	44	80	29–59	-0.31
Pipo – Quga, 40-60% canopy, light – medium burn intensity	175	134	147–203	0.16
Nighttime				
Grasslands, 0% canopy, unburned – extreme burn intensity	32	35	17–46	
Pinyon-juniper, <40% canopy, unburned – light burn intensity	302	324	263–339	
Pipo – Quga, 40-60% canopy, heavy – extreme burn intensity	2,013	1,681	1,956–2,072	0.25
Pipo – Quga, <40% canopy, unburned – light burn intensity	160	266	130–188	-0.27
Pipo – Quga, <40% canopy, medium burn intensity	136	171	110–162	-0.12
Pipo – Quga, 40-60% canopy, light – medium burn intensity	251	417	214–287	-0.28
Crepuscular				
Grasslands, 0% canopy, unburned – extreme burn intensity	24	45	13–32	-0.34
Pinyon-juniper, <40% canopy, unburned – light burn intensity	280	402	241–318	-0.20
Pipo – Quga, 40-60% canopy, heavy – extreme burn intensity	2,212	1,787	2,150–2,272	0.28
Pipo – Quga, <40% canopy, unburned – light burn intensity	192	357	161–222	-0.32
Pipo – Quga, <40% canopy, medium burn intensity	137	199	112–164	-0.19
Pipo – Quga, 40-60% canopy, light – medium burn intensity	369	424	328–411	-0.08
Daytime				
Grasslands, 0% canopy, unburned – extreme burn intensity	17	38	7–28	-0.38
Pinyon-juniper, <40% canopy, unburned – light burn intensity	200	355	169–231	-0.30
Pipo – Quga, 40-60% canopy, heavy – extreme burn intensity	2,151	2,044	2,085–2,216	0.07
Pipo – Quga, <40% canopy, unburned – light burn intensity	216	307	183–252	-0.19
Pipo – Quga, <40% canopy, medium burn intensity	188	238	159–221	-0.12
Pipo – Quga, 40-60% canopy, light – medium burn intensity	675	465	620–727	0.22

[1] Jacobs' D represents magnitude of selection or avoidance, from 1.0 to –1.0.
[2] Ponderosa pine (*Pinus ponderosa*)
[3] Gambel's oak (*Quercus gambelii*)

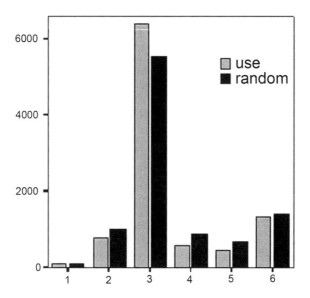

Figure 3. Annual use of habitat type (vegetation, canopy cover, burn intensity) by female elk (*n* = 9) compared to random sites, Apache-Sitgreaves National Forest, east-central Arizona, 2005-2007. Elk (*n* = 9) selected for habitat type number 3 and other habitat types were used according to availability.

fell within the perimeter of the burn, only 9,555 fell within the study area.

Habitat Classification

Based on remote sensing, we classified habitat of the 109.67 km² study area into six different vegetation-canopy cover-burn intensity classes. Ponderosa-oak, 40–60% CC, heavy to extreme burn intensity was the most common habitat type (Figure 2).

Habitat Selection

We developed five different *a priori* habitat selection models based on competing theories of factors affecting elk use (Table 1). The global model including all available habitat variables was the most parsimonious model describing annual and seasonal female elk habitat selection (Table 1).

Annually, female elk selected for the pine-oak, 40–60% CC, heavy to extreme burn intensity habitat type (Table 2). All other habitat types were avoided or used according to availability (Figure 3). This habitat selection pattern was consistent across the winter, spring, and summer seasons as well as during nocturnal and crepuscular times of day (Table 2). During autumn, female elk shifted their selection to the pine-oak, 40–60% CC, light to medium burn intensity habitat type, and all other habitat types were avoided or used according to availability (Table 2). During day female elk selected for the pine-oak, 40–60% CC, heavy to extreme burn intensity habitat type to a lesser degree than other times of day and selected the pine-oak, 40–60% CC, light to medium burn intensity habitat type more strongly. Female elk avoided all other habitat types during the day (Table 2).

Within the perimeter of the Rodeo-Chediski fire, female elk selected for areas that were classified as moderate to high burn severity, and avoided areas classified as unburned or low burn severity by the BAER team (Table 3). Average patch size of selected habitats used by elk was 31 km² (SD = 14.5).

Table 3. Annual use of habitat classified by burn severity[1] by female elk (n = 12) compared to random locations within the perimeter of the Rodeo-Chediski fire, Apache-Sitgreaves National Forest, east-central Arizona, 2005–2007.

Burn severity class	No. of locations observed	No. of locations expected	Bonferroni 90% CI	Jacobs' D[2]
High	7,030	5,583	11,035–11,336	0.23
Moderate	4,573	3,532	11,196–11,477	0.25
Low	3,381	4,049	9,005–9,286	-0.18
Unburned	5,116	6,936	8,402–8,683	-0.29

[1]Burn severity estimated from Landsat 7 TM Satellite imagery acquired 7 July 2002, from USGS EROS Data Center, Sioux Falls, SD by the U.S. Forest Service Burned Area Emergency Rehabilitation (BAER) team.

[2]Jacobs' D represents Magnitude of selection or avoidance, from 1.0 to –1.0.

DISCUSSION

We found female elk habitat use within the perimeter of the Rodeo-Chediski fire differed from availability with respect to habitat type (vegetation type, canopy cover, and burn severity). However, this selection and avoidance was less evident than that for many ungulates (Ockenfels et al. 1991, Ockenfels et al. 1994, Bristow et al. 1996). Elk habitat use is characteristic of availability and they can occupy and use a wider range of habitats than many ungulates (Hobbs and Hanley 1990). This low selectivity with regard to habitat is likely why our global habitat model best explained female elk habitat selection; the selection for any one habitat variable was not strong enough to accurately classify use and random sites by itself. Although elk seasonal habitat selection is affected by topography (Edge and Marcum 1991), slope (Zahn 1974), elevation (Beall 1974), and slope aspect (Mackie 1970), these effects are usually associated with forage availability, thermal factors, and cover.

We found that female elk selected the most intensively burned habitats with less canopy cover during crepuscular and evening hours, when elk usually feed. Many investigators have found that elk feed in more open canopy habitats, using denser canopy areas for resting and escape (Marcum 1976). After the Rodeo-Chediski fire, favorable precipitation within the intensively burned areas produced vigorous forb and shrub growth, improving forage conditions for elk and other ungulates. Because the selection for these areas was most pronounced during crepuscular and evening hours when elk primarily forage, this selection was likely due to improved forage conditions.

Improved forage conditions often increase elk use of burned areas in forested ecosystems (Bartos and Mueggler 1979, Leege and Godbolt 1985, Canon et al. 1987). Canon et al. (1987) interpreted increased use of burned aspen stands as evidence of improved forage palatability. Rowland et al. (1983) found no difference in nutritive quality of available forage in burned and unburned ponderosa pine forests in New Mexico. However, elk foraging in burned habitats had improved nutrient contents of their diet and greater body weights than elk foraging in unburned habitats (Rowland et al. 1983). Burned habitats may provide foraging advantages to elk through increased forage availability and diversity which would improve overall diet. Feeding site selection by elk optimizes nutrient intake while reducing energy expenditure (Wambolt and McNeal 1987, McCorquodale 1993). Consequently, improved forage conditions should influence foraging habitat selection and allow elk to optimize food intake.

Deposition and distribution of burned logs and debris (slash) can affect availability of forage to elk and impede movements (Dyrness 1965, Lyon and Jensen 1980). The Rodeo-Chediski fire burned with such intensity that some trees were entirely consumed. In some areas substantial accumulated slash, elimination of these obstacles may have affected elk use. Dyrness (1965) believed that accumulated slash in excess of 50% ground coverage precluded elk use of clear cuts in Oregon. Numbers of elk pellet groups decreased by more than 50% in Montana clear cuts when slash depth was >0.5 m (Lyon and Jensen 1980).

In summer, elk seek out thermal cover provided by shaded areas to bed during mid-day hours (Beall 1974, Brown 1994). We found female elk avoided areas with lower canopy cover throughout most seasons and times of day. During mid-day periods, elk selected areas that were burned less intensively which would ostensibly have more tree cover and consequently greater canopy cover. However, use of these areas during summer was not different from availability, suggesting a relationship to security cover, rather than thermal cover (shade) availability. Thomas et al. (1976) considered canopy cover >70% acceptable and 40–70% cover marginal in terms of providing thermal cover for elk. However, Cook et al. (1998) could not find evidence of a positive energetic benefit of thermal cover to elk body mass or condition. Additionally, Strohmeyer and Peek (1996) and Merrill (1991) have demonstrated that elk can exist in areas lacking any classically defined thermal cover. The benefit of thermal cover relative to elk habitat use warrants further investigation.

Elk habitat selection can also be affected by human activities (Irwin and Peek 1983). During autumn hunting seasons elk seek out areas with higher canopy cover and visual obstruction to avoid detection (Morgantini and Hudson 1979). We found elk use of more open intensively burned areas was reduced during daylight hours and autumn when human activity likely increased, especially in areas of high road density. Elk tend to avoid areas immediately adjacent (0.4–2.9 km) to roads, especially during periods of increased traffic (Marcum 1976, Morgantini and Hudson 1979). Black et al. (1976) recommended maintaining 20–30% hiding cover and 10–20% thermal cover where elk habitat occurred in patches of 2.6–10.5 ha. Conversely, we found elk used much larger habitat patches, averaging 31 km^2 (SD = 14.5). This difference in patch size may be more related to the scale of our habitat classification rather than reflecting elk habitat selection.

Kie et al. (2002) were able to explain 57% of the variation in mule deer (*Odocoileus hemionus*) home ranges based on spatial heterogeneity of habitat in eight discrete mule deer populations, associating better habitat quality with increased spatial heterogeneity. Spatial heterogeneity is a structural feature of landscapes that can be defined as the complexity of variability in space of the properties of the ecological system (Li and Reynolds 1994). Future research including a more advanced landscape analysis including edge density, patch shape, patch richness density, edge contrast index, and nearest suitable patch (McGarigal and Marks 1995) could provide more in depth information on how elk use habitat within their home ranges. This information could become extremely important in protecting or even enhancing elk habitat as forest restoration (thinning) projects continue in Arizona.

MANAGEMENT IMPLICATIONS

Forest treatments designed to improve elk habitat should consider canopy cover, habitat patch size, slash treatments, hiding cover availability, and road densities. We found that although elk used a variety of habitats with respect to canopy cover, within ponderosa pine communities, areas with 40–60% CC were selected during all times of day and seasons. As documented elsewhere, use of intensively burned areas by elk suggests that increased forage production within the perimeter of the Rodeo-Chediski fire likely attracted elk use. Prescribed fire is a common treatment to improve elk habitat quality that has been largely successful when properly applied (Leege and Hickey 1971).

Prescribed fire with intense burn severity similar to the Rodeo-Chediski fire improved sprouting of desirable forage plants in Idaho, and Lyon (1971) predicted increased forage availability would benefit mule deer and elk for 20 years. Forest treatments and prescribed fire designed to reduce canopy cover to 40–60% in a mosaic pattern would likely improve forage conditions for Arizona elk in ponderosa pine communities. Reseeding with preferred elk forage species and fertilization may further improve growth of forage species and increase elk use (Leege and Godbolt 1985). Slash accumulation should be kept below 50% ground coverage and <0.5 m depth (Dyrness 1965, Lyon and Jensen 1980). Efforts to limit roads and reduce vehicle traffic will improve access to available habitat for elk (Morgantini and Hudson 1979).

ACKNOWLEDGEMENTS

We thank L. D. Avenetti, M. L. Crabb, J. P. Ng, T. D. Rogers, and S. C. Sprague, for assistance in elk capture and telemetry. R. Birkeland and J. P. Greer provided logistical support. S. McAdams and W. H. Miller performed the habitat classification, and S. R. Boe assisted with geographic information systems. T. C. Atwood reviewed early drafts of the manuscript and provided valuable editorial comments. Funding for this study was provided by the Federal Aid in Wildlife Restoration Act through Project W-78-R.

LITERATURE CITED

Andren, H. 1994. Effects of habitat fragmentation on birds and mammal in landscapes with different proportions of suitable habitat: A review. Oikos 71:355–66.

Bartos, D. L., and W. F. Mueggler. 1979. Influence of fire on vegetation production in the aspen ecosystem in western Wyoming. In North American Elk: Ecology and management, edited by M. S. Boyce and L. D. Hayden-Wing, pp. 75–82. University of Wyoming, Laramie, Wyoming.

Beall, R. C. 1974. Winter habitat selection and use by a western Montana elk herd. Ph.D. Dissertation, University of Montana, Missoula, Montana.

Black, H., Jr., R. J. Scherzinger, and J. W. Thomas. 1976. Relationships of Rocky Mountain elk and Rocky Mountain mule deer habitat to timber management in the Blue Mountains of Oregon and Washington. In Proceedings of the Elk Logging-Roads Symposium, edited by S. R. Heib, pp. 11–31. University of Idaho, Moscow, Idaho.

Bowers, M. A., and S. F. Matter. 1997. Landscape ecology on mammals: Relationships between density and patch size. Journal of Mammalogy 788:999–1013.

Bristow, K. D., J. A. Wennerlund, R. E. Schweinsburg, R. J. Olding, and R. E. Lee. 1996. Habitat use and movements of desert bighorn sheep near the Silver Bell Mine, Arizona. Arizona Game and Fish Department Technical Report 25, Phoenix, Arizona.

Brown, R. L. 1994. Effects of timber management practices on elk. Arizona Game and Fish Department Technical Report 10, Phoenix, Arizona.

Burnham, K. P., and D. R. Anderson. 1992. Data-based selection of an appropriate biological model: The key to modern data analysis. In Wildlife 2001: Populations, edited by D. R. McCullough and R. H. Barret, pp. 16–30. Elsevier Science Publishers, London, United Kingdom.

Byers, C. R., R. K. Steinhorst, and P. R. Krausman. 1984. Clarification of a technique for analysis of utilization-availability data. Journal of Wildlife Management 48:1050–53.

Canon, S. K., P. J. Urness, and N. V. DeByle.

1987. Habitat selection, foraging behavior, and dietary nutrition of elk in burned aspen forest. Journal of Range Management 40: 433–38.

Carlson P. C., G. W. Tanner, J. M. Wood, and S. R. Humphrey. 1993. Fire in key deer habitat improves browse, prevents succession, and preserves endemic herbs. Journal of Wildlife Management 57:914–28.

Cook, J. G., L. L. Irwin, L. D. Bryant, R. A. Riggs, and J. W. Thomas. 1998. Relations of forest cover and condition of elk: A test of the thermal cover hypothesis in summer and winter. Wildlife Monographs 141:1–61.

Covington. W. W., and M. M. Moore. 1994. Southwestern ponderosa forest structure—changes since Euro-American settlement. Journal of Forestry 92:39–47.

Dills, G. G. 1970. Effects of prescribed burning on deer browse. Journal of Wildlife Management 34:540–45.

Dodd, N. L. 2003. Landscape-scale habitat relationships to tassel-eared squirrel population dynamics in north-central Arizona. Arizona Game and Fish Department Technical Guidance Bulletin 6, Phoenix, Arizona.

Dyrness, C. T. 1965. Soil surface condition following tractor and high-lead logging in the Oregon Cascades. Journal of Forestry 63:273–75.

Edge, W. D., and C. L. Marcum. 1991. Topography ameliorates the effects of roads and human disturbance on elk. In Proceedings of the Elk Vulnerability Symposium, compiled by A. G. Christiansen, L. J. Lyon, and T. N. Loner, pp. 132–37. Montana State University, Bozeman, Montana.

Gardner, R. H., B. T. Milne, M. G. Turner, and R. V. O'Neill. 1987. Neutral models for analysis of broadscale landscape pattern. Landscape Ecology 1:19–28.

Hayne, D. W. 1949. Calculation of size of home range. Journal of Mammalogy 30:1–18.

Hobbs, N. T., and R. A. Spowart. 1984. Effects of prescribed fire on nutrition of mountain sheep and mule deer during winter and spring. Journal of Wildlife Management 48:551–60.

Hobbs, N. T., and T. A. Hanley. 1990. Habitat evaluation: Do use/availability data reflect carrying capacity? Journal of Wildlife Management 54:515–22.

Hosmer, D.W., Jr., and S. Lemeshow. 1989. Applied logistic regression. John Wiley and Sons, New York, New York.

Hulbert, L. C. 1988. Causes of fire effects in tallgrass prairie. Ecology 69:46–58.

Irwin, L. L., and J. M. Peek. 1983. Elk habitat use relative to forest succession in Idaho. Journal of Wildlife Management 47:664–72.

Jacob, J. 1974. Quantitative measure of food selection. Oecologia 14:413–17.

Kie, J. G., T. Bowyer, M. C. Nicholson, B. B. Boroski, and E. R. Loft. 2002. Landscape heterogeneity at differing scales: Effects on spatial distribution of mule deer. Ecology 83:530–44.

Kolb, T. E., M. R. Wagner, and W. W. Covington. 1994. Concepts of forest health. Journal of Forestry 92:10–15.

Krohne, D. T. 1997. Dynamics of metapopulations of small mammals. Journal of Mammalogy 78:1014–26.

Kucera, T. E., and K. E. Mayer. 1999. A sportsman's guide to improving deer habitat in California. California Department of Fish and Game, Sacramento, California.

Leege, T. A., and W. O. Hickey. 1971. Sprouting of northern Idaho shrubs after prescribed burning. Journal of Wildlife Management 35:508–15.

Leege, T. A., and G. Godbolt. 1985. Herbaceous response following prescribed burning and seeding of elk range in Idaho. Northwest Science 59:134–43.

Li, H., and J. F. Reynolds. 1994. A simulation experiment to quantify spatial heterogeneity in categorical maps. Ecology 75:2446–55.

Lyon, L. J. 1971. Vegetal development following prescribed burning of Douglas-fir in south-central Idaho. U.S. Forest Service Research Paper INT-105, Ogden, Utah.

Lyon, L. J., and C. E. Jensen. 1980. Management implications of elk and deer use of clearcuts in Montana. Journal of Wildlife Management 44:352–62.

Mackie, R. J. 1970. Range ecology and relations of mule deer, elk, and cattle in the Missouri River breaks, Montana. Wildlife Monographs 20:1–79.

Main, M. B., and L. W. Richardson. 2002. Response of wildlife to prescribed fire in southwest Florida pine flatwoods. Wildlife Society Bulletin 30:213–21.

Marcum, C. L. 1976. Habitat selection and use during summer and fall months by a western Montana elk herd. In Proceedings of the Elk Logging-Roads Symposium, edited by S. R. Heib, pp. 91–96. University of Idaho, Moscow, Idaho.

Mast, J. N. 2003. Tree health and forest structure. In Ecological Restoration of Southwestern Ponderosa Pine Forests, edited by P. Freiderici. Island Press, Washington, D.C.

McCorquodale, S. 1993. Winter foraging behavior of elk in the shrub-steppe of Washington. Journal of Wildlife Management 57:881–90.

McGarigal, K., and B. J. Marks. 1995. FRAGSTATS: Spatial pattern analysis program for quantifying landscape structure. U.S. Forest Service GTR–PNW–351. Portland, Oregon.

Meneely, S. C., and S. D. Schemnitz. 1981. Chemical composition and in vitro digestibility of deer browse three years after a wildfire. The Southwestern Naturalist 26:365–74.

Merrill, E. H. 1991. Thermal constraints on use of cover types and activity time of elk. Applied Animal Behaviour Science 29: 251–67.

Morgantini, L. E., and R. J. Hudson. 1979. Human disturbance and habitat selection in elk. In North American Elk: Ecology and management, edited by M. S. Boyce and L. D. Hayden-Wing, pp. 132–39. University of Wyoming, Laramie, Wyoming.

Ockenfels, R. A., D. E. Brooks, and C. H. Lewis. 1991. General ecology of the Coues white-tailed deer in the Santa Rita Mountains. Arizona Game and Fish Department Technical Report 6, Phoenix, Arizona.

Ockenfels, R. A., A. Alexander, C. L. Dorothy Ticer, and W. K. Carrel. 1994. Home range, movement patterns, and habitat selection of pronghorn in central Arizona. Arizona Game and Fish Department Technical Report 13, Phoenix, Arizona.

Pearson, H. A., J. R. Davis, and G. H. Schubert. 1972. Effects of wildfire on timber and forage production in Arizona. Journal of Range Management 38:541–45.

Riggs, R. A., S. C. Bunting, and S. E. Daniels. 1996. Prescribed fire. In Rangeland Wildlife, edited by P. R. Krausman, pp. 295–315. The Society for Range Management, Denver, Colorado.

Rowland, M. M., A. W. Alldredge, J. E. Ellis, B. J. Weber, and G. C. White. 1983. Comparative winter diets of elk in New Mexico. Journal of Wildlife Management 47:924–32.

Singer, F. J., W. Schreier, J. Oppenheim, and E. O. Garton. 1989. Drought, fire, and large mammals. Bioscience 39:716–722.

Singer, F. J., A. Harting, K. K. Symonds, and M. B. Coughenour. 1997. Density dependence, compensation, and environmental effects on elk calf mortality in Yellowstone National Park. Journal of Wildlife Management 61:12–25.

Strohmeyer, D. C., and J. M. Peek. 1996. Wapiti home range and movement patterns in a sagebrush desert. Northwest Science 70:79–87.

Swetnam, T. W., and J. L. Betancourt. 1990. Fire: Southern oscillation relations in the southwestern United States. Science 304: 1017–20.

Thill, R. E., H. F. Morris, Jr., and A. T. Harrel. 1990. Nutritional quality of deer diets from southern pine-hardwood forests. American Midland Naturalist 124:413–17.

Thomas, J. W., R. J. Miller, H. Black, J. E. Rodiek, and C. Maser. 1976. Guidelines for maintaining and enhancing wildlife habitat in forest management in the Blue Mountains of Oregon and Washington. Transactions of the North American Wildlife and Natural Resource Conference 41:452–76.

Wambolt, C. L., and A. F. McNeal. 1987. Selection of wintering foraging sites by elk and mule deer. Journal of Environmental Management 25:285–91.

Wiens, J. A., N. C. Stenseth, B. Van Horne, and R. A. Ims. 1993. Ecological mechanisms and landscape ecology. Oikos 66:369–90.

Wilms, W., A. W. Bailey, A. McLean, and C. Kalnin. 1981. Effects of clipping or burning on the distribution of chemical constituents in bluebunch wheatgrass in spring. Journal of Range Management 34:267–69.

Wood, G. W. 1988. Effects of prescribed fire on deer forage and nutrients. Wildlife Society Bulletin 16:180–86.

Zahn, H. M. 1974. Seasonal movements of the Burdette Creek elk herd. Job Completion Report W–120–R–4. Montana Fish and Game Department, Helena, Montana.

SCENT-STATION SURVEYS: INDEXING RELATIVE ABUNDANCE OF MESOPREDATORS IN ARIZONA

Ted McKinney and Thorry W. Smith

ABSTRACT

We conducted scent-station surveys to index relative abundance of mesopredators (bobcat [*Lynx rufus*], coyote [*Canis latrans*], and common gray fox [*Urocyon cinereoargenteus*]) in five habitats across Arizona during 2004 and 2005. We developed scent-station transects in chaparral, pinyon-juniper (*Pinus* spp.–*Juniperus* spp.) woodland, ponderosa pine (*Pinus ponderosa*) forest, semi-desert grassland, and Upper Sonoran Desert scrub habitats. Bobcats visited scent stations only in chaparral and ponderosa pine forest habitats, and low visitation rates (≤3%) limit the usefulness of scent-station surveys to monitor long-term changes and trends for this species. Coyotes visited scent stations in all habitat types except ponderosa pine forest, but visitation rates consistently were <10%, a level that provides for limited usefulness of the survey method. Foxes visited scent-stations in all habitat types, but visitation rates were >10% only in chaparral and Upper Sonoran Desert habitats. Among habitats that we surveyed, scent-station surveys may provide more reliable survey indices for foxes than for coyotes, and we suggest that the technique may be of questionable merit for indexing relative abundance of bobcats in all habitat types. Scent-station surveys require large sample sizes, produce highly variable and imprecise results, and are potentially sensitive only to large changes in relative population abundance. Despite finding that scent-station visitation rates for mesopredators we studied often were <10%, we believe that scent-station surveys can be useful for management purposes. However, we believe that surveys must be habitat- and species-specific to be useful. We conclude that experimental research will be needed to evaluate the sensitivity and power of scent-station surveys for monitoring changes and trends of mesopredator populations in Arizona.

INTRODUCTION

Public involvement in management of mammalian mesocarnivores has increased dramatically during recent years, demanding biologically sound estimates of abundance and distribution of populations. One result of this trend in public attitudes is that professional resource managers face greater and more diverse management challenges regarding predator populations (Mech 1996, Weber and Rabinowitz 1996, Reiter et al. 1999, Casey et al. 2005). Population abundance is a dimensionally suspect concept in population biology because it is often based on dynamic indices useful only for small geographic areas (Adam 2003) that may not reflect abundance accurately at larger scale domains (Smallwood 1999). Importance of population size is sometimes overemphasized, but this parameter is often the currency by which success of management programs is judged (Lancia et al. 1994).

Wildlife managers increasingly recognize the need to address questions regarding population abundance and trends of mesocarnivores over large landscapes. Robust estimates of relative abundance, distribution, and trends based on noninvasive survey methods may enhance inferences about changes and trends in predator populations. Index surveys can be applied over relatively broad, habitat-specific landscapes (Sargeant et al. 1998), and thus can provide critical information about carnivore populations for wildlife conservationists, managers, and researchers.

Most mammalian carnivore species are cryptic, secretive, and occur at relatively low abundance. These characteristics make accurate, precise, and inexpensive estimations of absolute population abundance, size, and trend difficult. Researchers have used various methods to index relative abundance and trends for populations of diverse wildlife species, including scat abundance and deposition rates and scent-station surveys (Sargeant et al. 1998, Gese 2001, Gese 2004). Tracks detected at scent stations have been used for decades to index abundance and monitor distribution and trends of mammalian carnivore populations. Scent-station surveys have become more widely used during recent years by wildlife researchers and managers than other index techniques for monitoring relative abundance of mammalian carnivores (Sargeant et al. 1998, Gese 2001, Sargeant et al. 2003). Scent-station methodology has been used to index relative abundance of bobcats (*Lynx rufus*), coyotes (*Canis latrans*), common gray foxes (*Urocyon cinereoargenteus*), and many other mammalian carnivore species (Wood 1959, Linhart and Knowlton 1975, Brady 1979, Hon 1979, Lemke and Thompson 1960, Conner et al. 1983, Sargeant et al. 1998, 2003). Although survey indices do not unequivocally represent absolute abundance, scent-station surveys in particular likely index long-term trends in relative abundance of midsize and small mammalian predator (mesopredator) populations (Sargeant et al. 1998). Our objective was to index relative habitat-specific abundance of mammalian mesopredators in various major habitat types in Arizona using scent-station surveys.

STUDY AREAS

We conducted scent-station surveys during 2004 and 2005 throughout much of Arizona within five discrete habitat types (Figure 1). Vegetation types in these regions (Brown 1994) were semi-desert grassland, Arizona Upland subdivision of Sonoran Desert scrub, interior chaparral, ponderosa pine (*Pinus ponderosa*) forest, and pinyon (*Pinus* spp.)-juniper (*Juniperus* spp.) woodland. Elevations of semi-desert grassland ranged between about 1,100 and 1,700 m, and annual precipitation was between about 25 and 45 cm. Most of southwestern Arizona below 1,050 m elevation is in the Sonoran Desert scrub biome; elevations within Upper Sonoran Desert scrub ranged between about 500 and 1,050 m, and annual precipitation was about 25 cm. Elevations of chaparral ranged between about 1,050–2,000 m, and annual precipitation was between 35 and 50 cm. Elevations of pinyon-juniper woodlands ranged between about 1,555 and 2,300 m, and annual precipitation was between 25 and 50 cm. Elevations of ponderosa pine forest ranged between about 2,200 and 2,900 m, and annual precipitation was 64 cm (Sellers and Hill 1974, Brown 1994).

METHODS

We used consistent procedures to establish scent stations and scent-station transects along unpaved secondary roads during 2004 and 2005 following standardized survey designs (Sargeant et al. 1998, 2003). We conducted surveys between 8 November 2004 and 1 March 2005 in Upper Sonoran Desert scrub, semi-desert grassland, and chaparral habitats to reduce potential

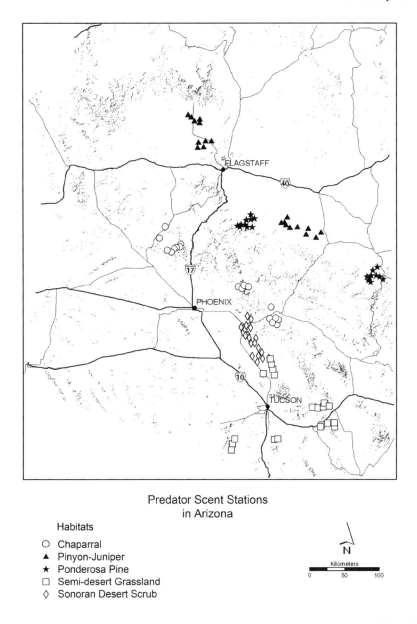

Figure 1. Locations of scent-station transects in five major habitat types in Arizona, 2004 to 2005.

conflicts with hunting seasons, and between 19 April and 6 May 2005 in pinyon-juniper woodlands and ponderosa pine forest habitats to avoid periods of adverse winter weather conditions. We placed 20 scent-station transects within each habitat type; each transect was designed to be 5 km long with 10 scent stations 0.5 km apart; all transects were located ≥5 km apart.

Each scent station consisted of a one-meter diameter circle of smoothed earth cleared of debris and covered with a ≤0.5

cm layer of lime as a tracking bed. We placed a single fatty acid scent tablet (olfactory lure; U.S. Department of Agriculture, Pocatello Supply Depot, Pocatello, Idaho) centrally on each tracking bed. We operated scent stations for one night and the following morning to document absence or presence of tracks (Murie 1974) on each scent station by bobcats, coyotes, and foxes. We defined scent-station visits by predators as ≥1 track by each species identified at each scent station, and calculated scent-station visitation rate as percent of operable stations visited per transect. We defined operable scent stations as those where accurate identification of tracks was possible (i.e., tracks were not obliterated by disturbance caused by humans, other animals, wind, or rain). We used 90% confidence intervals (CI) to compare mean scent-station visitation rates between habitats and mesopredator species because CIs provide more information than do statistical hypothesis tests (P values), provide estimates of uncertainty in estimating means, and address the question of whether between-mean differences are of practical significance (Yocoz 1991, McBride et al. 1993, Steidl et al. 1997, Johnson 1999, Vaske et al. 2002).

In the western United States, gray foxes inhabit areas with brushy vegetation in rugged, broken terrain (Fritzell 1987). In contrast, kit foxes (*Vulpes macrotis*) in southwestern North America inhabit desert shrub and shrub-grass habitats, typically in areas with little vegetative cover, relatively fine soils, and low topographical relief (McGrew 1979, Zoellick et al. 1989, Cypher 2003). We therefore assumed that scent-station visitations by foxes in semi-desert grassland represented kit foxes, whereas fox visitations in all other habitat types represented gray foxes.

RESULTS

We successfully established 200 scent-station transects and 982 scent stations, 975 of which were operable (99.3%; Table 1). Failure to establish 18 scent stations (1.8% of projected study design) resulted from limited length of suitable road available within a habitat type survey area or protracted storms (heavy rain, high winds). Scent-station visitation rates by bobcats, coyotes, and foxes were highly variable within and among habitats surveyed (Table 1). We recorded no visitations to scent stations by bobcats in semi-desert grassland, Sonoran Desert scrub, and pinyon-juniper habitats, and there was no clear difference in mean visitation rates by bobcats in chaparral and ponderosa pine forest habitats. We also recorded no visitations by coyotes in ponderosa pine forest habitat, and no differences in mean visitation rates by coyotes existed among other habitat types. Foxes visited scent stations in all habitat types, but mean visitation rates were higher in chaparral and Sonoran Desert scrub than in all other habitat types. Mean scent-station visitation by foxes tended to be lower in semi-desert grassland than in habitat types other than ponderosa pine forest (Table 1). Foxes had higher scent-station visitation rates than coyotes in chaparral, ponderosa pine forest, and Sonoran Desert scrub habitats, but visitation rates by coyotes were higher than for foxes in semi-desert grassland. Foxes had higher visitation rates than did bobcats in all habitat types. Coyote visitation rates were higher than those for bobcats in pinyon-juniper, semi-desert grassland, and Sonoran Desert scrub habitats, but were lower than for bobcats in ponderosa pine forest (Table 1).

The times required to set and run transects varied with distances of travel required to access survey sites and habitat types. In general, one person could set four or five transects per 8- to 10-hour day, and check on subsequent days about six transects per 8- to 10-hour day. While setting up and running nearly 1,000 scent-station transects, exclusive of commuting to sites, we drove nearly 2,400 km in Sonoran Desert scrub

Table 1. The number of operable scent stations (Operable) and mean (90% CI in parentheses) percent scent-station visits per transect by bobcats, coyotes, and foxes in five habitat types in Arizona, 2004–2005.

Habitat	Operable	Bobcat	Coyote	Fox
Interior chaparral	198	3.05 (0.82–5.29)	4.06 (0.86–7.25)	21.28 (14.67–27.88)
Pinyon-juniper woodland	190	0	2.00 (–0.02–4.02)	7.21 (2.64–11.79)
Ponderosa pine forest	199	0.50 (0.37–1.37)	0	2.50 (0.78–4.22)
Semi-desert grassland	192	0	8.34 (3.84–12.84)	0.50 (–0.37–1.37)
Upper Sonoran Desert scrub	196	0	7.27 (3.87–10.67)	23.56 (16.44–30.67)

habitat, >2,080 km in semi-desert grassland habitat, >1,600 km in chaparral habitat, about 960 km in pinyon-juniper habitat, and about 320 km in ponderosa pine forest habitat, for a total of about 7,360 km. At a current level of estimated vehicle costs ($0.278/km), this represented a vehicle cost of about $2,000. Assuming an employee wage of $30.00/hr, survey time for one person (2 days/6 transects) exclusive of commute time might cost $16,000. Thus, cost of operating 200 scent-station transects for one night each was >$18,000, exclusive of time and vehicle costs required for commuting to survey sites, equating to a cost of about $90/transect.

DISCUSSION

Scent-station visitation rates varied among mesopredator species and habitats. Foxes had higher mean visitation rates than did coyotes in chaparral, ponderosa pine forest, and Upper Sonoran Desert scrub habitats, and higher mean visitation rates than did bobcats in all surveyed habitats. In comparison, coyotes had higher mean visitation rates than did bobcats in pinyon-juniper woodlands, semi-desert grassland, and Upper Sonoran Desert scrub habitats. Foxes had higher mean visitation rates in chaparral and Upper Sonoran Desert scrub than in other habitats. We recorded no visitations by coyotes in ponderosa pine forest, but high variability resulted in no differences in mean visitation rates for coyotes among other habitat types. Bobcats did not visit scent stations in pinyon-juniper woodlands, semi-desert grassland, or Upper Sonoran Desert scrub habitats.

We distributed transects systematically along unpaved roads, a sampling approach widely used for scent-station surveys (Linhart and Knowlton 1975, Hatcher and Shaw 1981, Sargeant et al. 1998, Warrick and Harris 2001). We established transects within boundaries of extensive landscape areas for each habitat type during all surveys to reduce potential influence of adjacent habitat types. Although we used lime tracking beds, sifted soil is more commonly used, and we used lime primarily as a possible attractant for bobcats, based on research in Wyoming (M. Zornes, Arizona Game and Fish Department, personal communication). In Arizona, pilot studies to compare scent-station visitations on tracking beds of fine soil or lime ($n = 20$ transects and 200 stations for each type of tracking bed) in Lower Sonoran Desert scrub habitat during the same season found no differences in responses of bobcats, coyotes, and foxes (McKinney, unpublished data).

Attempts have been made to validate scent-station surveys by correlating visitation rates with direct estimates of carnivore abundance obtained using invasive methods (i.e., mark-recapture) or other noninvasive survey indices (e.g., Diefenbach et al. 1994, Warrick and Harris 2001, Schauster et al. 2002, Sargeant et al. 1998, Sargeant et al. 2003). Regardless, efforts by researchers to validate scent-station surveys have produced equivocal or conflicting results, and scent-station visitation indices have not been conclusively linked to absolute abundance of predators. Nonetheless, scent-station surveys have a long history of use, are widely employed to index relative abundance of mammalian predators, and may index long-term trends in relative abundance of midsize and small carnivores, particularly canids, if sampling designs are adequate (Sargeant et al. 1998, Gese 2001, Sargeant et al. 2003, Gese 2004).

Comparisons of survey results over short time periods indicated that between-year scent-station visitations by bobcats, coyotes, and foxes can be quite variable in different surveys (Linhart and Knowlton 1975, Davison 1980, Conner et al. 1983, Travaini et al. 1996, Windberg et al. 1997, Sargeant et al. 1998, White and Garrott 1999, Warrick and Harris 2001). Sample sizes required for scent-station surveys may increase dramatically when station visitation rates are lower than about 5–10 %, and visitation rates ≥10% might be critical to usefulness of surveys (Sargeant et al. 2003). Mean scent-station visitation rates of bobcats and coyotes in our study consistently were <10% in all habitat types, and visitation rates of foxes exceeded this level only in chaparral and Upper Sonoran Desert scrub habitats. Mean visitation rates in published studies often are <10 %, yet scent-station surveys have been widely used to index relative abundance of mammalian mesopredators (Sargeant et al. 1998, Gese 2001, Sargeant et al. 2003, Gese 2004). Based largely on scent-station surveys and modeling for red foxes (*Vulpes vulpes*) and striped skunks (*Mephitis mephitis*), our use of about 200 scent-stations in each habitat type might have allowed for detection of changes of 35–50% in visitation rates (Sargeant et al. 2003).

Thus, scent-station surveys require large sample sizes and small-scale surveys might lack utility, particularly when visitation rates are low (e.g., <10%; Sargeant et al. 2003). Scent-station visitation rates by bobcats typically are very low, making monitoring difficult (Sargeant et al. 1998, Sargeant et al. 2003, Harrison 2006), and ostensibly limiting reliability of surveys to index long-term changes and trends in relative abundance. Scent-station visitation rates for bobcats in our surveys and previously published research did not exceed about 5% (Knowlton and Tzilkowski 1979, Conner et al. 1983, Sargeant et al. 1998, Diefenbach et al. 1994). Scent-station visitation rates by coyotes and foxes in our studies also generally were within ranges found in previous research. Published visitation rates for coyotes (Linhart and Knowlton 1975, Roughton and Sweeny 1982, Windberg et al. 1997, Sargeant et al. 1998) and foxes (Wood 1959, Conner et al. 1983, Travaini et al. 1996, Sargeant et al. 1998, Warrick and Harris 2001, Schauster et al. 2002) have ranged from about 2–15 % and 2–65 %, respectively.

Some researchers have considered scent-station surveys to be a cost-effective and accurate approach for monitoring mesopredator populations (Travaini et al. 1996, Sargeant et al. 1998). A survey among 13 western states and Canadian provinces indicated that one agency used scent-station surveys to monitor bobcats, one used the method to monitor coyotes, and three used the method to survey foxes (Arizona Game and Fish Department, unpublished data). Although these and some eastern states monitor mesocarnivores with this technique, we are unaware of management decisions and recommendations made by agencies based on survey results.

Populations of a carnivore species of equal abundance in widely different habitat conditions might register widely different habitat-specific scent-station visitation rates (Roughton and Sweeny 1982), but possible effects of habitat on visitation rates have not been fully tested (Sargeant et al. 1998). We compared mean visitation rates by bobcats, coyotes, and foxes among habitat types to substantiate whether habitat-specific differences might occur. Our results demonstrated that scent-station visitation rates differed among mesopredator species and habitat types in Arizona, although we could not conclude if these differences dealt with mesopredator behavior or abundance. Consistent with others (Linhart and Knowlton 1975, Roughton and Sweeney 1982), we conclude that comparisons of scent-station indexes between habitats or species should be avoided or viewed with caution.

We conducted surveys from November through March in Upper Sonoran Desert scrub, semi-desert grassland, and chaparral habitats, and during April and May in pinyon-juniper woodlands and ponderosa pine forests, which may have influenced visitation rates. Among seasons, bobcats, coyotes, and gray foxes may exhibit different movement patterns and scent-station visitation rates (Wood 1959, Springer 1982, Conner et al. 1983, Martin and Fagre 1988, Chamberlain and Leopold 2000, Kolowski and Woolf 2002). We chose survey periods based on hunting seasons, weather, and human resource limitations, but recommend conducting surveys within Arizona from September through November to reduce the influence of seasonal animal behavior, hunting disturbance, and weather. Sources of bias in scent-station surveys might include habitat characteristics, human activities, predator behavior, vehicle traffic levels, wind, and precipitation (Fagre et al. 1981, Fagre et al 1983, Andelt 1985, Henke and Knowlton 1995, Sargeant et al. 1998).

Variable levels of rainfall influence abundance of small prey and diets of bobcats, coyotes, and foxes (Beasom and Moore 1977, Windberg 1995, White and Garrott 1999, McKinney and Smith 2007). However, we believe that precipitation during our studies did not bias our findings. Although rainfall locally is highly variable, statewide annual precipitation in Arizona was near normal during 2004 and 2005, and about half of 83 weather stations monitored throughout the state (National Weather Service, unpublished data) recorded at least average annual precipitation.

Intrinsic behavioral differences among mesopredators undoubtedly influence scent-station surveys. Felids rely on visual stimuli when hunting, whereas canids rely more on olfactory stimuli (Kleiman and Eisenberg 1973), contributing to differential responses to olfactory lures. Higher scent-station visitations by foxes than by coyotes may result from increased wariness by coyotes with regard to human activities and the tendency for foxes to investigate and deposit scat on novel objects (O'Farrell 1987, Schauster et al. 2002, Ralls and Smith 2004). Coyotes show aversion to novel objects (Heffernan et al. 2007), and exhibit wariness toward scent stations (Schauster et al. 2002) due to human activities associated with development of transects (Séquin et al. 2003), novel visual cues in a familiar environment (Windberg 1996, Harris and Knowlton 2001), or prior trap experience (Andelt 1985). Coyotes also may be more likely to visit scent stations located in unfamiliar areas when compared with areas in which they are familiar (Harris 1983).

Despite low visitation rates we observed on our surveys, we believe that scent-station surveys may offer a useful index for coyotes and foxes in some habitat types. Scent-station surveys appear to provide more questionable data for monitoring relative abundance of bobcat populations (Sargeant et al. 1998, Sargeant et al. 2003). Early studies lacking substantiation by statistical analysis suggested an optimal range of visitation rates of 40–60% for detecting change in scent-station visitation, well above the average visitation rate of 10% for coyotes in the western United States (Roughton and Sweeney 1982). However, little research has addressed species-specific minimum scent-station visitation rates that would make the use of scent-station surveys reliable, and visitation rates <10% may actually be meaningful for managing bobcat, coyote, and fox populations. For example, visitation rates ≤1.5% by bobcats, 2-5% by coyotes, and 10-16% by red foxes showed increasing trends over an 8-year period of scent-station surveys in Minnesota (Sargeant et al. 1998).

Analysis of statewide trends in scent-station indices of carnivore abundance likely sacrifices resolution that is useful to resource managers (Sargeant et al. 1998). Modern predator management requires more information than gross statewide trends (Roughton and Sweeny 1982). In comparison, small-scale scent-station surveys lack utility and provide questionable inference regarding broad landscapes, particularly when visitation rates are low (Linhart and Knowlton 1975, Davison 1980, Conner et al. 1983, Travaini et al. 1996, Windberg et al. 1997, Sargeant et al. 1998, 2003, White and Garrott 1999, Warrick and Harris 2001). Bias may be reduced in broad-scale scent-station surveys by partitioning a state into physiographic zones, distributing transects throughout these zones, and establishing transects in the same locations for successive surveys (Sargeant et al. 2003).

Long-term trends of scent-station visitation rates may reflect changes and trends in mesopredator populations, but scent-station surveys have poor spatial and temporal resolution, and susceptibility to confounding and low statistical power may limit usefulness of this method. Moreover, relationships between visitation rates and abundance of populations are unknown, but visitation rates are assumed to increase linearly with abundance (Sargeant et al. 1998). Despite these limitations, scent-station surveys have a long history of use that likely will continue because satisfactory alternatives are lacking (Sargeant et al. 2003).

We speculate that habitat-specific scent-station surveys in Arizona may offer a practical method to monitor long-term changes and trends in relative abundance for foxes and perhaps for coyotes, but are of questionable use in monitoring bobcats. Our study was not designed to predict long-term trends in scent-station visitation rates and trends in relative abundance of mesopredators (see Robinson and Wainer 2002). We believe that experimental research is needed to assess the sensitivity and power of scent-station surveys for monitoring changes and trends in bobcat, coyote, and fox populations in Arizona. Indexing relative abundance requires application of consistent, standardized techniques to detect population trends with some degree of accuracy, precision, and statistical power (Gese 2004). Recognizing limitations of scent-station surveys, we suggest that habitat-specific scent-station surveys, large sample sizes, and long-term analysis are essential for monitoring trends in relative abundance of mesocarnivores. Developing guidelines and consistent protocols is essential for long-term monitoring of mesopredator populations, and is critical for evaluating progress in meeting management objectives (Oakley et al. 2003).

ACKNOWLEDGMENTS

Funding for this research was provided by Project W-78-R of the Pittman-Robertson State Trust Fund Grant to the Arizona Game and Fish Department and the Heritage Fund (a portion of lottery monies approved for wildlife conservation by Arizona voters).

LITERATURE CITED

Adam, J. A. 2003. Mathematics in nature: Modeling patterns in the natural world. Princeton University Press, New Jersey.

Andelt, W. F. 1985. Behavioral ecology of coyotes in South Texas. Wildlife Monographs 94:1–45.

Andelt, W. F., C. E. Harris, and F. F. Knowlton. 1985. Prior trap experience might bias coyote response to scent stations. Southwestern Naturalist 30:317–318.

Beasom, S. L., and R. A. Moore. 1977. Bobcat food habit response to a change in prey abundance. Southwestern Naturalist 21:451–457.

Brady, J. R. 1979. Preliminary results of bobcat scent station transects in Florida. National Wildlife Federation Scientific and Technical Series 6:101–103.

Brown, D. E. 1994. Biotic communities–southwestern United States and northwestern Mexico. University of Utah Press, Salt Lake City.

Casey, A. L., P. R. Krausman, W. W. Shaw, and H. G. Shaw. 2005. Knowledge of and attitudes toward mountain lions: A public survey of residents adjacent to Saguaro National Park, Arizona. Human Dimensions of Wildlife 10:29–38.

Chamberlain, M. J., and B. D. Leopold. 2000. Spatial use patterns, seasonal habitat selection, and interactions among adult gray foxes in Mississippi. Journal of Wildlife Management 64:742–751.

Conner, M. C., R. F. Labisky, and D. R. Progulske, Jr. 1983. Scent-station indexes as measures of population abundance for bobcats, raccoons, gray foxes, and opossums. Wildlife Society Bulletin 11:146–152.

Cypher, B. L. 2003. Foxes. In Wild mammals of North America: Biology, management and conservation, edited by G. A. Feldhammer, B. C. Thompson, and G. A. Chapman, pp. 511-546. The Johns Hopkins University Press, Baltimore, Maryland.

Davison, R. P. 1980. The effect of exploitation on some parameters of coyote populations. Ph.D. dissertation, Utah State University, Logan.

Diefenbach, D. R., M. J. Conroy, R. J. Warren, W. E. James, L. A. Baker, and T. Hon. 1994. A test of the scent-station survey technique for bobcats. Journal of Wildlife Management 58:10–17.

Fagre, D. B., B. A. Butler, W. E. Howard, and R. Teranishi. 1981. Behavioral responses of coyotes to selected odors and tastes. In Worldwide Furbearer Conference Proceedings, edited by J. A. Chapman and D. Pursely, pp. 966–983. Frostburg, Maryland.

Fagre, D. B., W. E. Howard, D. A. Barnum, R. Teranishi, T. H. Schultz, and D. J. Stern. 1983. Criteria for the development of coyote lures. Vertebrate Pest Control and Management 4:265–277.

Fritzell, E. K. 1987. Gray fox and island fox. In Wild furbearer management and conservation in North America, edited by M. Novak, J. A. Baker, M. E. Offard, and B. Malloch, pp. 408–420. Ministry of Natural Resources, Ontario, Canada.

Gese, E. M. 2001. Monitoring of terrestrial carnivore populations. In Carnivore Conservation, edited by J. L. Gittleman, S. M. Funk, D. W. MacDonald, and R. K. Wayne, pp. 372–396. Cambridge University, Cambridge, United Kingdom.

Gese, E. M. 2004. Survey and census techniques for canids. In Canidae: foxes, wolves, jackals, and dogs. Status survey and conservation action plan, edited by C. Sillero-Zubiri, M. Hoffman, and D. W. Macdonald, pp. 273–279. International Union for Conservation of Nature and Natural Resources, Gland, Switzerland.

Harris, C. E. 1983. Differential behavior of coyotes with regard to home range limits. Ph.D. dissertation. Utah State University, Logan.

Harris, C. E., and F. F. Knowlton. 2001. Differential responses of coyotes to novel stimuli in familiar and unfamiliar settings. Canadian Journal of Zoology 79:2005–2013.

Harrison, R. L. 2006. A comparison of survey methods for detecting bobcats. Wildlife Society Bulletin 34:548–552.

Hatcher, R. T., and J. H. Shaw. 1981. A comparison of three indexes to furbearer populations. Wildlife Society Bulletin 9:153–156.

Heffernan, D. J., W. F. Andelt, and J. A. Shivik. 2007. Coyote investigative behavior following removal of novel stimuli. Journal of Wildlife Management 71:587–593.

Henke, S. E., and F. F. Knowlton. 1995. Techniques for estimating coyote abundance. In Symposium proceedings—Coyotes in the southwest: A compendium of our knowledge, edited by D. Rollins, C. Richardson, T. Blankenship, K. Canon, and S. Henke, pp. 71–78. Texas Parks and Wildlife Department, Austin.

Hon, T. 1979. Relative abundance of bobcats in Georgia: Survey techniques and preliminary results. National Wildlife Federation Scientific and Technical Series 6:104–106.

Johnson, D. H. 1999. The insignificance of statistical significance testing. Journal of Wildlife Management 63:763–772.

Kleiman, D. G., and J. F. Eisenberg. 1973. Comparisons of canid and felid social systems from an evolutionary perspective. Animal Behaviour 21:637–659.

Knowlton, F. F., and W. M. Tzilkoqaki. 1979. Trends in bobcat visitations to scent-station survey lines in western United States, 1972–1978. National Wildlife Federation Scientific and Technical Series 6:8–12.

Kolowski, J. M., and A. Woolf. 2002. Microhabitat use by bobcats in southern Illinois. Journal of Wildlife Management 66:822–832.

Lancia, R. A., J. D. Nichols, and K. H. Pollock. 1994. Estimating the number of animals in wildlife populations. In Research and management techniques for wildlife and habitats, edited by T. A. Bookhout, pp. 215–253. The Wildlife Society, Bethesda, Maryland.

Lemke, C. W., and D. R. Thompson. 1960. Evaluation of a fox population index. Journal of Wildlife Management 24:406–412.

Linhart, S. B., and F. F. Knowlton. 1975. Determining the relative abundance of coyotes by scent station lines. Wildlife Society Bulletin 3:119–124.

Martin, D. J., and D. B. Fagre. 1988. Field evaluation of a synthetic coyote attractant. Wildlife Society Bulletin 16:390–396.

McBride, G. B., J. C. Loftis, and N. C. Adkins. 1993. What do significance tests really tell us about the environment? Environmental Management 17:423–432.

McGrew, J. C. 1979. Mammalian species: *Vulpes macrotis*. American Society of Mammalogists 123:1–6.

McKinney, T., and T. W. Smith. 2007. Diets of sympatric bobcats and coyotes during years of varying rainfall in central Arizona. Western North American Naturalist 67:8–15.

Mech, L. D. 1996. A new era for carnivore conservation. Wildlife Society Bulletin 24:397–401.

Murie, O. J. 1974. A field guide to animal tracks, 2nd edition. Houghton Mifflin Company, New York.

Oakley, K. L., L. P. Thomas, and S. G. Fancy. 2003. Guidelines for long-term monitoring protocols. Wildlife Society Bulletin 31:1000–1003.

O'Farrell, T. P. 1987. Kit fox. In Wild furbearer management and conservation in North America, edited by M. Novak, J. A. Baker, M. E. Offard, and B. Malloch, pp. 423–431. Ministry of Natural Resources, Ontario, Canada.

Ralls, K., and D. A. Smith. 2004. Latrine use by San Joaquin kit foxes (*Vulpes macrotis mutica*) and coyotes (*Canis latrans*). Western North American Naturalist 64:544–547.

Reiter, D. K., M. W. Brunson, and R. H. Schmidt. 1999. Public attitude toward wildlife damage management and policy. Wildlife Society Bulletin 27:746–758.

Robinson, D. H., and H. Wainer. 2002. On the past and future of null hypothesis significance testing. Journal of Wildlife Management 66:263–271.

Roughton, R. D., and M. W. Sweeny. 1982. Refinements in scent-station methodology for assessing trends in carnivore populations. Journal of Wildlife Management 46:217–229.

Sargeant, G. A., D. H. Johnson, and W. E. Berg. 1998. Interpreting carnivore scent-station surveys. Journal of Wildlife Management 62:1235–1245.

Sargeant, G. A., D. H. Johnson, and W. E. Berg. 2003. Sampling designs for carnivore scent-station surveys. Journal of Wildlife Management 67:289–298.

Schauster, E. R., E. M. Gese, and A. M. Kitchen. 2002. An evaluation of survey methods for monitoring swift fox abundance. Wildlife Society Bulletin 30:464–477.

Sellers, W. D., and R. H. Hill. 1974. Arizona climate 1931–1972. University of Arizona Press, Tucson.

Sèquin, E. S., M. M. Jaeger, P. F. B. Russard, and R. H. Barrett. 2003. Wariness of coyotes to camera traps relative to social status and territory boundaries. Canadian Journal of Zoology 81:2015–2025.

Smallwood, K. S. 1999. Scale domains of abundance amongst species of mammalian carnivores. Environmental Conservation 26:102–111.

Springer, J. T. 1982. Movement patterns of coyotes in south central Washington. Journal of Wildlife Management 46:191–200.

Steidl, R. J., J. P. Hayes, and E. Schauber. 1997. Statistical power analysis in wildlife research. Journal of Wildlife Management 61:270–279.

Travaini, A., R. Laffitte, and M. Delibes. 1996. Determining the relative abundance of European red foxes by scent-station methodology. Wildlife Society Bulletin 24:500–504.

Vaske, J. J., J. A. Gliner, and G. A. Morgan. 2002. Communicating judgments about practical significance: Effect size, confidence intervals and odds ratios. Human Dimensions of Wildlife 7:287–300.

Warrick, G. D., and C. E. Harris. 2001. Evaluation of spotlight and scent-station surveys to monitor kit fox abundance. Wildlife Society Bulletin 29:827–832.

Weber, W., and A. Rabinowitz. 1996. A global perspective on large carnivore conservation. Conservation Biology 10:1046–1054.

White, P. J., and R. A. Garrott. 1999. Population dynamics of kit foxes. Canadian Journal of Zoology 77:486–493.

Windberg, L. A. 1995. Demography of a high-density coyote population. Canadian Journal of Zoology 73:942–954.

Windberg, L. A. 1996. Coyote responses to visual and olfactory stimuli related to familiarity with an area. Canadian Journal of Zoology 74:2248–2253.

Windberg, L. A., S. M. Ebbert, and B. T. Kelly. 1997. Population characteristics of coyotes (*Canis latrans*) in the northern Chihuahuan Desert of New Mexico. American Midland Naturalist 138:197–207.

Wood, J. E. 1959. Relative estimates of fox population levels. Journal of Wildlife Management 23:53–63.

Yocoz, N. G. 1991. Use, overuse, and misuse of significance tests in evolutionary biology and ecology. Bulletin of the Ecological Society of America 72:106–111.

Zoellick, B. W., N. S. Smith, and R. S. Henry. 1989. Habitat use and movements of desert kit foxes in western Arizona. Journal of Wildlife Management 53:955–961

MOUNTAIN LION DEPREDATION HARVESTS IN ARIZONA, 1976 – 2005

Ted McKinney, Brian F. Wakeling, and Johnathan C. O'Dell

ABSTRACT

We studied reported kills of mountain lions (*Puma concolor*) in Arizona related to livestock depredation events between 1976 and 2005 to determine if a relationship existed between mule deer (*Odocoileus hemionus*) abundance and livestock depredation. Depredation-related kills of mountain lions increased and contributed substantially to statewide hunter harvest of mountain lions when mule deer abundance waned. Depredation-related kills of mountain lions were negatively correlated with mule deer abundance. Depredation-related kills of mountain lions involved primarily adult males, but take of all age and sex classes of mountain lions increased concurrently. Cattle depredation initiated 90% of all reported mountain lion kills for depredation, and 98% of these reports involved depredation on calves. Mountain lions killed for depredation of cattle occurred in 12 of the state's 15 counties, although 5 counties accounted for 92% of all depredation kills. We believe that reduced relative abundance of mule deer contributes to increased depredation of cattle by mountain lions in Arizona.

INTRODUCTION

Predator reduction to mitigate domestic livestock depredation or benefit wildlife populations is a controversial action for wildlife management agencies (Ballard et al. 2001). Substantial and widespread conflicts between humans and large carnivores arise due to depredations of livestock, resulting in predator removals (Linnell et al. 1999). The public has widely disparate views on lethal removal of predators that threaten livestock or other domestic prey (Casey et al. 2005). Predator control efforts in Arizona between 1947 and 1969, when the state legislature offered a bounty on mountain lions (*Puma concolor*), resulted in 5,400 payments (Phelps 1989). In 1970, mountain lions were classified as big game by the Arizona legislature, and regulated take of mountain lions for reported depredations of livestock began in 1971 (Arizona Game and Fish Department 2006). Many western state wildlife management agencies consider depredation of livestock and other domestic animals by mountain lions an important management concern (Ballard et al. 2001, Torres et al. 1996, Barber 2005, Winslow 2005, Woolstenhulme 2005).

Depredations of cattle by mountain lions occurs in 11 western states, but is highest in Arizona (Shaw 1983, Cunningham et al. 1995, Cunningham et al. 2001, Mountain Lion Foundation 2007, www.pumaconservation.org/html/printable_version.html). About 850 livestock operators presently graze about 56,000 cattle on public lands in Arizona (Bureau of Land Management 2006, blm.gov/az/range.htm). Calves comprised an estimated 93% of cattle killed by mountain lions on a ranch in north-central Arizona (Shaw 1983). Calves comprised 44% of biomass eaten by mountain lions in southeastern Arizona (Cunningham et al.

1999). Other studies in Arizona reported cattle comprised 13% (Cashman et al. 1992), 14% (McKinney et al. 2006), and 26% (Shaw 1977) of mountain lion diets.

Hunter harvest is considered the primary cause of mountain lion mortality in hunted populations (Ruth et al. 1998, Logan and Sweanor 2001), but killing mountain lions that prey on livestock might account for a substantial portion of human-related mortality of the predator in Arizona (Cunningham et al. 1995, 1999). Depredation harvest has been the primary cause of mountain lion mortality in southeastern Arizona (Cunningham et al. 2001), and it accounted for 15% of all mountain lions harvested in Arizona between 1996 and 2004 (Barber 2005).

Depredation incidents involving mountain lions increased in California between 1972 and 1995 (Torres et al. 1996), and depredation harvest of mountain lions in Montana increased between 1971 and 1990 (Aune 1991). Depredation by mountain lions is affected by a complex interaction of factors (e.g., Shaw 1981), but not all have been adequately quantified. Our objectives were to describe statewide patterns, trends, and demographics of confirmed legal depredation harvest of mountain lions, quantify depredation of cattle and other prey by mountain lions, and determine relationships between kills of mountain lions for depredation, hunter harvest of mountain lion, and abundance of mule deer (*Odocoileus hemionus*).

METHODS

Nuisance and public safety issues associated with mountain lions are addressed differently than depredation by mountain lions (killing livestock or other domestic animals) in Arizona. Livestock operators are allowed to kill depredating mountain lions and must follow strict reporting requirements (Barber 2005). Records of mountain lion depredation kills and data on depredation of livestock have been maintained in Arizona since 1971, although data record consistency was incomplete through 1975 (Arizona Game and Fish Department 2006). We analyzed mountain lion depredation-related kill data between 1976 and 2005 only when depredation permits were issued, age and sex of depredating mountain lions were estimated, and depredated species and locations of depredation were were provided. We used data from depredation-related mountain lion kills where age and sex of mountain lions were provided by persons involved with mountain lion removal, based on necropsies, pelage, or tooth wear.

We also analyzed statewide data for hunter harvests of mountain lions and mule deer between 1976 and 2005 (Arizona Game and Fish Department 2006), and indexed mule deer abundance using harvest data (Marshal et al. 2002). We used Spearman's rank correlations to determine trends of mountain lion depredation and sport harvests, prey that were depredated, and abundance of mule deer. We used simple linear regression models to determine correlation between confirmed depredation harvest of mountain lions as the dependent variable and indexed abundance of mule deer as an independent variable. We then compared proportions of mountain lion adults, subadults, sexes, and mean sex ratio of adults and subadults in depredation harvests using Yate's adjusted chi-square and 2 x 2 contingency tables and *t*-tests, respectively (Zar 1996).

RESULTS

We documented 917 confirmed depredation incidents by mountain lions between 1976 and 2005 (Figure 1), with a mean of 30.6 depredation-related mountain lion kills per year (SD = 18.96, range = 2 to 66/year). Depredation-related kills of mountain lions during this period averaged 14.8% (SD = 10.33, range = 1.0 to 50.0%) of statewide hunter harvest of mountain lions. Hunter harvest of mountain lions was highly variable among years ($R = 0.1638$, $t_{28} = 0.8786$ P

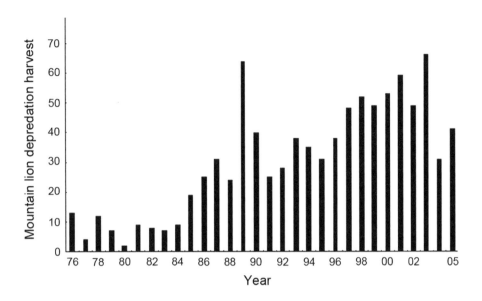

Figure 1. Confirmed number of mountain lions harvested for depredations in Arizona, 1976 to 2005.

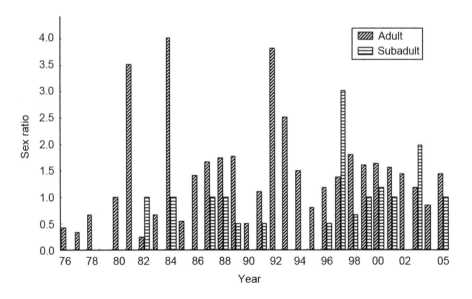

Figure 2. Sex ratios (M:F) of adult and subadult mountain lions harvested in confirmed depredation incidents in Arizona, 1976 to 2005.

= 0.387, SD = 43.83), but depredation-related kills as percent of hunter harvest of mountain lion increased after 1976 and was less variable than hunter harvest ($R = 0.7562$, $t_{28} = 6.115$, $P < 0.001$, SD = 10.33).

Of all depredation-related mountain lion kills between 1976 and 2005, 92.3% (847/917) occurred between 1985 and 2005; these kills comprised 17.9% (847/4,743) of hunter harvests. Between 1976 and 2005, depredation-related kills of all mountain lions—adult mountain lions (sexes combined), adult females, adult males, all females, and all males—increased ($R \geq 0.8108$, $t_{28} \geq 7.347$, $P < 0.001$). Similarly, depredation-related kills of subadult (sexes combined; $R = 0.6989$, $t_{28} = 5.171$, $P < 0.001$), subadult female ($R = 0.4480$, $t_{28} = 2.562$, $P = 0.013$), and subadult male mountain lions ($R = 0.5181$, $t_{28} = 3.206$, $P = 0.003$) increased between 1985 and 2005.

Females and males (combined age classes) were identified in 43.2% (356/825) and 56.8% (469/825) of depredation-related kills, respectively ($\chi^2 = 30.41$, $P < 0.001$). Adult and subadult mountain lions (combined sexes) were identified in 87.3% (729/835) and 12.7% (106/835) of depredation-related kills, respectively ($\chi^2 = 926.67$, $P < 0.001$). Among mountain lions classified as adults, females and males were identified in 42.1% (309/734) and 57.9% (425/734) of depredation-related kills, respectively ($\chi^2 = 36.64$, $P < 0.001$). Among mountain lions classified as subadults, females and males each were identified in 50% (47/94) of depredation-related kills. Mean sex ratio (M:F) of depredation-related kills (Figure 2) was 1.36 (SD = 0.89, range = 0.33 to 4.00). Mean sex ratios for adult and subadult mountain lions from depredation-related kills between 1976 and 2005 were 1.46 (SD = 0.96; range = 0.25 to 4.00) and 1.10 (SD = 0.67; range = 0.50 to 3.00), respectively, and did not differ ($t_{41} = 1.270$, $P > 0.211$). Sex ratios of adults ($R = 0.1967$, $t_{27} = 1.614$, $P > 0.118$) and subadults ($R = 0.3090$, $t_{12} = 1.126$, $P > 0.282$) from depredation-related kills showed no clear trends between 1976 and 2005.

Total reported depredation-related kills of mountain lions and depredations of cattle occurred most often between January and June, but reported depredations of other prey showed no clear seasonal pattern (Figures 3 and 4). Calves comprised 97.9% (573/585) of cattle depredations by mountain lions among events that specified relative ages of cattle killed. Other prey killed by mountain lions included chickens, colts, domestic goats and sheep, deer (*Odocoileus* spp.), domestic dogs, and ostriches (*Struthio camelus*). Depredation-related kills of mountain lions as a result of cattle depredation comprised 90.1% (826/917) of all reported depredation kills between 1976 and 2005 in Arizona. Mean harvest of mountain lions for depredation of cattle during this period was 26.7/year (SD = 20.71; range = 1 to 66/year), and mean harvest for depredation other than for cattle was 1.0/year (SD = 1.08; range = 0 to 3/year). Depredation of cattle ($R = 0.8328$, $t_{23} = 7.215$, $P < 0.001$) and calves ($R = 0.6958$, $t_{17} = 4.332$, $P < 0.001$) increased between 1976 and 2005, but depredation of other prey showed no annual trend ($R = 0.0693$, $t_{23} = 0.3330$, $P > 0.742$).

Statewide abundance of mule deer declined between 1976 and 2005 ($R = -0.6289$, $t_{28} = -4.280$, $P < 0.001$). Total depredation-related kills of all mountain lions (Figure 3; $F_{1,28} = 8.18$, $r^2 = 0.2261$, $P < 0.008$, $b = -0.48$), all adults ($F_{1,28} = 8.18$, $r^2 = 0.2261$, $P < 0.008$, $b = -0.48$), adult females ($F_{1,28} = 18.19$, $r^2 = 0.3938$, $P < 0.001$, $b = -0.63$), adult males ($F_{1,28} = 11.93$, $r^2 = 0.2788$, $P < 0.002$, $b = -0.55$), and subadults ($F_{1,28} = 9.44$, $r^2 = 0.2521$, $P < 0.005$, $b = -0.50$) were negatively correlated with indexed abundance of mule deer.

Mountain lions harvested for cattle depredations were reported in 12 of Arizona's 15 counties. Depredation harvests in 5 contiguous counties each comprised >

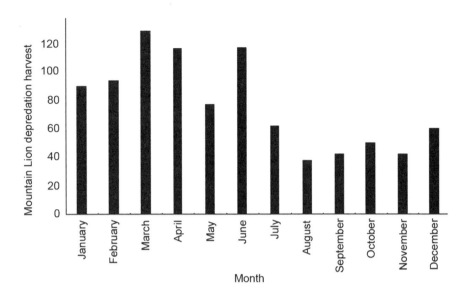

Figure 3. Monthly confirmed number of mountain lions harvested for depredations in Arizona, 1976 to 2005

6% of statewide cattle depredation harvests (Mohave [6.3%], Gila [7.7%], Graham [47.6%], Greenlee [24.2%], and Yavapai [6.5%]). These counties extend from northwestern to southeastern Arizona, and accounted for 92.4% (763/826) of reported depredation-related mountain lion kills for cattle depredations. The five counties encompass an estimated area of about 85,160 km^2, or about 28.9% of the area of Arizona. Depredation harvests in the five counties between 2000 and 2005 comprised 19.0% (292/1,533) of statewide sport harvest of mountain lions (Arizona Game and Fish Department, 2006). Remaining counties each comprised < 3% of mountain lions killed for cattle depredations (Apache [0.7%], Coconino [0.5%], Maricopa [2.4%], Cochise [2.5%], Pima [0.5%], Pinal [0.2%], Santa Cruz [0.7%]). Mountain lion harvests for depredations of domestic prey other than cattle (e.g., goats, horses) occurred in 10 counties (Cochise, Coconino, Gila, Graham, Greenlee, Maricopa, Mojave, Navajo, Santa Cruz, and Yavapai).

DISCUSSION

Depredation incidents by mountain lions have increased throughout much of western United States during the past few decades. Most depredations outside of Arizona involve domestic sheep, but mountain lions also have killed cattle, horses, alpacas (*Vicugna pacos*), llamas (*Lama glama*), emus (*Dromaius novaehollandiae*), and human pets and smaller animals such as chickens, geese, and domestic pigs (Cougar Management Guidelines Working Group 2005). Our results paralleled trends observed between 1972 and 1995 of mountain lion depredaton incidents in California, where hunting of mountain lions is prohibited (Torres et al. 1996). Mountain lion kills for depredation-related incidents also increased after 1970 in Montana, where hunting mountain lions is legal (Aune 1991). Consistent with previous studies in two limited areas of Arizona (Shaw 1977; Cunningham et al. 1995, 1999), we found that most mountain lion depredation-related kills were associated with predation of cattle, and 98% of depredation-related kills involved depredation of calves.

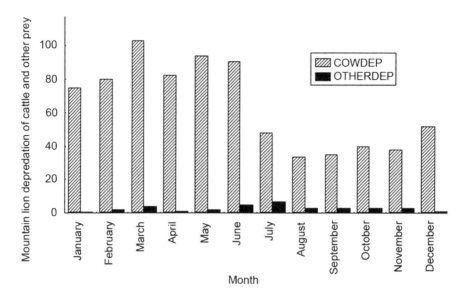

Figure 4. Monthly confirmed number of mountain lions harvested in response to depredations of cattle (COWDEP) and other prey (OTHERDEP) in Arizona, 1976 to 2005.

Depredation-related kills do not reliably index changes or trends in abundance of mountain lions (Cougar Management Guidelines Working Group 2005). Depredation-related kills contributed substantially to overall take of mountain lions in Arizona between 1976 and 2005, comprised on average 15% of hunter harvest, and ranged among years from <1–50% of hunter harvest.

About 92% of depredation-related kills from 1976 to 2005 occurred between 1985 and 2005, when they comprised 18% of sport harvest of mountain lions. The disproportionately high number of reports during 1985–2005 may have occurred at least in part because reporting requirements were better known and enforced than during 1976–1984. Adult males were killed for depredations more than adult females or subadults of either sex, consistent with previous findings (Aune 1991, Torres et al. 1996, Linnell et al. 1999, Cunningham et al. 2001, Woolstenhulme 2005). Mean sex ratio (M:F) of depredating mountain lions killed in our study (1.36) was higher than the average (1.14) for mountain lions harvested by hunters in Arizona between 1982 and 2002 (Zornes et al. 2006). Sex ratio also reportedly favored males (M:F = 1.48) for depredating mountain lions in California (Torres et al. 1996). The tendency for males to predominate in depredation-related kills has not been adequately explained, but males are not likely to be more vulnerable to methods of depredation take (Linnell et al. 1999).

Most mountain lion kills for depredation of cattle in Arizona occurred between January and June (Figure 4). In comparison, the number of calves and yearlings killed by mountain lions was highest between November and June in north-central Arizona (Shaw 1977). Estimated relative number of calves killed by mountain lions also increased during autumn and winter in southeastern Arizona (Cunningham et al. 1995). The higher predation of cattle by mountain lions in Arizona corresponds with the period of seasonal production of calves in free-ranging herds (Shaw 1977). In addition, a statewide survey in Arizona in 1984 indicated that the number of mountain lions killed for cattle depredations underrepresents the number of cattle killed by mountain lions (Shaw 1984).

Mountain lions occupy various habitat types throughout most of Arizona, and are believed to inhabit about 187,000 km^2 of suitable habitat that includes about 31,000 km^2 classified as high quality habitat (Barber 2005). Ninety-two percent of depredation harvests for predation of cattle occurred within five counties that comprise about 35% of habitat occupied by mountain lions. These counties are essentially contiguous with the northwest-southeast distribution of the chaparral zone in Arizona (Swank 1958). Vegetation consisting of Great Basin conifer and Madrean evergreen woodlands, Rocky Mountain and Madrean montane conifer forests, and Arizona Upland Sonoran Desert scrub also is contiguous with chaparral in much of the region (Brown 1994). Most reports of depredations of cattle by mountain lions in Arizona originate from mid-elevation chaparral and pine-oak (*Pinus* spp.-*Quercus* spp.) woodlands, with few documented in high-elevation or low desert areas. Vegetation types and topography within these five counties probably increase the likelihood that livestock will suffer from mountain lion predation. Further, the elevations are relatively low, temperatures are moderate, and habitat in these counties are conducive to yearlong stocking of cow-calf livestock operations, whereas higher elevation ranges may not allow for yearlong livestock operations that include age classes of cattle especially vulnerable to mountain lion predation.

Relationships between husbandry practices and mountain lion depredations of livestock have not been adequately demonstrated (Cougar Management Guidelines Working Group 2005). Nonetheless, depredation of cattle by mountain lions is often higher if free ranging cow-calf herds are grazed in areas of rugged terrain and dense vegetation cover, and if abundance of prey other than cattle is comparatively low (Shaw 1983; Cunningham et al. 1999; Bueno-Cabrera et al. 2005).

Yearlong cow-calf operations predominate in mountainous areas of Graham and Greenlee counties with the highest depredation rates, coincident with historically intensive mountain lion depredation control efforts that have contributed to low survival rates of mountain lions in some regions of Graham county (Dodd and Brady 1986, Cunningham et al. 2001). Between 1988 and 1993, hunters and depredation-control efforts removed 32 and 26 mountain lions, respectively, from one area of Graham county (Cunningham et al. 2001). Another 46 and 52 mountain lions were removed from the area between 2000 and 2005 by hunters and depredation control, respectively (Arizona Game and Fish Department 2005, 2006). Despite high mountain lion removal in these areas, large numbers of livestock suffer from mountain lion depredation annually.

Depredations of livestock concern ranchers and wildlife managers, but killing depredating mountain lions may provide only a short-term solution for preventing or reducing losses of cattle (Cougar Management Guidelines Working Group 2005, Graham et al. 2005). Intensive levels of hunter harvest of mountain lions may alter demographics and reduce populations of the predator, but mountain lion populations recover relatively rapidly if hunting pressure is not maintained over time (Lindzey et al. 1992, Ross and Jalkotzy 1992, Cunningham et al. 2001, Anderson and Lindzey 2005, Stoner et al. 2006). Longer-term solutions to depredation may require significant reductions in mountain lions over broad areas, reduction in the number of adult female mountain lions in an area, or modification of present husbandry practices, such as grazing cattle in low quality mountain lion habitat when calves are present (Shaw 1977, 1988,

Cougar Management Guidelines Working Group 2005). Research is needed to evaluate the effectiveness of different animal husbandry practices and hunt structures in reducing livestock depredations.

Potential explanations for increased depredations by mountain lions in western United States are speculative, but include factors such as changes in land use, elimination of bounties for mountain lions, increasing abundance of mountain lions, and declining abundance of deer (Cougar Management Guidelines Working Group 2005). Mule deer are the primary prey of mountain lions in North America, and are widely distributed in Arizona (Hoffmeister 1986; Lindzey et al. 1994). Abundance of mule deer may influence the abundance of mountain lions (Hemker et al. 1984; Lindzey et al. 1994, Pierce et al. 2000, Riley and Malecki 2001), but relationships between predator abundance and depredation are complex. Availability of natural prey influences depredation of cattle by mountain lions (Polisar et al. 2003), and our findings suggest that decline in abundance of mule deer corresponded with increased depredation-related kills of mountain lions in Arizona.

MANAGEMENT IMPLICATIONS

Fundamentally, reducing livestock depredation primarily requires avoiding the placement of livestock within good mountain lion habitat. Changing the type and age of livestock (e.g., switching to adult cattle, avoiding calves) grazed within an area with demonstrated depredation history may reduce the likelihood and frequency of depredation events. Improving habitat and encouraging healthy wildlife communities, including deer, on ranges grazed by livestock may provide predators with a natural prey base, thereby reducing depredation on cattle. Changes to the mountain lion populations, such as reductions in overall abundance or abundance of female mountain lions through hunter harvest or targeted removal may reduce depredation events, although little effect has been shown to occur through standard management practices. Research to determine the effectiveness of these various management alternatives should include a financial costs assessment.

ACKNOWLEDGMENTS

We thank M. L. Zornes for reviewing an earlier draft of this paper and A. A. Munig for assistance in data procurement. The manuscript benefited from reviews from H. G. Shaw and an anonymous referee. Unexpectedly, Ted McKinney (senior author) passed away on October 22, 2008, before the manuscript was revised and accepted. His career spanned more than 40 years, and he left quite a legacy in scientific study.

The Heritage Fund, a portion of lottery monies approved for wildlife conservation by Arizona voters, Wildlife Conservation Fund; monies generated by tribal gaming approved by Arizona voters; and Pittman-Robertson funds, monies generated by an excise tax on hunting and shooting equipment approved by the United States Legislature and President in 1937, provided funding for this research.

LITERATURE CITED

Anderson, C. R., and F. G. Lindzey. 2005. Experimental evaluation of population trend and harvest composition in a Wyoming cougar population. Wildlife Society Bulletin 33:179–188.

Arizona Game and Fish Department. 2005. Hunt Arizona–2005 edition. Arizona Game and Fish Department, Phoenix.

Arizona Game and Fish Department. 2006. Hunt Arizona–2006 edition. Arizona Game and Fish Department, Phoenix.

Aune, K. E. 1991. Increasing mountain lion populations and human-mountain lion interactions in Montana. In C. S. Braun, editor. Mountain lion-human interaction symposium and workshop, pp. 86–94. Colorado Division of Wildlife, Denver.

Ballard, W. B., D. Lutz, T. W. Keegan, L. H. Carpenter, and J. C. deVos, Jr. 2001. Deer-predator relationships: A review of recent North American studies with emphasis on mule and black-tailed deer. Wildlife Society Bulletin 29:99–115.

Barber, S. P. 2005. Arizona mountain lion status report. Mountain Lion Workshop 8:65–69.

Brown, D. E. 1994. Biotic communities–southwestern United States and northwestern Mexico. University of Utah Press, Salt Lake City.

Bueno-Cabrera, A., L. Hernandez-Garcia, J. Laundré, A. Contreras-Hernandez, and H. Shaw. 2005. Cougar impact on livestock ranches in the Santa Elena Canyon, Chihuahua, Mexico. Mountain Lion Workshop 8:141–149.

Casey, A. L., P. R. Krausman, W. W. Shaw, and H. G. Shaw. 2005. Knowledge of and attitudes toward mountain lions: A public survey of residents adjacent to Saguaro National Park, Arizona. Human Dimensions of Wildlife 10:29–38.

Cashman, J. L., M. Peirce, and P. R. Krausman. 1992. Diets of mountain lions in southwestern Arizona. Southwestern Naturalist 37:324–326.

Cougar Management Guidelines Working Group. 2005. Cougar management guidelines. 1st ed. WildFutures, Bainbridge Island, Washington.

Cunningham, S. C., L. A. Haynes, C. Gustavson, and D. D. Haywood. 1995. Evaluation of the interaction between mountain lions and cattle in the Aravaipa-Klondyke area of southeast Arizona. Technical Report No. 17, Arizona Game and Fish Department, Phoenix. 64 pp.

Cunningham, S. C., C. R. Gustavson, and W. B. Ballard. 1999. Diet selection of mountain lions in southeastern Arizona. Journal of Range Management 52:202–207.

Cunningham, S. C., W. B. Ballard, and H. A. Whitlaw. 2001. Age structure, survival, and mortality of mountain lions in southeastern Arizona. Southwestern Naturalist 46:76–80.

Dodd, N. L., and W. W. Brady. 1986. Cattle grazing influences on vegetation of a sympatric desert bighorn range in Arizona. Desert Bighorn Council Transactions 20:8–13.

Graham, K., A. P. Beckerman, and S. Thirgood. 2005. Human-predator-prey conflicts: Ecological correlates, prey losses and patterns of management. Biological Conservation 122:159–171.

Hemker, T. P., F. G. Lindzey, and B. B. Ackerman. 1984. Population characteristics and movement patterns of cougars in southern Utah. Journal of Wildlife Management 48:1275–1284.

Hoffmeister, D. F. 1986. Mammals of Arizona. University of Arizona Press, Tucson.

Lindzey, F. G., W. D. Van Sickle, S. P. Laing, and C. S. Mecham. 1992. Cougar population response to manipulation in southern Utah. Wildlife Society Bulletin 20:224–227.

Lindzey, F. G., W. D. Van Sickle, B. B. Ackerman, D. Barnhurst, T. P. Hemker, and S. P. Laing. 1994. Cougar population dynamics in southern Utah. Journal of Wildlife Management 58:619–624.

Linnell, J. D. C., J. Odden, M. E. Smith, R. Aanes, and J. E. Swenson. 1999. Large carnivores that kill livestock: Do "problem individuals" really exist? Wildlife Society Bulletin 27:698–705.

Logan, K. A., and L. L. Sweanor. 2001. Desert puma. Island Press, Washington, D.C.

McKinney, T., T. W. Smith, and J. C. deVos, Jr. 2006. Evaluation of factors potentially influencing a desert bighorn sheep population. Wildlife Monographs 164:1–34.

Marshal, J. P., P. R. Krausman, V. C. Bleich, W. B. Ballard, and J. S. McKeever. 2002. Rainfall, El Niño, and dynamics of mule deer in the Sonoran Desert, California. Journal of Wildlife Management 66:1283–289.

Phelps, J. S. 1989. Status of mountain lions in Arizona. In R. H. Smith, editor. Proceedings of the third mountain lion workshop, pp. 7–9. Arizona Chapter of the Wildlife Society and Arizona Game and Fish Department, Phoenix.

Pierce, B. M., V. C. Bleich, and R. T. Bowyer. 2000. Social organization of mountain lions: Does a land-tenure system regulate population size? Ecology 81:1533–1543.

Polisar, J., I. Maxit, D. Scognamillo, L. Farrell, M. E. Sunquist, and J. F. Eisenberg. 2003. Jaguars, pumas, their prey base, and cattle ranching: Ecological interpretations of a management problem. Biological Conservation 109:297–310.

Riley, S. J., and R. A. Malecki. 2001. A landscape analysis of cougar distribution and abundance in Montana, USA. Environmental Management 28:317–323.

Ross, P. I., and M. G. Jalkotzy. 1992. Characteristics of a hunted population of cougars in southwestern Alberta. Journal of Wildlife Management 56:417–428.

Ruth, T. K., K. A. Logan, L. L. Sweanor, M. G. Hornocker, and L. J. Temple. 1998. Evaluating cougar translocation in New Mexico. Journal of Wildlife Management 62:1264–1275.

Shaw, H. G. 1977. Impact of mountain lion on mule deer and cattle in northwestern Arizona. In R. L. Phillips and C. J. Jonkel, editors. Proceedings of the 1975 Predator Symposium, pp. 17–32. University of Montana, Missoula.

Shaw, H. G. 1981. Comparison of mountain lion predation on cattle on two study areas in Arizona. In Proceedings of the wildlife-livestock relations symposium, pp. 17–32. University of Idaho, Moscow.

Shaw, H. G. 1983. Mountain lion field guide. Special Report, Arizona Game and Fish Department, Phoenix. 9:1–37.

Shaw, H. 1984. Cattle growers and lions. In J. Roberson and F. Lindzey, editors. Proceedings of the Second Mountain Lion Workshop, pp. 119–128. Zion National Park, Utah.

Shaw, H. G., N. G. Woolsey, J. R. Wegge, and R. L. Day, Jr. 1988. Factors affecting mountain lion densities and cattle depredation in Arizona. Final Report, Pittman-Robertson Project W-

78-R, Arizona Game and Fish Department, Phoenix.

Stoner, D. C., M. L. Wolfe, and D. M. Choate. 2006. Cougar exploitation levels in Utah: Implications for demographic structure, population recovery, and metapopulation dynamics. Journal of Wildlife Management 70:1588–1600.

Swank, W. B. 1958. The mule deer in Arizona chaparral. Wildlife Bulletin No. 3, Arizona Game and Fish Department, Phoenix.

Torres, S. G., T. M. Mansfield, J. E. Foley, T. Lupo, and A. Brinkhaus. 1996. Mountain lion and human activity in California: Testing speculations. Wildlife Society Bulletin 24:451–460.

Winslow, R. 2005. New Mexico mountain lion status report. Mountain Lion Workshop 8:70–72.

Woolstenhulme, R. 2005. Nevada mountain lion status report. Mountain Lion Workshop 8:49–56.

Zar, J. H. 1996. Biostatistical analysis. Prentice Hall, Upper Saddle River, New Jersey.

Zornes, M. L, S. P. Barber, and B. F. Wakeling. 2006. Harvest methods and hunter selectivity of mountain lions in Arizona. In J. W. Cain and P. R. Krausman, editors. Managing Wildlife in the Southwest, pp. 85–89. Southwest Section of the Wildlife Society, Tucson, Arizona.

Synthesis

SHAPING CONSERVATION THROUGH THE INTEGRATION OF RESEARCH WITH RESOURCES MANAGEMENT ON THE COLORADO PLATEAU: A SYNTHESIS

Charles van Riper III, Brian F. Wakeling, and Thomas D. Sisk

In published reviews of the previous eight volumes of the Colorado Plateau series, book-reviewers suggested that a synthesis chapter would be a useful addition in subsequent volumes (Fleming 2006, Quartaroli 2007). Because all of the Colorado Plateau books present widely disparate facets of research, and every volume has chapters that span aspects of cultural, physical, and biological sciences, these reviewers felt that a closing summary chapter would help readers to better synthesize the disparate topics. This fourth volume from the University of Arizona Press (the ninth in the full Colorado Plateau series) has the same broad span of material as previous volumes, and this summary chapter will attempt to tie together the other chapters of the book.

The sixteen chapters within this book represent the results of fifteen studies and two panel discussions from the Biennial Conference; they are presented in four sections according to overarching topics: (1) conservation visions and frameworks, (2) assessing monitoring frameworks and systems, (3) wildlife surveys as a conservation framework, and (4) large mammal conservation and management. Many of the chapters address conservation and management issues of multiple resources; as a group, they address diverse resources from various geographic regions over the Colorado Plateau. Some chapters explain integrative tools that can be used to manage resources at broader geographic scales, which we now recognize is important for sustaining many ecological functions. This book, drawn from the Ninth Biennial Conference of Research on the Colorado Plateau, focuses more heavily on conservation than did the previous eight volumes, with an emphasis on assessing monitoring frameworks and wildlife research, in concert with a presentation of conservation success stories.

CONSERVATION VISIONS AND FRAMEWORKS

The book begins with a section that speaks to the broadest conceptual scope—conservation visions and frameworks. These chapters set a foundation for directions that future conservation can take on the Colorado Plateau, particularly in light of ongoing climatic change. The first chapter, by Jones et al., provides a summary of the Conservation Biology Panels that were held 30–31 October 2007 during the Ninth Biennial Conference. Drawing on presentations and discussions in these panels, Jones and several co-authors convey the broad range of the participants' impressions and suggestions on how conservationists need to "carry on" in an era of explosive population growth, habitat loss and fragmentation, and climate change that collectively place increasing stress on the vulnerable ecosystems and biodiversity of the Colorado Plateau. Panel members— among the most active and experienced scientists working in the region—were asked to evaluate how their individual research and related efforts had served the Colorado Plateau, and assessments were modest and often equivocal. From these exchanges, a general consensus emerged that successes had not fully matched early expectations, and that future successes would demand novel approaches and great creativity. Panel members advocated for greater integration of the natural and social sciences, for more attention to garnering and sustaining political support, and for a renewed pursuit of aesthetic and spiritual appreciation that can help maintain focus and motivation among conservationists. With few exceptions, the panelists, mostly research scientists, deemphasized new research needs while highlighting the importance of improved public communication, engagement of broader constituencies, and social processes to inform conservation goals and objectives. If these perceptions represent the thoughts, feelings, and aspirations of the broader community of scientists and land managers concerned with conservation, this chapter may well define a turning point in conservation action over the Colorado Plateau.

Chapter 2, by Garfin et al., presents the results of down-scaled projections of climatic models for projecting possible climatic changes over the Colorado Plateau region. As climate change exerts a stronger influence on Colorado Plateau ecosystems, it becomes increasingly important to incorporate insight from climate models into future management planning. The authors address the issue of down-scaled projections of climatic models, demonstrating approaches for managers that highlight how down-scaling regional climate models can be integrated with the ecological process models, and how they can be directly related to conservation and management of resources. By exploring the suite of models currently available to researchers and providing techniques for model integration, the authors take a big step forward in bringing climate change research into a management-relevant framework. They anticipate that, in the future, linking downscaled climate models to a host of management applications will allow a more seamless and repeatable way to project the effects of change and the efficacy of management responses to changing resources.

The conservation visions and framework section of this book is tied together with Chapter 3, by Sisk et al., who describe the acquisition and conservation-based management of the Kane and Two Mile Ranches north of the Colorado River in Arizona. The purchase of these ranches by the Grand Canyon Trust and The Conservation Fund, and the initiation of management-driven research efforts, provides an example "par excellence" of how landowners, managers, and scientists can work with the public to integrate restoration and conservation at broad geographic scales by developing novel public/private partnerships. These authors illustrate how rangeland ecologists, soils scientists, remote-sensing scientists, fire ecologists, wildlife biologists, silviculturalists, and livestock producers can use science to articulate objectives, integrate efforts, and guide adaptive management. Four case studies describe this approach. The first study addresses post-fire management of forested habitat following the Warm Fire in 2006. The second study delves into the intricacies of restoring winter range for mule deer, an effort complicated by historical wildfires, livestock grazing, and the spread of undesirable invasive plant species. The third study focuses on challenges in restoring native cool-season grasses, which on paper may seem to be a fairly straightforward procedure, but is in fact subject to many environmental vagaries that demand careful planning, with attention to climate, timing, and soil disturbance. The final case study describes the use of geospatial modeling to identify risk for cheatgrass invasion and how managers might use that information to prioritize landscapes for restoration efforts. Preliminary models were appraised for overall accuracy using accepted evaluation criteria and these models were employed to evaluate possible impacts of planned management treatments, with respect to the likelihood of cheatgrass spread. By documenting the employment of a spatial decision-support system for vegetation- and ranch-planning efforts, Chapter 3 outlines a foward-thinking demonstration and test of how science can provide an analysis of land cover for the management of lands on the Colorado Plateau. It also highlights the capability of private (nongovernmental) organizations to facilitate this kind of collaboration, blending disciplines to integrate research with adaptive management on the Colorado Plateau. This conservation approach sets a new benchmark that could be applied broadly within the field of natural resources.

ASSESSING MONITORING FRAMEWORKS AND SYSTEMS

Chapter 4, by Garman et al., transitions into the "assessing monitoring frameworks" portion of this book. Monitoring has been a perennial topic of discussion among scientists, resource managers, and participants of every previous Biennial Conference (e.g., Sisk et al., 1999). The authors incorporate protocols from the National Park Service (NPS) Inventory and Monitoring Program (I&M) specifically related to soil inventory surveys. The initial data were collected in Canyonlands National Park, and then tested in Capital Reef National Park.

The authors note that many NPS areas have existing soil maps, but most maps are at a coarse resolution, and this lack of detail is not conducive to choosing sites for soil study, thus the necessity for a finer detailed soil study at the ecological site resolution. In order to refine the selection of locations for soils inventories, they implemented two predictive mapping methods: (1) feature-extraction maps and (2) decision tree models.

In their modeling efforts, Garman and his co-authors found that the number, and even types, of ecological sites within a map-unit polygon can differ considerably from a described average. Using existing soil maps to define the spatial extent of targeted ecological sites can lead to over-estimation of the number of targeted ecological sites and the selection of numerous monitoring samples in non-targeted ecological sites. This result led to the modeling efforts, with the decision tree being identified as the most efficient approach, overall. Garman et al. propose that the advantages of predictive mapping of ecological site types outweigh the costs of model-development efforts. They argue that it is reasonable to assume that feature-extraction is the method of choice where spectral and contextual features are largely different among targeted ecological sites. The authors further suggest that with the planning horizon (80+ yrs) for monitoring programs such as the NPS Northern Colorado Plateau Network I&M, developing an effective survey design has long-term benefits. Moreover, they note that in remote areas over the Colorado Plateau where travel time to sampling locations can constitute the bulk of monitoring costs, increased efficiency of modeling can bring a substantial benefit to long-term monitoring.

In Chapter 5 Lauver et al. continue the theme of monitoring and habitat modeling, using a case study at Bandelier National Monument where they examined the implications of modifying statistically robust designs to account for logistical feasibility. By taking a quantitative look at the extent to which changes in ecological indicators affect soil and vegetation monitoring results, they provide a sound footing for future consideration of the way in which scientific robustness and practical limitations can be negotiated between parties cooperating to achieve effective and efficient monitoring, the results of which will guide adaptive management. By incorporating an experimental component to applied work at Bandelier National Monument, they also demonstrated that incisive science can be combined with programmatic activities, another requirement for any successful adaptive management program. They suggest that these new monitoring programs will allow land managers, policy makers, and planners to make better-informed land-use decisions, especially throughout northwestern New Mexico. It is clear that this sort of practical approach to monitoring is essential for the implementation of adaptive management, and this chapter offers insight into issues that will need to be examined and resolved during the critical design stage of any monitoring effort.

The "Assessing Monitoring Frameworks and Systems" section continues with a chapter by Yamamoto et al., who discuss the development, construction, and deployment of a wireless cyberinfrastructure for environmental monitoring. The authors suggest that with the increasing complexity of ecological questions, it is imperative that scientists have the correct tools to collect environmental data. They argue that the WiSARD-Net (Wireless Sensing and Relay Device Network) that they have developed will fit the needs of scientists who need environmental information. The WiSARD-Net system that Yamamoto et al. have developed is simple, yet elegant. It can collect meteorological data (e.g., air and soil temperature, humidity) and also photosynthetically active radiation measurements. The environmental information is collected continuously and downloaded directly to a computer. Chapter #6 presents a test that was run on the WiSARD-Net system as a pilot study at the Flagstaff Arboretum and details the lessons learned on this initial pilot study. A conceptual model is provided for the system, as well as information on how to make WiSARD-Net run in "real time" for scientists. The authors also provide an overview of WiSARD-Net, showing how to integrate the *in situ* wireless sensor network and data-center hardware to provide a complete solution for sampling the environment, transferring the information to a networked server, and then displaying the information, via the Web, anywhere in the world. This is certainly a tool that managers and scientists on the Colorado Plateau should seriously consider employing in future environmental monitoring efforts.

Vegetation studies are introduced with Chapter 7, where Walter Fertig examines the use of vascular plant taxa to determine existing gaps in protected areas over the Colorado Plateau. Using a National Gap Analysis Program (GAP) approach (Scott et al. 1991, 1993), Fertig conducts a comprehensive examination of native plant distributions across the Utah portion of the Colorado Plateau. He provides an original and innovative look at the sufficiency of our current conservation areas network. By consulting botanical databases and examining the existing network of protected areas, he demonstrates that nearly one third of the region's native plant species are currently unprotected. He argues that this is a sizeable proportion, given extensive federal ownership and the existing broad interagency mandate for protecting native species and the ecosystems upon which they depend. Furthermore, Fertig identifies those areas in which the biggest gaps in the protection network occur, concluding that 70 percent of the unprotected taxa occur in just 12 plant diversity hotspots, demonstrating that future expansion of the region's network of conservation areas could be pursued with great efficiency, using the existing "gaps" identified in this chapter.

Chapter 8 concludes the set of chapters of the "Assessing Monitoring Frameworks and Systems" section. In this chapter Poff et al. analyze how fire and fire surrogate treatments impact soil moisture conditions in southwestern ponderosa pine forests. The authors argue, as did Garret et al. (1997) in an earlier book in the Colorado Plateau series, that ponderosa pine forests since at least the early 1900s have exhibited substantial increases in small-diameter pines and decreases among the largest classes of trees, a change largely due to selective logging and reduced fire frequencies since the late 1800s. Garret and Soulen (1999) also suggested that declines in large-diameter trees were due to competition for limited moisture with burgeoning smaller trees. Working under the USDA/USDI Fire and Fire Surrogates Program, Poff et al. had two major objectives: (1) to determine the small-scale hydrologic responses of ponderosa pine in Arizona to operational fuels treatments and (2) to gather information on the soil moisture that will be available to plants in different rooting depths as well as for soil microbiology processes on similar soils. The authors found no statistically significant differences between aspects, treatments, or stand density (basal area), as related to soil moisture, a finding that they attribute to the high variability among plots in the same treatment areas. However, it may be significant that the "cut-and-burn" or "burn-only" treatment units had the highest soil moisture content throughout the year at all depths. Considering that fire used to be a natural part of the southwestern ponderosa pine ecosystem, this is an important observation. The authors suggest that future studies should look at fire and fire surrogate treatments at other locations in southwestern ponderosa pine forests.

WILDLIFE SURVEYS AS A CONSERVATION FRAMEWORK

The next quartet of chapters covers management aspects of using wildlife surveys as a framework for conservation on the Colorado Plateau. In Chapter 9, Nowak and Persons use long-term monitoring data for milksnakes from Petrified Forest National Park (Holbrook, Arizona) to describe survey efforts for this cryptic reptile. Despite the fact that milksnakes are widely distributed throughout much of the United States and South America, they are difficult to detect and, thus, information regarding their status is difficult to acquire. Building on previous work (Drost et al. 2001), Nowak and Persons employed a variety of survey methodologies, including nighttime and daylight driving, walking and bicycling surveys, pitfall traps and cover arrays, box traps and drift fences, and radio-telemetry monitoring of captured animals. They suggest that nighttime driving surveys may be the single most efficient and effective method for determining long-term trends in milksnake abundance.

In Chapter 10 Graham et al. provide an assessment of the affects of the pesticide Diflubenzuron, a treatment applied for Mormon Cricket control, on non-target arthropod communities. This chapter provides a wealth of information on insect species present within habitats of northwestern Utah. In a unique application of inventory techniques, documenting invertebrate species assemblages along a pesticide disturbance gradient, Graham et al. examine the potential impact of Diflubenzuron to all arthropod communities both proximately and ultimately, thus the importance of this study. The authors found a wide variety of Coleoptra, Diptera, Hemiptera, non-ant Hymenoptera, Lepidoptera, Orthoptera, and Scorpiones susceptible to Diflubenzuron, with some differences attributed to specific aspects of the three sampled locations. The authors demonstrate, through the use of principal components analysis, that there may be a lag effect of up to one year on assessing the true impact of Diflubenzuron on arthropod communities. They conclude this chapter urging continued monitoring, which would provide managers with a better understanding of the long-term potential impacts of spraying to invertebrate species within the Desert Southwest.

In Chapter 11, the survey theme shifts to the examination of birds, as Johnson et al. examine the distribution and habitat associations of the Yellow-billed Cuckoo throughout the state of Arizona. By examining historical records, the authors determined that Yellow-billed Cuckoos and their breeding habitat are likely substantially less common in Arizona today than they were in the late 1800s and early 1900s. In their statewide survey in 1998 and 1999, Yellow-billed Cuckoos were primarily detected around the San Pedro River, Cienega Creek, Verde River, Sonoita Creek, and the Agua Fria River. The authors affirm that changes to Yellow-billed Cuckoo habitat and distribution can be linked largely to the well-documented changes to historical flows in Arizona's rivers and periodic flood-dependant riparian habitat, along with habitat fragmentation due to agricultural development, invasive exotic vegetation, and urbanization.

In a second chapter on monitoring birds, van Riper et al. examine differences in population trends of birds in the Camp Verde region of central Arizona. Chapter 12 provides the results of an inventory and long-term monitoring effort documenting birds reported at the Montezuma Castle National Monument complex and the surrounding Camp Verde region of Central Arizona. The authors compared historical records in the published literature, detections by qualified volunteer observers, and recent surveys they had conducted to determine trends in species at the national monument and throughout the Camp Verde region. This chapter provides a detailed account of the status of more than 200 species, giving information on residency status, dates present, breeding status, abundance, and preferred habitat type. Of those bird species that have been detected within Montezuma Castle National Monument, the authors found that 14 species have increased, whereas six have declined in abundance. They conclude that much of the changes observed in avian community composition are likely due to large-scale changes (e.g., urbanization) outside of the region. Only a few changes in bird status can be tied directly to management actions taken within Montezuma Castle National Monument or the Camp Verde region.

LARGE MAMMAL CONSERVATION AND MANAGEMENT

The final four chapters of the book move from large biological surveys to the management and conservation of large mammals on the Colorado Plateau. In Chapter 13 Wakeling presents the current status and historical changes in management for trophy deer on the Kaibab Plateau. This chapter provides a much needed summary of numerous years of biological inventory undertaken on the Kaibab deer herd. The author compares results of hunter check-station information from Jacob Lake prior to and after the implementation of a 2001 change in hunting quotas set by the Arizona Fish and Game Department. Wakeling found that deer maximum-mean outside antler spread was reached at five years of age. The new 2001 regulations had little effect on the size of deer taken, except that older deer were taken later in the hunting season. The author concludes that state management of the Unit 12A Kaibab deer hunt is successful, largely meeting the needs of the hunting public, while properly managing for trophy-class bucks.

In Chapter 14, Bristow and Cunningham report on the use of habitat by female elk subsequent to the Rodeo-Chediski fire that occurred in east-central Arizona in 2002. This stand-replacing fire consumed almost 185,000 ha of ponderosa pine forest below and above the Mogollon Rim. While unplanned, the fire provided a valuable opportunity to learn how elk use altered habitats. The authors note that in this case favorable precipitation patterns that immediately followed the fire probably enhanced forage growth. They found that after the fire, female elk selected ponderosa pine-oak habitats with 40–60 percent canopy closure in the heavy-to-extreme burn intensity regions. This trend was consistent through the winter, spring, and summer seasons during both daylight and nighttime hours. In autumn, female elk shifted to light-to-medium burn intensity regions, although still using ponderosa pine-oak habitat with 40–60 percent canopy closure. Bringing this study to bear on land management, Bristow and Cunningham recommend prescribed fire that retains mosaics of moderate canopy closure and forage in order to improve elk habitat.

Chapters 15 and 16 both look at managing and conserving predators on the Colorado Plateau, as well as specific survey methodology. In Chapter 15 McKinney and Smith report on the efficacy of scent-station transects for indexing mesopredators. The authors investigated visitation by bobcats, coyotes, and foxes in chaparral, pinyon-juniper woodland, ponderosa pine forest, semi-desert grassland, and Upper Sonoran Desert scrub habitats throughout Arizona. They conclude that this technique, while ineffective as a bobcat population index, does provide a reliable method for tracking relative abundance of coyote and fox populations in most habitat types, with the exception of coyotes in ponderosa pine habitats, where they did not visit scent stations. The authors caution against drawing conclusions about differences in bobcat numbers among habitat types, but argue that scent-station transects can provide a reliable method of comparing relative abundance of foxes and coyotes among time periods (or locations) within specific habitat types.

In the final chapter of the large mammal conservation and management section, McKinney et al. provide a summary and analysis covering 29 years of mountain lion depredation harvests in Arizona. For the period from 1976 through 2005, they examined existing records to see if mule deer abundance was inversely related to livestock depredation. The authors found that more than 90 percent of livestock depredation by mountain lions targeted cattle, and almost all of those cases involved calves. Mountain lions that were killed as a result of livestock depredation were taken in 12 of the Arizona's 15 counties, but 92 percent of those lions were taken within five counties. The authors also note that most lions taken after killing livestock were adult males. After comparing the statistics, the authors concluded that a reduced relative abundance of mule deer does contribute to greater depredation of cattle, and consequently to more lions being killed for livestock depredation. Studies such as this one and the others in this section provide invaluable data that both land managers and biologists can draw upon in evaluating both harvest and conservation programs proposed in coming years.

The scientific works published in this series of books, based on the biennial conferences of research on the Colorado Plateau, present the results of collaborative efforts between scientists and land managers and contribute significantly to the peer reviewed literature. We have found that many of the protocols and management techniques presently being used in land management units over the Colorado Plateau are a result of past collaborative works published in this series, including those of federal scientists working jointly with university and other partner agency scientists, and with land managers from various states (e.g., the region's Wildlife and Game and Fish Departments). It has been clearly demonstrated that techniques that work in one management unit of the Colorado Plateau are applicable to a number of other areas (e.g. Drost et al. 2001), a fact due primarily to the similarity of habitat and climatological conditions across the region.

Like other research compilations that are centered on a particular theme, this book should help to focus attention on research and conservation efforts presently being conducted within diverse habitats over the Colorado Plateau. In particular, we hope that private land owners and state land stewards of

Arizona, Utah, Colorado, and New Mexico, as well as managers of National Parks, National Monuments, U.S. National Forests, Fish and Wildlife Service refuges, Bureau of Land Management districts, Bureau of Reclamation waters, and tribal lands will be able to make good use of the ideas and concepts presented within this book. We hope that the new information presented in these chapters will allow land stewards to undertake more effective and efficient efforts in the management, conservation, and stewardship of southwestern lands and natural resources. Finally, if the material in this volume can stimulate support for future research in the conservation and management of cultural, natural, and physical resources over the Colorado Plateau, it will make the organizational and editorial work of the past two years a worthwhile and productive effort.

LITERATURE CITED

Drost, C. A., T. B. Persons, and E. M. Nowak. 2001. Herpetofauna survey of Petrified Forest National Park, Arizona. In Proceedings of the Fifth Biennial Conference of Research on the Colorado Plateau, edited by C. van Riper III, K. A. Thomas, and M. A. Stuart, pp. 83–102. U. S. Geological Survey/Forest and Rangeland Ecosystem Science Center USGSFRESC/COPL/2001/21 Rep. Ser., Flagstaff, Arizona.

Fleming, B. 2006. Review of: The Colorado Plateau II: Biophysical, socioeconomic, and Cultural Resources. New Mexico Historical Review 81(1): 155–117.

Garret L. D., and M. H. Soulen. 1999. Changes in character and structure of Apache/Sitgreaves forest ecology: 1850-1990. In Proceedings of the Fourth Biennial Conference of Research on the Colorado Plateau, edited by C. van Riper III and M. A. Stuart, pp. 25–29. U.S. Geological Survey/Forest and Rangeland Ecosystem Science Center CPFS Rep. Ser., 99/16, Flagstaff, Arizona.

Garret L. D., M. H. Soulen, and J. R. Ellenwood. 1997. After 100 years of forest management: "The north Kaibab." In Proceedings of the Third Biennial Conference of Research on the Colorado Plateau, edited by C. van Riper III and E. Deshler, pp. 129–149. U.S. Department of the Interior National Park Service Transactions and Proceedings Series NPS/NRNAU/NRTP-97/12.

Quartaroli, R. D. 2007. Review of: The Colorado Plateau II: Biophysical, socioeconomic, and Cultural Resources. Journal of Arizona History 48(1):104–106.

Scott, J. M., B. Csuti, K. Smith, J. E. Estes, and S. Caicco. 1991. Gap analysis of species richness and vegetation cover: An integrated biodiversity conservation strategy. In Balancing on the Brink of Extinction: The Endangered Species Act and Lessons for the Future, edited by K. Kohm, pp. 282–297. Island Press, Washington, D.C.

Scott, J. M., F. Davis, B. Csuti, R. Noss, B. Butterfield, C. Groves, H. Anderson, S. Caicco, F. D'Erchia, T.C. Edwards, Jr., J. Ulliman, and R. G. Wright. 1993. Gap Analysis: A geographic approach to protection of biological diversity. Wildlife Monographs. 123:1-41.

Sisk, T. D., T. E. Crews, R. T. Eisfeldt, M. King and E. Stanley. 1999. Assessing impacts of alternative livestock management practices: Raging debates and a role for science. In Proceedings of the Fourth Biennial Conference of Research on the Colorado Plateau, edited by C. van Riper III, and M. A. Stuart, pp. 89–103. U.S.Geological Survey/Forest and Rangeland Ecosystem Science Center CPFS Rep. Ser., 99/16, Flagstaff, Arizona.

CONTRIBUTORS

Christine Albano
Grand Canyon Trust
2601 N. Fort Valley Rd.
Flagstaff, AZ 86001-8341
calbano@grandcanyontrust.org

Terence Arundel
U.S. Geological Survey
Southwest Biological Science Center
Colorado Plateau Research Station
Northern Arizona University
P.O. Box 5614
Flagstaff, AZ 86011
terry_arundel@usgs.gov

Ethan Aumack
Grand Canyon Trust
2601 N. Fort Valley Rd
Flagstaff, AZ 86001-8341
Eaumack@grandcanyontrust.org

Jan Balsom
Grand Canyon National Park
2601 N. Fort Valley Rd.
Flagstaff, AZ 86001-8341
jan_balsom@nps.gov

Paul Beier
Northern Arizona University
South San Francisco St.
Flagstaff, AZ 86011
Paul.Beier@nau.edu

Jayne Belnap
U.S. Geological Survey
Biological Resources Division
2290 S. West Resource Blvd.
Moab, UT 84532
janye_belnap@usgs.gov

Eli J. Bernstein
Center for Sustainable Environments
Northern Arizona University
South San Francisco St.
Flagstaff, AZ 86011
eliljbernstein@gmail.com

Anne M. D. Brasher
U.S. Geological Survey
Utah Water Science Center
121 West 200 South
Moab, UT 84532
abrasher@usgs.gov

Kirby Bristow
Arizona Game and Fish Department
5000 W. Carefree Highway
Phoenix, AZ 85086
kbristow@azgfd.gov

James Catlin
Wild Utah Project
68 S. Main St., Suite 400
Salt Lake City, UT 84101
Jim@wildutahproject.org

Rebecca N. Close
U.S. Geological Survey
Utah Water Science Center
121 West 200 South
Moab, UTAH 84532
in transit: current email
becky.close@gmail.com

Neil Cobb
Merriam-Powell Center
    for Environmental Research
Northern Arizona University
South San Francisco St.
Flagstaff, AZ 86011
Neil.Cobb@nau.edu

Kenneth L. Cole
U.S. Geological Survey
Southwest Biological Science Center
Colorado Plateau Research Station
P.O. Box 5614
Northern Arizona University
Flagstaff, AZ 86011
ken_cole@usgs.gov

Timothy E. Crews
Center for Sustainable Environments
Northern Arizona University
South San Francisco St.
Flagstaff, AZ 86011
and
Environmental Studies Program
Prescott College
220 Grove Avenue
Prescott, AZ 86301
tcrews@prescott.edu

Stan Cunningham
Arizona State University
Polytechnic Campus
7001 E. Williams Field Rd.
Mesa, AZ 85212
Stanley.Cunningham@asu.edu

Jim DeCoster
Southern Colorado Plateau Network
National Park Service
Northern Arizona University
1298 S. Knoles Dr.
Flagstaff, AZ 86011
Jim_DeCoster@nps.gov

Brett G. Dickson
Center for Sustainable Environments
and Merriam-Powell Center for
    Environmental Research
Northern Arizona University
South San Francisco St.
Flagstaff, AZ 86011
and
Grand Canyon Trust
2601 N. Fort Valley Rd.
Flagstaff, AZ 86001-8341
Brett.Dickson@nau.edu

Jon K. Eischeid
NOAA Earth Systems Research
    Laboratory
325 Broadway
Boulder, CO 80305
jon.k.eischeid@noaa.gov

Walter Fertig
Moenave Botanical Consulting
1117 W Grand Canyon Dr.
Kanab, UT 84741
walt@kanab.net

Thomas L. Fleischner
Prescott College
220 Grove Ave.
Prescott, AZ 86301
tfleischner@prescott.edu

Paul Flikkema
Wireless Networks Research Laboratory
Department of Electrical Engineering
Northern Arizona University
South San Francisco St.
Flagstaff, AZ 86011
Paul.Flikkema.nau.edu

Steve Fluck
Grand Canyon Trust
2601 N. Fort Valley Rd.
Flagstaff, AZ 86001-8341
sfluck@grandcanyontrust.org

Gregg M. Garfin
Institute for Study of Plant Earth
University of Arizona
715 N. Park Ave.
P.O. Box 210156
Tucson, AZ 85721
gmgarfin@email.arizona.edu

Steven L. Garman
National Park Service
Northern Colorado Plateau Inventory
P.O. Box 848
Moab, UT 84532
and
Monitoring Program
Department of Biology
Mesa State College
1175 Texas Avenue
Grand Junction, CO 81501
Steven_garman@nps.gov

Tim B. Graham
U.S. Geological Survey
Southwest Biological Science Center
Canyonlands Research Station
2290 S. West Resource Blvd.
Moab, UT 84532
current address:
1701 Murphy Lane
Moab, Utah 84532
lasius17@gmail.com

Ed Grumbine
Prescott College
220 Grove Avenue
Prescott, AZ 86301
egrumbine@prescott.edu

Yuxin He
Wireless Networks Research Laboratory
Department of Electrical Engineering
Northern Arizona University
South San Francisco St.
Flagstaff, AZ 86011
hoocolin@gmail.com

Paul Heinrich
Merriam-Powell Center for
    Environmental Research
Northern Arizona University
South San Francisco St.
Flagstaff, AZ 86011
Paul.Heinrich@nau.edu

Kirsten Ironside
Environmental Sciences and Public Policy
Northern Arizona University
P.O. Box 6077
Flagstaff, AZ 86011
Kirsten.Ironside@nau.edu

Matthew J. Johnson
U.S. Geological Survey
Southwest Biological Science Center
Colorado Plateau Research Station
Northern Arizona University
P.O. Box 5614
Flagstaff, AZ 86011
mjjohnson@usgs.gov

Allison L. Jones
Wild Utah Project
68 South Main St.
Salt Lake City, UT 84101
Allison@wildutahproject.org

Chris L. Lauver
Southern Colorado Plateau Network
National Park Service
Northern Arizona University
1298 S. Knoles Dr.
Flagstaff, AZ 86011
Chris_lauver@nps.gov

Melanie T. Lenart
Institute for Study of Plant Earth
University of Arizona
715 N. Park Ave.
P.O. Box 210156
Tucson, AZ 85721
mlenart@email.arizona.edu

Robert T. Magill
Arizona Game and Fish Department
5000 West Carefree Highway
Phoenix, AZ 85086
and
Wildlife Specialties L.L.C.
P.O. Box 1314
Palisade, CO 81526
rob.magill@wildlifespecialtiesllc.com

David J. Mattson
U.S. Geological Survey
Southwest Biological Science Center
Colorado Plateau Research Station
Northern Arizona University
P.O. Box 5614
Flagstaff, AZ 86011
david_mattson@usgs.gov

Ted McKinney, now deceased
Arizona Game and Fish Department,
    Research Branch
5000 West Carefree Highway
Phoenix, AZ 85086

Melissa McMaster
Grand Canyon Trust
2601 N. Fort Valley Rd
Flagstaff, AZ 86001-8341
and
Department of Politics and International
   Affairs
Northern Arizona University
South San Francisco St.
Flagstaff, AZ 86011
mam492@nau.edu

Daniel G. Neary
USDA Forest Service
Rocky Mountain Research Station
2500 S. Pine Knoll Drive
Flagstaff, AZ 86001
dneary@fs.fed.us

Jodi Norris
Southern Colorado Plateau Network
National Park Service
Northern Arizona University
1298 S. Knoles Dr.
Flagstaff, AZ 86011
Jodi_Norris@nps.gov

Erika M. Nowak
Northern Arizona University
Department of Biological Sciences
P.O. Box 5640
Flagstaff, AZ 86011
erika.nowak@nau.edu

Johnathan C. O'Dell
Arizona Game and Fish Department,
   Game Branch
5000 West Carefree Highway
Phoenix, AZ 85086
jodell@azgfd.gov

Alex Orange
Wireless Networks Research Laboratory
Department of Electrical Engineering
Northern Arizona University
South San Francisco St.
Flagstaff, AZ 86011
ajo6@nau.edu

Trevor B. Persons
206 Bigelow Hill Rd.
Norridgewock, ME 04957
trevor.persons@nau.edu

Boris Poff
National Park Service
Mojave National Preserve
2701 Barstow Rd.
Barstow, CA 92311
boris_poff@nps.gov

Andi S. Rogers
Arizona Game and Fish Department
5000 West Carefree Highway
Phoenix, AZ 85086
arogers@azgfd.gov

Steven S. Rosenstock
Arizona Game and Fish Department
5000 West Carefree Highway
Phoenix, AZ 85086
srosenstock@azgfd.gov
or Steven.Rosenstock@nau.edu

Bill Ruggeri
Wireless Networks Research Laboratory
Department of Electrical Engineering
South San Francisco St.
Flagstaff, AZ 86011
Bill.Ruggeri@nau.edu

David Schlosberg
Center for Sustainable Environments
   and Department of Politics
   and International Affairs
Northern Arizona University
South San Francisco St.
Flagstaff, AZ 86011
david.schlosberg@nau.edu

Ron Sieg
Arizona Game and Fish Department
5000 West Carefree Highway
Phoenix, AZ 85086
rsieg@azgfd.gov

Thomas D. Sisk
Center for Environmental Science
  and Education
Northern Arizona University
P.O. Box 5694
Flagstaff, AZ 86001
Thomas.Sisk@nau.edu

Thorry W. Smith
Arizona Game and Fish Department
Research Branch
5000 West Carefree Highway
Phoenix, AZ 85086
tsmith@azgfd.gov

Mark K. Sogge
U.S. Geological Survey
Southwest Biological Science Center
Colorado Plateau Research Station
Northern Arizona University
P.O. Box 5614
Flagstaff, AZ 86011
mark_sogge@usgs.gov

Aregai Tecle
Northern Arizona University
School of Forestry
P.O. Box 15018
Flagstaff, AZ 86011
Aregai.Tecel@nau.edu

Andrea Thode
School of Forestry
Northern Arizona University
P.O. Box 15018
Flagstaff, AZ 86011
Andrea.Thode@nau.edu

Lisa Thomas
Southern Colorado Plateau Network
National Park Service
Northern Arizona University
1298 S. Knoles Dr.
Flagstaff, AZ 86011
Lisa_Thomas@nps.gov

Charles van Riper III
U.S. Geological Survey
Southwest Biological Science Center
Sonoran Desert Research Station
University of Arizona
125 Biological Sciences East
Tucson, AZ 85721
charles_van_riper@usgs.gov

Brian F. Wakeling
Arizona Game and Fish Department,
  Game Branch
5000 West Carefree Highway
Phoenix, AZ 85086
bwakeling@azgfd.gov

Aneth Wight
National Park Service
Northern Colorado Plateau Inventory
  and Monitoring Program
P.O. Box 848
Moab, UT 84532
aneth_wight@nps.gov

Holland Wilberger
Department of Systems and Industrial
  Engineering
University of Arizona
Tucson, AZ 85721
hmw@email.arizona.edu

Dana Witwicki
National Park Service
Northern Colorado Plateau Inventory
  and Monitoring Program
P.O. Box 848
Moab, UT 84532
dana_witwicki@nps.gov

Kenji Yamamoto
Wireless Networks Research Laboratory
Department of Electrical Engineering
Northern Arizona University
South San Francisco St.
Flagstaff, AZ 86011
kry3@nau.edu

INDEX

A

Abert's Towhee 256
acacia 219
accessibility 76, 88, 93, 95
Accipiter
　cooperii 226
　gentiles 227
　striatus 226
accountability 11
accounting space 12
accuracy 12, 27, 104, 282, 300
　soil-mapping 67-86
　chance-corrected 76
ACEC (Areas of Critical Environmental Concern) 111, 117
Achnatherum hymenoides 57
Actitus macularis 230
adaptive management 4, 48, 52, 55, 64, 316, 318
advocacy 14
Aegolius acadicus 234
aerial 235
　foragers 235
　photography 74
　spray 152
Aeronautes saxatalis 234
aerosol 25
Agelaius phoeniceus 234
agencies 1, 3, 9, 10, 17, 46, 59, 64, 66, 115, 146, 271, 303
agriculture 18, 55, 116, 206, 261, 265, 296
Agua Fria River 197, 201, 205, 320
AIC (Akaike's Information Criterion) 60, 62, 282
　weight 60, 61, 63
Aimophila ruficeps 256
Aix sponsa 220
Akaike's Information Criterion (*see also* AIC) 282
all terrain vehicle (*see also* ATV) 133
alluvial seedbeds 206
Alnus oblongifolia 199
Alpacas 307
Alydidae 183
American Birds 209
American Coot 230
American Crow 243
American Goldfinch 263

American Kestrel 228
American Ornithologists' Union 198, 217
American Pipi 250
American Robin 249
American Tree Sparrow 257
American Wigeon 221
amphibians 117, 136
Amphispiza 258
　belli 258
　bilineata 258
Anabrus simplex 152
Anas
　acuta 220
　americana 221
　clypeata 221
　crecca 220
　cyanoptera 220
　discors 220
　penelope 221
　platyrhynchos 220
　strepera 221
Anasazi forts 8
ancient forests 46
animosity 16
Anna's Hummingbird 234
Anser albifrons 219
Anthus rubescens 250
antler 271-276
　spread 320
ants 159, 160, 162, 170, 180, 183, 189, 190, 193, 196
AOGCMs (ocean atomosphere models) 25-40
Apache-Sitgreaves National Forest 278-287
Aphelocoma
　californica 242
　ultramarine 243
Aphididae 183
aphids 183
Aquila chrysaetos 228
AR4 (Fourth Assessment Report) 25, 27, 29, 33, 41
Araneae 160, 172-175
ArcGIS 60, 93, 282
archeological resources 8
Ardea vivescens 224
Areas of Critical Environmental Concern (ACEC) 111, 117

330 Index

area-weighting 80
Arizona alder 199
Arizona Bird Records Committee 237
Arizona Board of Regents 9
Arizona Breeding Bird Atlas 213
Arizona elegans 142
Arizona Game and Fish Commission (*see also* AZGFD) 271
Arizona Game and Fish Department 272, 278
Arizona legislature 303
Arizona Strip 45, 64
Arizona sycamore 199
Arizona Upland Sonoran Desert scrub 309
Arizona walnut 199
arroyos 256
Artemesia 142
arthropods 151-194
   community structure 153
   relative abundance 160
   target and nontarget 84
   ground-dwelling 152
Ashdown Gorge 111
Ash-throated Flycatcher 240
Asio flammeus 234
astronomers 9
Atriplex canescens 53
ATV 133, 136-139, 146-147
Auriparus flaviceps 245
avian community 213-265, 320
Aythya
   affinis 222
   americana 221
   collaris 222
   valisineria 221
AZGFD (*see also* Arizona Game and Fish Department) 53, 55, 65

B

Baccharis salicifolia 199
Baeolophus
   ridgwai 245
   wollweberi 245
Baird's Sandpiper 231
Bald Eagle 264
Bandelier National Monument 89-93, 98, 317
   topography 90
Bank Swallow 244
Barberry 256
bark beetle 24
Barn Owl 233
Bar-T-Bar 8

basal area 121-127, 319
baseline assessment 48
   Grand Canyon Trust 45
beach-habitat-building flow 5
Beaver Creek 215-264
bedrock geology 75
bees 160, 171, 190
beetles 55, 160, 166-71, 190-92
Bell's Vireo 241
Belted Kingfisher 237
best-fit solution 160
Bewick's Wren 247
Bicknell's Thrushes 208
Bidahochi Formation 8
big game 303
Bill Williams River 197, 202, 203, 206
biomass, herbaceous 50
bird list of Camp Verde region 219-263
Black Hills 232
Black Phoebe 240
Black Ridge Canyons 111
Black Vulture 225
Black-bellied Plover 230
Blackbrush 72, 73
Black-chinned Sparrow 257
Black-crowned Night-Heron 224
Black-headed Grosbeak 255
Black-throated Blue Warbler 252
Black-throated Gray Warbler 252
Black-throated Sparrow 258
BLM (*see also* Bureau of Land Management) 8-10, 18
Blue Grosbeak 255
Blue-gray Gnatcatcher 248, 265
Blue-throated Hummingbird 234
Blue-winged Teal 220
bobcat 293-300
bole scorch 50
Bombycilla cedrorum 250
bootstrapping 77
bottlebrush squirreltail 57
bottomland 207
Box-Death Hollow 111
box elder 242
bran bait 152
Branta canadensis 219
Brewer's Blackbird 260
Brewer's Sparrow 257
Bridger Knoll Fire 53
Bridled Titmouse 245
Brigham Young University Herbarium 111
broadcast seeding 56,

broadleaf deciduous vegetation 199
Broad-tailed Hummingbird 237
Bromus tectorum 45
Bronzed Cowbird 260
brood parasite 207, 261
brood patch 255
Brown Creeper 246
Brown Thrasher 249
Brown-crested Flycatcher 240
Brown-headed Cowbird 241, 251, 254, 261
Bryce Canyon National Park 110-112, 118-19
Bubo virginianus 233
Bubulcus ibis 224
Bucephala
   albeola 222
   clangula 222
buck 271-275, 320
   -to-doe ratio 272
   older age class 271-273
Bufflehead 222
bulldozer pushes 53
Bullock's Oriole 261
Bulrushes 219
bunchgrass 56, 58, 142
Bureau of Land Management (*see also* BLM) 64, 111, 194, 303
Bureau of Reclamation 4, 75, 78, 322
burn 319
   intensity 277
   prescribed 321
   severity class 49
burn severity geographic information system 281
Burned Area Emergency Rehabilitation 278
Bushtit 246
Buteo
   jamaicensis 227
   lagopus 228
   regalis 228
   swainsoni 227
Buteogallus anthracinus 227
Butorides striatus 224

C

Cactus Wren 246
Calamospiza melanocorys 258
Calidris bairdii 231
California Condor 52
Calypte
   anna 234
   costae 234
Camp Verde (region) 213
   bird list 219-263
Campbell Scientific 101

Camptostoma imberbe 238
Campylorhynchus brunneicapillus 246
Canada Goose 264
candidate species 198, 232
   Yellow-billed Cuckoo 197-211, 232, 265, 319, 320
Canis latrans 293
canopy 123, 200, 205-06, 278-89, 320
Canvasback 221
Canyon Towhee 256
Canyon Wren 247
Capitol Reef National Park 69-86, 110, 112
Caprimulgids 235
Carbaryl 152, 189, 192
carbon footprint 13
   sequestration 46
Carduelis
   pinus 262
   psaltria 263
   tristis 263
Cardinalis cardinalis 254
Carpodacus
   mexicanus 262
   purpureus 262
   cassinii 262
Cassin's Finch 262
Cassin's Kingbird 241
Catoptrophorus semipalmatus 230
caterpillars 198
Cathartes aura 225
Catharus
   bicknelli 208
   guttatus 248
   ustulatus 248
Cathedral-Rock Springs 72
Catherpes mexicanus 247
Cattails 219
cattle 18, 53, 56, 195, 261, 290, 303, 311
   trampling 56, 58
   stocking levels 18
Cattle Egret 224
CBC (Community-Based Collaboratives) 8, 17
Cedar Breaks National Monument 110
Cedar Waxwing 250
Celtis reticulata 199
cementum 271-75
centroid-assignment scheme 79
Certhia americana 246
Cervus elaphus 277
Ceryle alcyon 237
Chaetura vauxi 233
Champurrado Wash 202, 203
chaparral 293-99, 309, 321
Charadrius vociferous 230

cheatgrass 45, 49, 52-53, 55, 59-64, 316, 317
Check-list of North American Birds 209, 266
Chen caerulescens 219
Chenopodeaceae 185
chi square 95, 138, 273, 304
chickens 306
China 13
Chinle Formation 8
Chino Valley 133, 257
Chipping Sparrow 257
chitin 151
Chondestes grammacus 257
Chordeiles
  acutipennis 235
  minor 233
Christmas bird counts 243
Chyrsothamnus 141
cicadas 198
Cicadellidae 183
Cicadidae 198
Cienega Creek 197, 201, 202-205, 320
Cinnamon Teal 220
Circus cyaneus 226
Cistothorus palustris 247
Citellus spilosoma 140
citizen-activists 17
Clarkdale 216
Cliff Swallow 244
cliffrose 53
climate 57, 60, 70, 96, 107, 147, 216, 264, 315, 316
  projections 14, 21
  historic records 21
  variability 21
climate models 26, 30
  22-model 25, 26, 29
  CMIP3 25, 26
  CNRM 26
  CSIRO 26, 29, 30-35
  ensemble mean 25
  HAD 26, 28-41
  NCAR 26, 30, 33
climate parameters 21, 24
  statistical downscaling 21
CO_2 level 6
Coccothraustes vespertinus 263
Coccyzus americanus occidentalis 197, 210, 232
Coconino National Forest 123, 124
Colaptes auratus 238
Coleogyne ramosissima 73
collaboration 1, 3, 6, 9, 29, 59, 90
Colorado River 4, 5, 40, 73, 199, 206, 316
Colorado River Compact 4

colts 306
Columba livia 231
Columbina passerine 232
commission error 96, 117
Common Black-Hawk 227
common gray fox 293
Common Goldeneye 222
Common Ground Dove 232
common kingsnake 142
Common Merganser 222
Common Moorhen 229
Common Nighthawk 233
Common Poorwill 235
Common Raven 243
Common Yellowthroat 253
Community-Based Collaborative 6, 8, 17
competing science 17
complementarity 115, 118
conifer forests 234, 309
conifer species 40
connectivity 2, 3, 47
conservation successes 5, 7, 18, 316
continental divide 198
Contopus
  cooperi 239
  sordidulus 239
Convention on Biological Diversity 6
Cooper's Hawk 226
Coos 208
Coragyps atratus 225
Cordilleran 239
Corvus brachyrhynchos 243
Corvus corax 243
Cornville 228, 234
Costa's Hummingbird 236
cost-distance analysis 93
cottonwood 197-255
  canopy 200, 205, 206
Cougar Management Guidelines Working
  Group 307
counties
  Apache 307
  Cochise 134
  Coconino 47
  Gila 307
  Graham 307
  Greenlee 307
  Maricopa 307
  Mohave 307
  Pima 307
  Pinal 307
  Santa Cruz 307

Yavapai 307
Coupled Model Intercomparison Project version 3 25
cowbird host 252, 255
coyote 293-300, 321
Crematogaster 176, 180, 181
creosote bush 219
Crissal Thrasher 249
Crotalus viridis viridis 142
Crotaphytus collaris 147
Crotophaga sulcirostris 233
cultural barriers 4
Cyanocitta stelleri 242
cyberinfrastructure 101, 108, 318

D

Dark Canyon 111
dark skies movement 10
Dark-eyed Junco 259
data collection 13, 59, 101
　layers 59
　log-transformed 170
　remotely sensed 48
data analysis 51, 105, 160
　equal variance 160, 162
　normality 160, 162
database system 106
dataloggers 101-103
Davis Dam 206
DDT 229, 264
Decagon ECH_2O Probes 104
deer management 121
　trophy quality 272
DEM (Digital Elevation Model) 75
Dendroctonus 24
Dendroica
　caerulescens 252
　coronata 252
　graciae 252
　nigrescens 252
　petechia 252
　townsendi 252
Denver Museum of Natural Sciences 187
Department of Energy 42
desert shrew 140
Diablo Trust 8
Diflubenzuron 151-196, 319
　treatments 152-194
digital elevation model 75
Digital Orthophoto Quads 74
digital soil-data layers 59
Dinosaur National Park 110

Diptera 160, 166, 171-75, 189, 191-94, 319
DISTANCE software 159
distraction behavior 231
diversity (plant)
　alpha 115
　beta 118
dogs 306
dorsal blotch patterns 138
downscaling 21-41
Dromaius novaehollandiae 307
drought 22, 46, 96, 123, 145, 198, 278
Duke Forest 107

E

ear tags 281
eastern collared lizards 147
ECHAM5 26-39
ecological degradation 70
ecological indicators 90, 317
Ecological Restoration Institute 9
ecological site 69,-86, 92-98, 317
ecosystem types
　forest 96, 103, 107, 114
　grassland 122, 135, 147, 156, 223, 227, 234
　pinyon-juniper 96
　riparian 13, 21, 22, 46, 198, 206
　shrubland 17, 46, 70
　upland 69, 70, 90, 294, 309
　woodland 46, 53, 82, 96, 294, 309, 321
eddy covariance 107
edge contrast index 288
edge density 288
eggs 225, 229
Egretta thula 224
El Niño-Southern Oscillation 29
Eleagnus angustifolia 199
Elf Owl 234
elk 121, 277-289,320
Elymus elymoides 57
emissions 5, 6, 18, 25, 29
Empidonax
　difficilis 239
　occidentalis 239
　traillii 239
　wrightii 239
emus 307
endangered species
　Bald Eagle *(Haliaeetus leucocephalus)* 226
　Willow Flycatcher *(Empidonax traillii)* 239
Endangered Species Act (ESA) 198
ENSO (El Nino Southern Oscillation) 29
ENVI (image processing software) 282

environmental-sensor-network experiments 102
environments, "rock and ice" 115
eolian deposits 72, 80
Ephedra torreyana 142
ERDAS Imagine (image processing software) 282
Eremobates 187
Eremophila alpestris 243
ERI (Ecological Restoration Institute) 9
error
   matrix 76, 81
   rates 117
   rectification 84
ethical stance 11
Euclidean distance measure 185
Eugenes fulgens 236
Euphagus cyanocephalus 260
Eurasian Wigeon 221
European Starling 250
evapotranspiration 23, 40-42, 123
   potential 24
Evening Grosbeak 263
exotic species 52
   cheatgrass (*Bromus tectorum*) 45, 49, 52-53, 55, 59-64, 316, 317
   tamarisk (*Tamarix* spp.) 5, 197, 199, 200, 204, 205, 207, 219
   Russian olive (*Eleagnus angustifolia*) 199, 205, 207
expertise-based prerogatives 12
extinction 2, 146
extraction rules 69, 74, 77, 84

F

Falco
   columbarius 228
   peregrinus 229
   sparverius 228
false negatives 117
false positives 117
fawn-to-doe ratio 272
Feature Analyst Package 74
feeders
   sugar 234, 264
   seed 242, 265
Ferruginous Hawk 228
field efficiency 79
Field-to-Desktop 101-08
fire 45-65, 93, 95, 121-28, 277-89, 316, 319-20
   community-level effects 121
   high-severity 122
   low-severity 122
   prescribed 122
   public controversy 49
   regimes 123
   stand-replacing 277
   suppression 277
   Warm 49-65, 316
fire surrogate 122, 123, 128, 318, 319
Flagstaff City Council 9
flies 160, 162, 166, 70-72, 190
flocks 217
flood flow 122
flux corrections 29
Flying M 8
forage conditions 277, 287-89
Forelius 151
Forest Ecosystem Restoration Analysis 59
forest
   mortality 22, 40
   restoration 52
   thinning 121
forests
   ancient 46
   spruce-fir/aspen 114-15
foresummer, arid 21, 34
Formica 152
Formicidae 159
Fort Mohave 207
Fort Verde 232
Fort Yuma 207
Fossil Creek 257
Fourth Assessment Report 21
four-wing saltbush 141
Fraxinus pennsylvanica 199
Fremont cottonwood 199
frosts 40
fuel(s)
   accumulation 277
   load 9, 49
   reduction strategies 121
   post-fire 278, 281
Fulica americana 230

G

Gadwall 221
Gallinago delicate 231
Gallinula chloropus 229
Game Management Unit 12A 271
gap analysis 109-10, 318
   ReGAP 59
   Regional Gap analysis 153
GCM (Global Climate Model) 21-42
   fidelity 24

performance metrics 28
projections 28
ranking procedure 28
selection 29
Grand Canyon National Park 2, 4, 47, 49
Grand Canyon Trust 2, 4, 45, 48, 316, 323-28
geese 219, 307
Generalized Random-Tessellation Stratified design 94
Geococcyx californianus 232
Geocoridae 185
Geothlypis trichas 253
GeoTiff 282
germination 41, 56
   of native grasses 56
Gila River 201-02, 205
Gila Woodpecker 237
GIS (*see also* geographic information system) 60, 76, 78, 93, 95, 111, 194, 281-282, 289
Glaucidium gnoma 234
Glen Canyon Dam 4
Glen Canyon Environmental Studies 5
Glen Canyon National Recreation Area 111
global climate change 147
Global Climate Models (GCMs) 21-42
Global Positioning System (*see also* GPS) 281
glossy snakes 142
goats 306
Golden Eagle 228
Golden-winged Warbler 251
Goodding willow 199
gophersnake 142
GPS (Global Positioning System) 74, 281
   spectrum 282
   averaging option 282
Grace's Warbler 252
Grand Canyon Monitoring and Research Center 5
Grand Canyon Protection Act 4
Grand Canyon Trust 45
Grand Staircase Escalante National Monument 8, 110
Granier method 104
grasses, cool-season 316
grasshopper 151
grassland, arid 58, 256
Gray Flycatcher 239
Gray Vireo 241
grazing
   allotments 9
   livestock 17, 45
   permits 45
Grazing Land Conservation Initiative 55

greasewood 72, 82, 185
Great Plains 151
Great Blue Heron 224
Great Horned Owl 233
Greater Roadrunner 232
Greater White-fronted Goose 219
Great-tailed Grackle 260
Green Heron 224
greenhouse gas 25
Green-tailed Towhee 255
Green-winged Teal 220
grid resolutions 27
Groove-billed Ani 233
Grouse Creek 151
growth
   urban 206, 261
   exurban 14
growth processes, temperature limited 40
GRTS (Generalized Random-Tessellation Stratified design) 94, 98
Gymnogyps californianus 52
Gymnorhinus cyanocephalus 243

H

habitat
   class 282
   classification 281
   exotic 200
   fragmentation 14
   improvement 55
   landscape-scale 14
   mixed exotic 200, 205
   mixed native 204
habitats
   aquatic 216, 219
   high elevation 109, 115
   montane 116
   oak 320
   riparian 198, 232, 261, 320
   Sonoran 235, 250
   upland 215, 242
   yucca 262
hackberries 250
Haliaeetus leucocephalus 226
Harris's Hawk 227
Hartnet allotment 72-84
Hassayampa River 201, 202, 205
hazard-rate model 159
Heber Ranger Station 280
Heber-Overgaard 280
Hemerotrecha 187
Hemiptera 152, 160, 182

immature 183
Hepatic Tanager 254
herbicide 53, 55
herbivore pressure 41
Heritage Fund 300, 310
Hermit Thrush 248
herpetological inventory 136, 146
Hesperostipa comata 57
hierarchical learning 74
hierarchical re-learning 69, 76
highways 2
Holohil Systems 137
Hooded Merganser 222
Hooded Oriole 261
Hooded Warbler 253
Hoover Dam 206
Horned Lark 243
hotspots (plant diversity) 112, 318
House Finch 262
House Rock Valley 45, 56, 64
House Sparrow 263
House Wren 247
Hovenweep National Monuments 110
HRV (House Rock Valley) 56-58
hunt
   composition 275
   harvest 271
   quality 271
   success 272
husbandry practices 309
Hutton's Vireo 242, 265
hydroclimatic implications 40
hydrologic condition 122
hydrologic function 90
hydrological systems 4
Hyla 198
Hymenoptera 160, 162
Hypsiglena torquata 142

I

Ibapah, Utah 151-193
Icteria virens 253
Icterus
   bullockii 261
   cucullatus 261
   parisorum 262
image processing software 282
Imazapic 53
Imperial Dam 206
incisors 272, 275
India 13
Indian ricegrass 57, 72

Indigo Bunting 255
infestations, Orthoptera 152
in-holding 56
insect life cycles 22
insecticide 151-194
institutional barriers 4
instream flow 13
Intergovernmental Panel on Climate Change 6
   Fourth Assessment Report 21
Ips 24
island biogeographic theory 115
Island in the Sky 74
Ixobychus exilis 223

J

jackknife methods 77
Jacob Lake 271, 320
James' galletta 141
Jerome 261
jet streams 28
Joint Fire Sciences Program 121
Juglans major 199
Junco hyemalis 259
Juniper 52, 73, 90, 114, 239, 296
Juniper Hill 216
Juniper Ridge 280
juniper titmouse 245
Juniperus 52, 73, 293
   deppeana 280
   monosperma 90, 142

K

Kaibab National Forest 49, 64, 123
Kaibab Plateau 49, 63, 272, 320
Kane Ranch 45-49, 56-64
Kappa 76
katydids 198
Killdeer 230
kills
   depredation-related 303
kit foxes 296
knapweed 123
kowlp call 200, 208
Krascheninnikovia lanata 53
kuks 208

L

Lacey Point 142
Ladder-backed Woodpecker 238
Laguna Dam 206

Lake Mead 5, 201
Lake Montezuma 219, 233
Lama glama 307
LAMP (Linux Apache MySQL PHP) 106
Lampornis clemenciae 236
Lampropeltis 133, 142, 144, 148, 149
  caelanops 133
  getula 142
  taylori 133
  triangulum 133, 148
Land Use History of North America (LUHNA) 10
landcover types 59
Lanius ludovicianus 241
Landsat 86
  satellite imagery 281, 287
  thematic mapper 282
landscape
  cell 79
  planning 45
  scale 14, 45, 60, 279
lapse-rate correction 27
large carnivores 303
Lark Bunting 258
Lark Sparrow 257
Larrea tridentata 219
Lazuli Bunting 255
Le Conte's Thrasher 250
Least Bittern 209, 224
Lepidoptera 160, 166, 193, 319
Leptothorax 175, 181, 183
Lesser Goldfinch 263
Lesser Nighthawk 235
Lesser Scaup 222
lighting ordinance 9
lightning 49
lime tracking beds 297
Lincoln's Sparrow 258
linear regression models 24, 304
Linux 105
  Apache MySQL PHP 106
litter 278
livestock
  allotments 56
  depredation 303, 310, 321
  grazing 9
  husbandry practices 309
  management 19, 48, 322
  operations 309
  operators 303
lizards 134, 147
llamas 307
Loggerhead Shrike 241

logging 45, 121
  salvage 50
Lolium perenne 49
Lophodytes cucullatus 222
lottery draw system 272
Loxia curvirostra 262
Lower Sonoran Desert 297
Lucy's Warbler 251
LUHNA (Land Use History of North America project) 10
Lygaeidae 152, 183, 191
Lynx rufus 293

M

MacGillivray's Warbler 253
Magnificent Hummingbird 236
malathion 152, 189
Mallard 220
Mann-Whitney rank sum tests 160, 162
maple 242
mapping
  predictive 69, 84, 317
  spatial extent 70, 78, 317
  object recognition 71, 74
Marble Canyon 56
marsh 219, 223, 260
Marsh Wren 247
Masticophis taeniatus 142
Maxim iButton 101
MCNM (Montezuma Castle National Monument) 213-266
MCP (minimum convex polygons) 281
mechanical grinding 53
Mediterranean climate 31
Megascops kennicottii 233
Melanerpes
  erythrocephalus 237
  lewis 237
  uropygialis 237
Melospiza
  lincolnii 258
  melodia 258
Mergus
  merganser 222
  serrator 223
Merlin 228
Merriam Powell Center for Environmental Research 107
mesocarnivores 293, 298
mesopredators 293, 321
mesquite bosque 199, 215, 248, 259
Mexican Jay 243

Micrathene whitneyi 234
microchip 138, 148
Microcoryphia 159, 169-175
milksnake 133-149
   animal processing 138
   artificial cover arrays 133
   collection 134
   detection 145
   diet 144
   mortality rates 144
   New Mexico 133
   range 134
   recaptures 139
   roundup weekends 136
   secretive nature 147
   surveys (night driving) 133, 147
   Utah 133
Mimus polyglottus 249
mineral development 117
Mingus Mountain 238, 257
minimum convex polygons 281
Miridae 183-186
mixed desert upland shrub 219
model
   logistic regression 50, 282
   ocean-atmosphere 25
   vegetation-change 21, 41
MODIS ASTER 282
Mogollon Rim 123, 236, 277, 320
moisture deficit 24, 40
Molothrus aeneus 260
   ater 261
Monomorium 175, 179, 183
monsoon 21, 29, 43, 126
Monte Carlo simulations 40, 69, 79
   test 160
Montezuma Castle National Monument (*see also* MCNM) 213-265, 320
Montezuma Well 213-219, 220-264, 265
Mormon cricket 151-194, 319
   study sites 189
   study zones 153
Mormon tea 72, 142
motion sensors 281
Mountain Bluebird 248
Mountain Chickadee 245
mountain lion
   bounties 310
   habitat 309
   hunter harvest 303, 306, 310
   removal 303, 309
   sex ratio 304, 308
   survival rates 309
Mourning Dove 232
Muhlenbergia 141
muhly 141
mulch 53
mule deer 45-64, 271-276
   home ranges 52
   management 52, 271
   migration 52
   nutritional requirements 53
   population level 275
multivariate analysis 170, 195
Myadestes townsendi 248
Mycteria americana 225
Myiarchus
   cinerascens 240
   tyrannulus 240

N

Nabidae 185-187
NASA (National Aeronautics and Space Administration) 6
Nashville Warbler 251
National Elevation Data 60
National Forest Foundation 65
National Park Service (see also individual parks) 4, 70, 89, 117, 140
National Park Service Inventory and Monitoring Program 10, 87
National Vegetation Classification 112
National Wildlife Refuges 111, 206
Natural Resources Conservation Service 59, 92
NAU (*see also* Northern Arizona University) 1, 9, 53, 59, 107, 108, 147
Navajo Nation 198
NCPN (*see also* Northern Colorado Plateau Inventory and Monitoring Network) 69-71, 73, 74, 77, 78, 86
needle and thread grass 57, 72
Needles, Utah 74, 205
neotropical migrants 197, 218
nest predators 207
nesting 198, 218, 224, 250, 261, 265
nestlings 229
NetBridge 104
   queue 106
netleaf hackberry 199
network hub 102
New York Botanical Garden 111
NGOs (nongovernmental organizations) 3, 14
Nightsnakes 142, 145

NMS (nonmetric multidimensional scaling) 160, 170, 179
nodes 77
nonmetric multidimensional scaling 160, 169, 179
North American Birds 209, 214
North American Weed Management Association 49
North Kaibab Plateau 272
Northern Arizona Audubon Society 214, 265
Northern Arizona University (also see NAU) 1, 9, 53, 59, 107, 108, 147
Northern Beardless-Tyrannulet 238, 265
Northern Cardinal 254
Northern Colorado Plateau Inventory and Monitoring network 69, 87, 317
Northern Flicker 238, 251
Northern Goshawk 227
Northern Harrier 226
Northern Mockingbird 249
Northern Pintail 220, 265
Northern Pygmy-Owl 234
Northern Rough-winged Swallow 244
Northern Saw-whet Owl 234
Northern Shoveler 221
Northern Waterthrush 253
Notiosorex crawfordi 140
Nycticorax nycticorax 224
nymphs 183

O

Oak Creek Canyon 232, 236
occupancy models 208
Odocoileus hemionus 45, 271, 288, 303
O-horizon density 123
olfactory lure 296, 299
Olive-sided Flycatcher 239
one-seed juniper 142
Onset Computer Hobo 101
Oporornis tolmiei 253
Orange-crowned Warbler 251
ordination space 171, 181
Oreoscoptes montanus 249
Orthoptera 151-194
    control efforts 152
    infestations 153
Osprey 225, 264
ostriches 306
overgrazing 121
overstory 50, 90, 279
    density 126
owls 211, 233

Oxyura jamaicensis 223

P

Pacific-slope Flycatcher 239
Page Springs 224, 234, 237
paleontological resources 8
Pandion haliaetus 225
PAR (photosynthetically active radiation) 101, 103
Parabuteo unicinctus 227
Parameter-elevation Regressions on Independent Slopes Model 21, 24
parasitism 241, 258
    cowbirds 266
Paria Canyon-Vermilion Cliffs 111
Parker Dam 206
Passer domesticus 263
Passerina
    amoena 255
    caerulea 255
    cyanea 255
patch richness 288
pattern template 74
Payson 272, 280
PCMDI (Program for Climate Model Diagnosis and Intercomparison) 25, 27
PC-ORD (software) 160, 170, 185, 195
Peck Lake 216
Peregrine Falcon 229, 264
permit
    allocations 273
    hunting 275
    levels 275
person-hour of effort 133
Program for Climate Model Diagnosis and Intercomparison 25, 27
Petrified Forest National Park 8, 133, 142, 147
Petrochelidon pyrrhonota 244
Phainopepla 250
Phainopepla nitens 250
Phalaenoptilus nuttallii 235
Phalaropus tricolor 231
Phasianus colchicus 223
Pheidole 174, 190
phenology 159, 174, 183, 191
pollinator 41
Pheucticus melanocephalus 255
photosynthetically active radiation (*see also* PAR) 101, 318
Picoides scalaris 238
Pied-billed Grebe 223
pigs 307

Pine Siskin 262
Pine Valley Mountains 113, 117
pine-oak woodlands 286, 309, 320
Pinus
   edulis 52, 73, 90, 280
   ponderosa 24, 49, 65, 277, 293
pinyon 52, 55, 73, 90
pine mortality 22
Pinyon Jay 243
pinyon-juniper woodland 53, 90, 294
Pipilo
   aberti 256
   fuscus 256
   maculatus 256
   chlorurus 255
Piranga
   flava 254
   ludoviciana 254
   rubra 254
PIT tag 138, 140, 144
pitfall traps 136, 151, 157, 180, 319
Pituophis catenifer 142
pixel brightness values 281
plant communities 45, 49, 63, 92
plants
   Colorado Plateau endemics 109, 115, 117
   invasive 14, 58, 123, 156, 316
   native 22, 52
   non-native 50, 52
   rare 114, 116, 119
   vascular 109, 112
Platanus wrightii 219
playback-listen 200
Plegadis chihi 225
Pleuraphis jamesii 141
Plumbeus Vireo 242
Pluvialis squatarola 230
Poa secunda 57
Podilymbus podiceps 223
Poecile gambeli 245
Pogonomyrmex 174, 179, 181
Polioptila caerulea 248
pollination 193
polygon 92, 281, 317
   map unit 71
ponderosa pine 24, 50, 96, 121, 123, 287, 297, 318, 321
   crown mortality 49
Pooecetes gramineus 257
population abundance 146, 293, 301
Populus fremontii 199, 219
Porzana carolina 229

power generation 4
Powerline site 124
precipitation
   amplitude 28
   bimodal distribution 23
   Mediterranean climate 31
   monsoon 21, 24, 28, 29, 33, 43, 126, 128
   observed-average 35
   projections SCP 21, 31, 41, 90
   seasonal 26, 32, 56
predator abundance 310
   control efforts 303, 309
   reduction 303
predictive capability 46, 80
predictive science 14
Prescott 2, 255, 272
Prescott Valley 133
Presidential Proclamation # 6920 8
PRISM (Parameterelevation Regressions on Independent Slopes Model) 21, 24, 27, 60
Progne subis 244
Program for Climate Model Diagnosis and Intercomparison 25
Prosopis 197, 219
protected areas 109, 114, 117, 318
Psaltriparus minimus 246
Psyllidae 185
public
   land 18, 45, 55, 64, 199
   meetings 272
   -private partnership 45, 64
Puma concolor 303
Purple Finch 262
Purple Martin 244
Purshia mexicana var. *stansburiana* 53
Pygmy Nuthatch 246
Pyrocephalus rubinus 240

Q

Quercus gambellii 280
Qusicalus mexicanus 260

R

R statistical computing 94
rabbitbrush 82, 141
radiation, photosynthetically active 101, 318
radio 102
radio frequency band 103
   900-928 MHz ISM 103
radiocollar 281, 282
   store-on-board 282

radiotracking 139
Rainbow Bridge National Monument 110
Rainbow Forest 144
rainshadows 24
Rallus limicola 229
range expansion 214, 224, 231, 261
rangeland
  degradation 58
  management, sustainable 8
rangeland-site concept 70
rank correlations 304
ranking procedure 28
raptor 226, 264
rare species
  sage grouse (*Centrocercus urophasianus*) 152
  spotted frogs *(Rana luteiventris)* 152
rattles 208
Rattlesnake Butte 141
rattlesnakes 142
real-time displays 103
receiver operating characteristic 62
Red Crossbill 262
Red-breasted Merganser 223
Redhead 221
Red-headed Woodpecker 237
Red-naped Sapsucker 238
Red-tailed Hawk 227
Red-winged Blackbird 259
Redwoods 107
reflectance 74
  properties 76
reforestation 50
Regulus calendula 247
remote sensing 14, 19, 86, 281, 286, 316
reproductive capability 271
resampling 40
Research Natural Area 111
reseeding 56, 58, 289
reserve network 110, 116, 117
resilience 11, 70
resolution, spatial 22, 76, 87, 102
resource extraction 3, 4
restoration
  landscape-level 45, 48, 52, 55, 63
Rhynchopsitta pachyrhyncha 232
Rim Rock 233
Ring-necked Duck 222
Ring-necked Pheasant 223
Riparia riparia 244
riparian
  drainages 197
  habitat 197, 206, 208, 241, 254, 259, 265, 320

restoration 209
riparian-obligate 232
RNA (*see also* Research Natural Area) 117
ROC (receiver operating characteristic curve) 62, 66
rock outcrop 60, 71, 84
Rock Pigeon 231
Rock Wren 246
Rocky Mountain Research Station 53, 65, 128
Rocky Mountains 123
rodenticide 265
Rodeo Chediski fire 277-291
rooting 121, 319
Rough-legged Hawk 228
Ruby-crowned Kinglet 242
Ruddy Duck 223
Rufous Hummingbird 237
Rufous-crowned Sparrow 256
runoff 40, 43, 206
Russian olive 199, 205, 207
Russian thistle 58
ryegrass 49, 52, 65

S

sage grouse 152, 192-193
Sage Sparrow 258
Sage Thrasher 249
sagebrush 53, 55, 72, 82, 109, 114, 116, 142, 147, 156-157, 185, 192
Salix 197-199, 219
  gooddingii 199
Salpinctes obsoletus 246
Salsola tragus 58
Salt River 201, 202, 206
saltbush 53, 72, 82, 141, 142, 185
salvage logging 50
sampling 40, 50-52, 69-71, 89-98, 101-103, 133, 225, 235, 297, 298,
  design 48, 69-71, 85, 89-90, 317-318
  frames 71, 85, 89-98
  probability-based 90
  sample bias 89-98, 216
  milk snakes 136-147
  Orthoptera populations 151-194
San Bernardino Valley 133, 201, 202
San Francisco Peaks 22, 23
San Pedro River 197, 201-206, 320
sand sage 142
sandbars 5
Sandberg bluegrass 57
sandy loam sagebrush grasslands 147
Santa Cruz River 201, 202, 205

sap flux 101, 104
SAS statistical analysis software 61, 124
SAS Institute 61
Say's Phoebe 240
Sayornis nigricans 240
Sayornis saya 240
scenarios, climate change 21
  A1B 25, 29, 34
  business as usual 26
  climate change 21
  medium non-mitigation 25
scenario, management 19
scent tablet 296
scent-station
  surveys 293-300
  visitation rates 293, 296-300
science advisory council 47
Science Advisory Council, Grand Canyon Trust 59
scientific rigor 3
Scirpus 219
Scorpiones 166, 172-175, 187, 193, 319
scorpions 152, 168, 187-188, 191-193
Scott's Oriole 262, 265
SCP (*see also* Southern Colorado Plateau) 21-23, 31-35, 38-41
SCPN (Southern Colorado Plateau Network) 89-90
scripts 106
  security 106
sediment 4, 5, 122
Sedona 228, 234, 238, 262
seed predation 193
seed-harvester 174
seeding methods 56
  aerial 50
  broadcast 56
  cattle trampling 56
  mechanical drill 56
seedling recruitment 50, 52, 58
seedling survival 41
seep willow 199, 219
Seiurus noveboracensis 253
Selasphorus
  platycercus 237
  rufus 237
Seligman 133
semi-desert grassland 156, 293-299, 231
sensitive species 136, 146
  on National Forests 198
  spotted tree frog (*Rana luteiventris*) 152
  Western Yellow-billed Cuckoo (*Coccyzus americanus occidentalis*) 198
servers, computer 102-108
  Apache 106
  LAMP 106
  MySQL 105, 106
sewage treatment facilities 265
shared purpose 15
Sharp-shinned Hawk 226
sheep 306-307
shorebirds 230
Short-eared Owl 234
Show Low 280
shrub
  restoration 45, 49, 63
  establishment 53, 55
shrub steppe communities 59
shrubland 17, 46, 70-74, 82, 142, 156-157
Sialia
  currucoides 248
  mexicana 248
SigmaStat 160
silt banks 5
silverfish 170
silvicultural practices 121
silvicultural treatments 121, 123, 124, 126, 127, 279
simultaneous confidence interval 282
sinkhole, limestone 216
site selection 78, 89, 90, 94-96, 98, 199, 288
Sitta
  carolinensis 246
  pygmaea 246
slash 288
  treatments 289
slick rock 73, 75-76
Slide Fire 53
slope 23, 60, 76, 78-79, 84, 89, 93-97, 282-83, 287
Snake River 207
snakes 140, 142-143, 145
Snow Goose 219
snowmelt 40, 126
snowpack 40
Snowy Egret 224
social and cultural context 15
social desires 271, 275
societal myths 11
Society for Conservation Biology 1
  Colorado Plateau Chapter 2, 7, 18
software, sample-selection 71
soil 46-60, 69-82, 89-98, 121-128
  analysis 48

and bobcats 297
and fire 46, 49
and kitfoxes 296
and milksnakes 134, 142
crust 80
disturbance 53, 58, 63, 316
fertility 46, 63
erosion 58, 74
map unit 69, 71, 74-82, 89-98
microbiology 123, 319
moisture 22, 40, 101, 103, 104, 121-128, 318
monitoring 89-98
nutrients 277
properties 69-71, 74
stability 50, 58, 59, 70, 90
surveys 69, 71, 74, 89, 92-93, 96, 316
temperature 101, 278, 318
texture 60
water-repellent 123
Soil Survey, NRCS 60, 75, 78
solar array 106
Solenopsis 152, 176, 179-180
Solifugae 172-175, 187, 194
Song Sparrow 258, 259
Sonoita Creek 197, 201-205, 320
Sora 229
South America 133, 197, 319
South Coast Wildlands 3
Southern Colorado Plateau 21-25, 30-41, 64, 146
Southern Colorado Plateau Network 89, 98
Southwest Strategy 10
spatial allocation method, GRTS 98
spatial
  congruence coefficient 28
  correlation coefficient 28
  heterogeneity 288
  modeling 59, 64
  geospatial 316
spatio-temporal sensing 108
Special Report on Emissions Scenarios A1B 25
species (animals)
  avian 213-266
  crepuscular 235, 282
  endangered 198, 226, 239
  cliff-nesting 244
  bank-nesting 244
  secretive 134, 144, 208, 217, 224, 229, 232, 238, 249, 250, 254, 256, 259
  cavity nester 251
  passerine 262, 265
  waterfowl 221, 264-265
  neotropical migrants 197, 218, 219-263

species (plants)
  browse 52-53, 121, 277
  exotic 52, 53, 56, 63, 197, 199, 200, 207
  imperiled 46
  invasive 316
  loss 5
  native 7, 46, 58, 113, 114, 197, 200, 251, 318
  non-native 14, 109, 112-114, 147
  "pest" 251
  winter shrub 53
Species of Special Concern 232
spectral
  bands 281
  classification 281, 282
  properties 76, 282
Sphyrapicus
  nuchalis 238
  thyroideus 238
spiders 160, 162, 166, 168, 172, 192, 194
Spizella
  arborea 257
  atrogularis 257
  breweri 257
  passerina 257
spotted frogs 152
spotted ground squirrel 140
Spotted Owl 234
Spotted Sandpiper 230
Spotted Towhee 256, 265
springs 46, 193,
SPSS statistical analysis software 124
squirreltail grass 57
SRES (*see also* Special Report on Emissions Scenarios) 25
SSURGO soil surveys 71, 75, 78
St. Johns 133
stakeholders 9, 17, 18
stand density 121, 124, 126-127, 319
station clustering 24
statistical significance
  soil moisture 125
  milksnake sampling 138
  arthropod data 160, 166
statistics, spatial 29
Stelgidopteryx serripennis 244
Steller's Jay 242
stem density 121
storms 23, 28, 296
stream flow 13, 122
stress,
  on vegetation 22, 24, 41, 46, 70, 278
  on ecosystems 47, 89, 315

values 160, 170, 179
striped whipsnake 12
Strix occidentalis 234
students 12
Sturnella neglecta 260
Sturnus vulgaris 250
Struthio camelus 306
Sulphur Springs Valley 201
summer 21, 23, 24, 30-42, 123, 133, 140, 144, 216, 218, 283-284, 288
summer grazing allotments 9
Summer Tanager 254
support
   political 8, 11, 316
   financial 8, 11
survey
   methods 133, 146, 199, 294
   protocol 199, 204
   reliability with Yellow-billed Cuckoo 208
   with milksnakes 145
   with Yellow-billed Cuckoo's 200
sustainability
   of land or ecosystems 11
   of sage grouse populations 193
   of societies 11
sustainable agriculture 18
   land uses 46
   muledeer management 271
Swainson's Hawk 227
Swainson's Thrush 2 48
Swallows 243, 245
Swifts 234, 235
sycamore 199, 219, 220, 226, 232, 238, 239, 241, 245, 246, 250, 251,
   balls 262
systems of governance 12

T

Tachycineta thalassina 244
take-home lessons 1
tamarisk 5, 197, 199, 200, 204, 205, 207, 219
Tamarix 197, 199
   ramosissima 219
Tapco 237
tape playback 208, 233, 234
Tapinoma 152, 175-176, 179-181, 183, 190
target population 7 0, 79, 86, 90, 93-95, 98, 151
Tavasci Marsh 220, 221, 224, 225, 229, 233
taxa richness, reduction of 193
T_{ds} 27
Teec Nos Pos Wash 200
telemetry 133, 135-137, 141, 144, 208, 289, 319

Telonics 137, 281
temperature,
   ambient, sampled by radiocollar 281
   and ponderosa pine 123
   data gathering 101, 103, 104
   global 30
   historic 24-25, 30
   in cottonwood-willow galleries 207
   mean across North America 40
   observed 31
   projections for Southern Colorado Plateau 21-42
   seasonality 23-24, 30, 33, 41
   sensitivity of models 34, 39
   simulated 28, 30, 31
   soil 101, 278, 318
temperature-limited growth processes 40
temporal patterns 140
Terrestrial Ecosystem Survey 60
Tettigoniidaen 198
The Arboretum at Flagstaff 101, 107
The Conservation Fund 45, 316
The Nature Conservancy 17, 111
thermocouples 104
Thick-billed Parrot 232, 265
three-axis solution 185
Thryomanes bewickii 247
Thysanura 159
timber management 277
Time Domain Reflectometry 124
Tonto Creek 201-202, 206
Tonto National Forest 278
tooth eruption, mule deer 272, 275
topographic
   diversity 22, 23, 32
   facet 24
   heat load 60
   roughness 60, 63
topographical
   error 96
   maps 199
topography
   in elk habitat selection 287
   in mountain lion predation 309
   Bandelier National Monument 90
   San Francisco Peaks 22
   Vernon, Utah 189-190
tourism 45
Townsend's Solitaire 248
Townsend's Warbler 252
Toxostoma
   crissale 249
   lecontei 250

rufum 249
training samples 69, 77-79, 84
transducers 101, 103-105
    PAR (photosynthethically active radiation) 101, 103, 104
transects 133, 136-137, 299
    artificial cover 133
    box trap 139, 140, 142, 147
    coverboard 139, 145
    cover object 136, 137
    drift fence 133, 137, 146
    scent-station 293-300, 321
    walking 146
transmitters 137, 141, 145
transparent incremental deployment 103
transparent process 9
transponder 138
trap
    box 133, 136, 137, 139, 140, 142, 146, 147
    coverboard 139
    clover traps 281
    funnel 140
    pitfall 133, 136, 137, 139, 151-194
    sampling seed 101
TRASE System I 124
treatment
    vegetation management 93
    zone 151
tree
    cavities 228, 233
    diseases 121
    frogs 198
    mortality 40, 50
    pathogens 121, 278
    removal 50, 55
    seedling recruitment 50
    trunk diameter 101
tree models
    classification and regression 71
    decision 69, 71, 77-86, 317
tribal relations 18
Troglodytes aedon 247
true bugs 160, 171
trust 3, 9, 17, 18
Turdus migratorius 249
Turkey, wild 121
Turkey Vulture 225
Tuzigoot Bridge 261
Two Mile Ranch 45-65
two-axis solution 170
two-tailed statistical significance 138

Typha 219
Tyrannus verticalis 241
Tyrannus vociferans 241
Tyto alba 233

U

understory
    multi-storied 198
    multistructure 206
    vegetation 49-52, 64, 123, 249, 253, 254, 256, 258, 265
ungulates 277-278, 287
uniform random number 80
United Nations 6, 7
United States 10, 13, 133, 152, 197
    FFS study sites 122
    government 8, 16
    western 16, 21-44, 55, 121, 123, 198, 277, 296, 300, 307, 310
Universal Transverse Mercato 200
University of Arizona 42, 137, 209, 315
Upper Sonoran Desert scrub 293-299, 321
urban areas 2, 123
urbanization 206, 231, 320
Urocyon cinereoargenteus 293, 294
u.r.v. (uniform random number) 80
USDA (*see also* U.S. Department of Agriculture) 22, 77, 78, 82, 198
U.S. Department of Agriculture (*see also* USDA) 55, 152, 296
U.S. Fish and Wildlife Service (*see also* USFWS) 198, 209
U.S. Forest Service (*see also* USFS) 9, 53,198, 277, 278, 319
    Burned Area Emergency Rehabilitation 287
    Research Natural Areas (RNA) 111, 117
USFS (*see also* U.S. Forest Service) 53, 59, 60, 64, 123, 278, 281
USFWS (*see also* U.S. Fish and Wildlife Service) 199
U.S. Highway 260 280
U.S. Geological Survey (*see also* USGS) 86, 151, 194, 199, 266, 261, 318
USGS (*see also* U.S. Geological Survey) 1, 2, 5, 59, 86, 141, 147
    National Elevation Data 60. 75, 78
    satellite imagery 287
Utah 18, 45, 71, 74, 75, 78, 109-118, 123, 134, 198
    Colorado Plateau flora 113
    governor 17
    west desert 151-194

Utah Basin 111
Utah Department of Food and Agriculture 194
Utah Gap and Southwest ReGap programs 111, 116, 117
Utah High Plateaus 110
Utah juniper 73, 74, 80
Utah milksnake 133
Utah Natural Heritage Program 109, 112-115
Utah State University 111
Utah Water Science Centers 194

V

Vaisala WXT-520 104
values (people's) 5, 8, 9, 12, 15, 109
variability
    climate 29-42, 59
    ecosystem 47, 122, 288
    landscape 70
Vaux's Swift 235
vegetation 21, 42, 109-118, 309, 320
    alpine 114
    and fire 59, 121, 278
    change 21, 39, 40, 41
    classes in Utah 114
    GAP communities 153, 156-157, 185
    and habitat types 284-286, 294
    in bird surveys 199, 200, 204-206, 216
    landscape-scale 48
    life zones 23
    management 92-93, 317
    modeling 21, 22, 24, 25, 28, 39 41, 42
    monitoring 64, 69, 70, 71, 75, 89-92, 96, 111-112, 122, 281, 317
    maps 77, 78, 80, 111
    riparian 207
    stress 24, 40, 41
velvet ash 199
Verde Hot Springs 231
Verde River 197, 201-205, 215, 219-221, 224, 227, 229, 231, 233, 237-239, 246, 254, 264, 320
Verde Valley 213-264
Verdin 245, 265
Vermilion Cliffs 56, 111
Vermilion Flycatcher 240, 265
Vermivora
    celata 251
    chrysoptera 251
    luciae 251
    ruficapilla 251
    virginiae 251

Vernon, Utah 151-194
Vesper Sparrow 257, 265
VHF (very high frequency) beacons 281
Vicugna pacos 307
Violet-green Swallow 244
Vireo
    bellii 241
    gilvus 242
    huttoni 242, 265
    plumbeus 242
    vicinior 241
Virginia Rail 229
Virginia's Warbler 251
VLS 2002 (Feature Analyst Package) 74
Vulpes macrotis 296

W

Walker Basin 233
Walnut Canyon National Monument 8
Warbling Vireo 242
Warm Fire 49-52, 59, 64, 316
wasps 160, 171, 190
water
    and fire 122
    as ecosystem service 46, 63, 122
    availability 13, 278
    diversion 206-207
    ground- 216
    in the Colorado River 4, 206
    loss 123
    management 17, 64
    open bodies of 193, 219
    purification 63
    rights 1
    surface 207, 216
    supplementation 58, 206
    supply 122, 123
    vapor pressure 24
    yield 122
watersheds 45, 56, 122, 264
    condition 122
    hydrology 122
    management 122
wealth 11, 319
wear patterns, in mule deer teeth 272
weather 49, 103, 124, 137, 138, 295, 299
    stations 299
    systems 27
web-based information 10
WebSim 159, 194
weed treatment 55
weighting 27, 80

function 24
Western Area Power Administration 4
Western Bluebird 248
Western Flycatcher 239
 Pacific-slope or Cordilleran 239
Western Kingbird 241
Western Meadowlark 260
Western Screech-Owl 233
Western Scrub-Jay 242
Western Tanager 254
Western Wood-Pewee 239
Wet Beaver Creek 201, 202, 219, 220, 227, 231, 232, 237, 238, 242, 243, 263
White Mountain Apache Reservation 277, 278
White-breasted Nuthatch 246
White-crowned Sparrow 259
White-faced Ibis 225, 264
White-throated Sparrow 259
White-throated Swift 234, 235
White-winged Dove 231
Wilburforce Foundation 65
Wilderness Areas 109, 111, 116, 117
wildfire 14, 25, 41-64, 93-95, 120, 122, 277-289
 anthropogenic influences 277, 291, 316
 ecological functions 121, 128, 278
 frequency 41
 mitigation treatments 122
 suppression 49, 59, 121, 124, 277, 278
wildlife
 corridors 2, 19, 53, 201-203, 234, 248
 forage 235, 243, 236, 276-279, 287
 management 52, 271, 276, 303
Wildlife of Special Concern 198
Willet 230
Williamson's Sapsucker 238
Willow Flycatcher, Southwestern 209-211, 239
willow 197-199, 204
Wilsonia
 citrina 253
 pusilla 253
Wilson's Phalarope 231
Wilson's Snipe 231
Wilson's Warbler 253
Wingfield Mesa 216
winterfat 72
Wireless Networks Research Lab 107
Wireless Sensing And Relay Device Network 102, 318
wireless sensor networks 101, 102, 108
 battery life 103, 108
WiSARD
 sensor node 102

hub 102
server 105
Wisardnet 103-108, 318
witness tree 124
Wood Duck 220, 264
Wood Stork 225, 265
woodlands
 Madrean evergreen 309
 pine-oak 309
 pinyon-juniper 52-53, 90, 96, 295-299
 riparian 219-263
World-Wide-Web 102
Wupatki National Monument 145, 147, 149
Wyoming Basins 111

X

Xanthocephalus xanthocephalus 260

Y

Yellow Warbler 252, 261
Yellow-billed Cuckoo 197-211
 breeding pairs 207
 detections 204
  historical 197, 200-206
 food availability 198
 habitat type association 200, 204
 nesting 200
 population
  declines 198
  trends 198
  historical estimates 197-199
 survey methods 199-200
 vocalization (call) 208
Yellow-breasted Chat 253, 254
Yellow-headed Blackbird 260
Yellow-rumped Warbler 252
Yellowstone 44, 278

Z

Zenaida
 asiatica 231
 macroura 232
Zion National Park 110-113, 117
Zonotrichia
 albicollis 259
 leucophrys 259